Elliptic Partial Differential Equations from an Elementary Viewpoint

What the Laplacian Is Good For

Elliptic Partial Differential Equations from an Elementary Viewpoint

What the Laplacian Is Good For

Serena Dipierro
Enrico Valdinoci

University of Western Australia, Australia

World Scientific

EW JERSEY · LONDON · SINGAPORE · BEIJING · SHANGHAI · HONG KONG · TAIPEI · CHENNAI · TOKYO

Published by

World Scientific Publishing Co. Pte. Ltd.

5 Toh Tuck Link, Singapore 596224

USA office: 27 Warren Street, Suite 401-402, Hackensack, NJ 07601

UK office: 57 Shelton Street, Covent Garden, London WC2H 9HE

Library of Congress Cataloging-in-Publication Data

Names: Dipierro, Serena author | Valdinoci, Enrico, 1974- author
Title: Elliptic partial differential equations from an elementary viewpoint :
    what the laplacian is good for / Serena Dipierro, Enrico Valdinoci,
    University of Western Australia, Australia.
Description: New Jersey : World Scientific, [2026] | Includes bibliographical references and index.
Identifiers: LCCN 2025012312 | ISBN 9789819813063 hardcover |
    ISBN 9789819813940 paperback | ISBN 9789819813070 ebook |
    ISBN 9789819813087 ebook other
Subjects: LCSH: Differential equations, Elliptic | Differential equations, Partial
Classification: LCC QA377 .D6425 2025 | DDC 515/.3533--dc23/eng/20250717
LC record available at https://lccn.loc.gov/2025012312

British Library Cataloguing-in-Publication Data
A catalogue record for this book is available from the British Library.

Copyright © 2026 by World Scientific Publishing Co. Pte. Ltd.

For any available supplementary material, please visit
https://www.worldscientific.com/worldscibooks/10.1142/14311#t=suppl

Desk Editors: Nambirajan Karuppiah/Lai Fun Kwong

Typeset by Stallion Press
Email: enquiries@stallionpress.com

The Big Crunch was her nickname for the mythical result that the Niah had aspired to reach: a unification of every field of mathematics that they considered significant.

Greg Egan, Glory.

Preface

There is more to life than partial differential equations.

But not much more.

Partial differential equations capture and understand the rules of the universe. They detect the optimal strategy for an animal to search for food. They teach us to optimally take advantage of environmental resources. They tell us how to process images and sounds for better performances. They describe how different biological species can share an ecological niche. They figure out how natural crystals grow. They predict the weather and foresee climatic changes. They quantify the melting of ice and glaciers. They find out how fast bushfires spread. They describe the shape of the stones in the rivers. They can tell us (maybe) why an airplane flies. They explain how cells and tumors diffuse in a living organism. They can explain social phenomena (despite mathematicians being chronically antisocial). They are the cornerstone of artificial intelligence (and they try to prevent natural stupidity). They can be used for nerd sniping.

They do this, and much more, because partial differential equations are the language spoken by Mother Nature. But, more importantly, partial differential equations are beautiful and fun.

The notes collected here and in the companion book [DV23] are the outcome of some courses taught to undergraduate and graduate students at the University of Western Australia, Pontifícia Universidade Católica do Rio de Janeiro, Indian Institute of Technology Gandhinagar, Ukrainian Catholic University (Український Католицький Університет) and Politecnico di Milano in 2021–2023.

Far from aiming to be all-encompassing, the following pages wish to shed light on a number of selected topics in the theory of elliptic partial differential equations in a style accessible to third-year undergraduate students, possibly under an inspired mentorship, but may also pique the interest of more advanced students and professional researchers. While all the topics presented are of classical flavor, the exposition and the way the material is organized are perhaps rather original, with the intention of addressing several quite difficult points in a style that is as self-contained as possible and rigorous, as well as intuitive and approachable without major prerequisites. (Indeed, we only assume prior knowledge of the "basic" analysis, providing references to useful results and theorems as well whenever the less advanced readers may need to consolidate their background.)

The focus of this book is on models and motivations. Namely, before diving into the more sophisticated theory of partial differential equations, which will be presented in the companion book [DV23], we offer an intuitive approach to a variety of situations in which partial differential equations play a decisive role. Not only do we believe that this is the only possible approach to a truly cross-disciplinary perspective of science and mathematics and that a unified vision is essential from the cultural point of view, but also we believe that many readers may find it helpful to underpin the deep understanding of advanced analytical theories with a concrete grasp on the essence of things, which uses intuition as a foundational keystone for a high-rise mathematical building.

So, without aiming at being exhaustive, we present here some classical circumstances in which partial differential equations, especially (but not only) those relying on the Laplace operator, naturally arise to model interesting natural phenomena. The objective is not merely to use mathematics to find the "Answer to the Ultimate Question of Life, the Universe, and Everything" (which is well known to be 42 anyway, see [Ada80]) but rather to develop familiarity with some of the basic features of partial differential equations, to see them in action and to cultivate an intuition and instinctive feeling of the problems, which will turn out handy in our voyage through the mighty jungle of technical and seemingly abstract mathematics.

On the one hand, every line of this set of notes is intimately intertwined with any other line, like droplets flowing within a variegated ocean. On the other hand, every part of this book can be read independently. Cross-references are given when some pages are strictly linked, but in general readers are free to choose the direction they prefer and plan their journey through the knobs and gullies of this multiform mathematical landscape.

Though a list of classical and modern references is given at the end, this is not a book focused on the history of mathematics; hence, we will not address topics such as priority of discoveries or progress of this subject over time.

As usual, these notes may contain errors or inconsistencies; if you find any, please let us know. In general, we will be happy to receive comments and criticisms, which may help us improve this work. After all, scientific knowledge is based on a dynamic flow of information, and we will certainly cherish and treasure readers' feedback and advice.

All right, enough chitchat, please! It's time to start our journey together. Let's watch the Laplacian coming onto the scene.

Serena and Enrico

Courtesy of Burbuqe Shaqiri.

About the Authors

Serena Dipierro is Professor of Mathematics, Fellow of the Australian Mathematical Society, and Australian Mathematical Society Hanna Neumann Lecturer. She carried out her academic career in Santiago de Chile, Edinburgh, Magdeburg, Berlin, Melbourne, Milan, and Perth.

She has obtained prestigious awards, including the Australian Mathematical Society Medal, the Mahony-Neumann-Room Prize, the Bartolozzi Prize of the Italian Mathematical Union, the Christopher Heyde Medal of the Australian Academy of Science, the Book Prize of the Italian Mathematical Union, and the Prize of the Bulletin of the Brazilian Mathematical Society.

Enrico Valdinoci is Professor of Mathematics and Australian Laureate Fellow. He carried out his academic career in Pisa, Rome, Milan, Berlin, Melbourne, and Perth.

He is a highly cited researcher, Ladyzhenskaya Lecturer and has been awarded the James S. W. Wong Prize, the Mahony-Neumann-Room Prize, the Orazio Arena Prize, the Book Prize of the Unione Matematica Italiana, the Prize of the Bulletin of the Brazilian Mathematical Society, and the Amerio Gold Medal Prize.

Contents

Chapter 1

The Heat Equation

Among the many occurrences in which elliptic equations naturally arise in nature, one of the most widely popularized is the stationary state of the heat equation (see Figure 1.1 for a popular application of the heat equation).

In its most basic formulation, the setting of this equation is due to Jean-Baptiste Joseph Fourier who[1] presented for the first time a

---

[1] Interestingly, while changing the course of mathematics and physics, Fourier was acting as a full-time politician, since in 1801 Napoléon Bonaparte decided to appoint him as prefect of the Department of Isère, in the Alps. And perhaps it was not a completely trivial task to upset the First Consul (Emperor of the French from 1804) and dedicate oneself to mathematical investigations; in any case, playing around with the heat equation at that point was not really part of Fourier's workload model.

See Figure 1.2 for a watercolor caricature of Fourier (by Julien-Léopold Boilly).

By the way, Napoleon had already included Fourier as the leader of the "Legion of Culture", which was sent to Egypt in 1798. The 150 renowned scholars of this selected group had the (perhaps debatable) honor of traveling to Egypt with Napoleon himself on the flagship of the French fleet "L'Orient" and entertaining him every evening with a scientific lecture. Allegedly, Fourier enjoyed the desert weather so much that, upon returning to Paris, he kept his rooms unreasonably hot and would wrap himself in multiple layers of very warm clothing.

The fate of the flagship "L'Orient" was instead unfortunate since it exploded spectacularly at 10:30 PM on August 1, 1798, during the Battle of the Nile, see Figure 1.3. Someone reported that the ship was repainted slightly before the battle, which was not a very wise choice since the inflammable contents of the fresh paint acted as an energizer for the fire produced on the ship by Horatio Nelson's cannonballs.

1

Figure 1.1.   The heat equation applied in a restaurant in Texas (photo by vxla, image from Wikipedia, licensed under the Creative Commons Attribution 2.0 Generic license).

Figure 1.2.   Caricature of Joseph Fourier (work by Julien-Léopold Boilly, 1820, Public Domain image from Wikipedia).

partial differential equation to describe conductive diffusion of heat (and, to study this equation, he also introduced a marvelous instrument that was going to change forever the history of the universe, namely, the Fourier series).

The *ansatz*[2] made by Fourier is that, at some[3] point $x \in \mathbb{R}^n$, at an instant of time $t \in \mathbb{R}$, the variation of temperature $u(x,t)$ of some body is determined by the heat flow around the point under

---

[2] *Ansatz* (plural: *Ansätze*) is a masculine German name. As is quite customary in German, this name is obtained by a preposition ("an", meaning "at", or "on", or "to", or "against", or pretty much whatever one likes) and another name ("Satz", plural: "Sätze", meaning "sentence", or "statement"). And, as is quite customary in German, a precise translation goes well beyond the expertise of modest mathematicians of our caliber. The "*ansatz*" could be literally a "starting point". In jargon, it is frequently used to denote an "educated guess" about a problem, typically an additional simplifying assumption set at the beginning of an argument, possibly to be verified later on (if it produces any result worthy of further consideration).

Maybe mathematics is not quite a deductive discipline, after all. Perhaps it is mostly the art of making good guesses. Of course, anybody can make guesses. But for a good guess, one needs to become acquainted with the problem and to have developed that sort of familiarity that permits separating the essential aspects from the minor details (and, in trying to do this, sometimes a little bit of luck doesn't hurt).

[3] The space variable x here is supposed to be in $\mathbb{R}^n$. Though the models presented refer to physical spaces in which one can focus on the case $n = 3$ (or even $n = 2$, when dealing with plates, or $n = 1$, when dealing with ropes of negligible thickness), whenever possible we prefer to work in (Euclidean) spaces of arbitrary dimension. When one does so, the notation becomes usually more effective, and the essential traits of the main ideas are quite often more transparent. Also, when it will be needed to focus our attention on the special case of particular dimensions, one will be able to understand why and to what extent these dimensions are different from the general case. In dealing with the arbitrary dimension case, we can't help but quote the first "general working principle" in the Preface of [Eva98]: "PDE theory is (mostly) not restricted to two independent variables. Many texts describe PDE as if functions of the two variables (x, y) or (x, t) were all that matter. This emphasis seems to me misleading, as modern discoveries concerning many types of equations, both linear and nonlinear, have allowed for the rigorous treatment of these in any number of dimensions. I also find it unsatisfactory to classify partial differential equations: this is possible in two variables but creates the false impression that there is some kind of general and useful classification scheme available in general".

Figure 1.3.   The explosion of "L'Orient" at the Battle of the Nile, as depicted by artist George Arnald (Public Domain image from Wikipedia).

consideration, plus possibly additional heat[4] sources. Let us denote by $B(x,t)$ the heat flux vector and by $f(x,t,u(x,t))$ the scalar intensity of the external[5] heat sources.

A precise measure of the temperature at a point x is certainly rather difficult from a practical point of view, so it is useful sometimes

---

[4]For the sake of precision, we should probably say here "heating or cooling sources", e.g., fireplaces, bonfires, stoves, radiators, gas heaters, infrared heaters, but also refrigerators, air conditioners, swamp boxes, chilled beams, or whatever heating or cooling device Fourier might have possessed at that time. With this understanding, positive values of f would correspond to heating systems and negative values to cooling systems.

[5]To make things simpler, one can suppose that f depends only on x, as a heat source which is permanently switched on or off, or on x and t, in case the source is turned on and off or its intensity is modified over time. The additional dependence on u itself models the interesting case of a thermostat controlling a heating or cooling system.

to consider instead an average[6] temperature, measured in some region of the space Ω, say

$$U(t; \Omega) := \fint_\Omega u(x, t)\, dx.$$

In this setting, up to physical constants that we omit, we can equate the variation of U with the heat flux through $\partial\Omega$, possibly adding to it the effect of the heat sources (if any) in Ω; namely, we write that

$$\partial_t U(t; \Omega) = -\frac{1}{|\Omega|} \int_{\partial\Omega} B(x, t) \cdot \nu(x)\, d\mathcal{H}_x^{n-1} + \fint_\Omega f(x, t, u(x, t))\, dx.$$
(1.1)

In terms of notation, we denote here (and we will mostly retain this notation throughout the book) by $\nu(x)$ the unit normal vector at $x \in \partial\Omega$ pointing outward from Ω and by $d\mathcal{H}_x^{n-1}$ the surface element[7] on $\partial\Omega$. The minus sign appearing on the right-hand side of (1.1) is due to the fact that the normal ν points toward the exterior, while we compute there the flux coming into the region Ω.

By differentiating under the integral sign[8] in (1.1), we find that

$$\fint_\Omega \partial_t u(x, t)\, dx = -\frac{1}{|\Omega|} \int_{\partial\Omega} B(x, t) \cdot \nu(x)\, d\mathcal{H}_x^{n-1} + \fint_\Omega f(x, t, u(x, t))\, dx.$$

---

[6]As usual, we use here (and repeatedly over this set of notes) the integral notation for averages, that is

$$\fint_\Omega u(x, t)\, dx := \frac{1}{|\Omega|} \int_\Omega u(x, t)\, dx,$$

where $|\Omega|$ the Lebesgue measure of Ω.

[7]In many textbooks, the surface element is denoted by ds, or dS, or $d\Sigma$. The notation $d\mathcal{H}_x^{n-1}$ comes from the fact that we are denoting by $\mathcal{H}^k$ the k-dimensional Hausdorff measure, and for smooth $(n-1)$-dimensional surfaces, the surface measure is itself precisely equal to $\mathcal{H}^{n-1}$. See e.g. [EG15, Chapter 2] for a thorough presentation of the Hausdorff measure.

The notation $d\mathcal{H}_x^{n-1}$ used here is perhaps a bit heavier than ds, or dS, or $d\Sigma$, or similar ones, but it has the benefit of being clearer and more explicit. Also, it allows us to consider surface integrals along less regular objects without having to introduce a new notation on a case-by-case basis.

[8]In this chapter about motivations, the arguments are developed at a formal level, we feel free to exchange derivatives and integrals, we do not keep track of lower order terms in the expansions, we do not discuss convergence issues and the existence of limits, etc. This is a rather customary approach when one deals with providing convincing, but not necessarily circumstantial, motivations for a problem with the objective of developing some intuition about it. We promise to try to be more rigorous from the following chapter on.

Hence, by the divergence[9] theorem,

$$\fint_\Omega \partial_t u(x,t)\,dx = -\fint_\Omega \operatorname{div} B(x,t)\,dx + \fint_\Omega f(x,t,u(x,t))\,dx. \quad (1.2)$$

From this, Fourier took a further step by realizing that to make this identity manageable, one needs to take a constitutive law about the flux vector B, relating it to the temperature. Fourier's new *ansatz* was to suppose that the flux of heat between two adjacent (infinitesimal) regions is proportional to the (infinitesimal) difference of their temperatures, namely

$$B(x,t) = -\kappa(x,t)\,\nabla u(x,t). \quad (1.3)$$

The proportional coefficient κ is sometimes called[10] the "heat conduction coefficient". We take κ to be positive: with respect to this, we stress that the minus sign in (1.3) is motivated by the fact that heat flows from hotter regions to colder ones (hence in the opposite direction of the growth of the temperature function u). The relation in (1.3) is also sometimes[11] called "Fourier's law".

---

[9]Typically, we reserve the notation "∇" for the vectors of the derivatives with respect to the space variables and "div" for the corresponding divergence. No confusion should arise with respect to derivatives with respect to the time variable, which is usually denoted here by "∂_t".

[10]See Chapter 38 for its counterpart in electrostatics.

[11]Actually, the same presentation here could have been used to describe the transport of mass through diffusive means. In this setting, one argues that the flux goes from regions of high concentration to regions of low concentration. The *ansatz* corresponding to (1.3) would be that the magnitude of the flux is proportional to the concentration gradient. In the context of mass diffusion, κ is sometimes called "diffusion coefficient", and (1.3) is sometimes referred to as "Fick's law", after the German physician and physiologist Adolf Eugen Fick.

Alternative choices for the constitutive relation are possible. For instance, one can replace (1.3) with a more general equation in which the flux depends possibly in a nonlinear way on the gradient of temperature (in the case of Fourier's model, or transported mass in the case of Fick's model); that is, one could suppose that $B = -\Phi(\nabla u)$, for some function $\Phi : \mathbb{R}^n \to \mathbb{R}^n$. A typical example is the case in which $\Phi(\nabla u) = |\nabla u|^{p-2}\nabla u$, that is $B = -\kappa|\nabla u|^{p-2}\nabla u$, meaning the flux is proportional to a power of the gradient of u (more precisely, the vector B has the same direction of the gradient of u, and its magnitude is proportional to a power

By substituting (1.3) into (1.2), we find that

$$\fint_\Omega \partial_t u(x,t)\,dx = \fint_\Omega \mathrm{div}(\kappa(x,t)\,\nabla u(x,t))\,d\mathcal{H}_x^{n-1} + \fint_\Omega f(x,t,u(x,t))\,dx.$$
$$(1.4)$$

Since this identity holds true for all regions Ω, we find a pointwise counterpart of (1.4) by writing

$$\partial_t u(x,t) = \mathrm{div}(\kappa(x,t)\,\nabla u(x,t)) + f(x,t,u(x,t)). \qquad (1.5)$$

The case of homogeneous media in which κ is constant (say, equal to 1 up to a renormalization of units of measure) is of particular interest: in this case, (1.5) boils down[12] to

$$\partial_t u(x,t) = \Delta u(x,t) + f(x,t,u(x,t)). \qquad (1.6)$$

---

of the magnitude of the gradient of u): this setting would lead to the so-called "p-Laplace equation".

Another setting of interest is the one in which the flux is related to a pressure drop (a framework which is usually related to the so-called "Darcy's law"), say $B = -\kappa\nabla P$, where P has some physical meaning of pressure. In this sense, the case $P = u$ reduces to (1.3), but other situations are of interest. For instance, one could assume that P and u are related by a state equation of the form $P(x) = p(u(x))$. A natural possibility is to take the function p to be a power of u, e.g. $p = u^m$. This choice would lead to the so-called porous medium equation.

These notes are of an elementary nature, and hence we will not address the cases of the p-Laplace equation, the porous medium equation, or other types of "anomalous diffusions", such as the type of diffusion that takes into account mass transfer from remote regions due to long-range interactions. The reader interested in these more advanced topics may look to e.g. [ZBLM85, DiB93, HKM06, Váz07, BV16] and the references therein.

For the readers interested in using Darcy's law to have a proper cup of coffee, see e.g. [Kin08] and the references therein.

[12]We exploit the standard notation for the Laplacian, given by

$$\Delta u := \mathrm{div}(\nabla u) = \sum_{j=1}^{n} \frac{\partial^2 u}{\partial x_j}.$$

In some textbook, the use of "Δ" is replaced by "∇^2" or by "$|\nabla|^2$". Of course all notations are good. Personally, we have a preference for the Δ notation since we find it simpler to read. Also, in a sense, it highlights the fact that the Laplace operator has a "dignity" which is "independent from the one of the gradient". This "philosophical" point will perhaps be clarified by the geometric interpretation of the Laplacian, which we will discuss in Theorem 1.1.1 of the companion book [DV23].

Figure 1.4.   Portrait of Pierre-Simon Laplace by Johann Ernst Heinsius (image from Wikipedia, licensed under the Creative Commons Attribution-Share Alike 4.0 International license).

The name of the Laplace operator comes from Pierre-Simon, marquis de Laplace, who introduced it while studying celestial mechanics, in connection with the gravitational potential (an approach that will be exploited here when dealing with the fundamental solution in Section 2.7 of the companion book [DV23]. See Figure 1.4 for a portrait of Laplace (by Johann Ernst Heinsius) when he was 35 years old, with an easel that quite resembles a blackboard – a working instrument and sign of distinction of every passionate mathematician.

Let us remark that a slightly different approach to the derivation of the heat equation is possible by considering "heat" instead of "temperature" as the main building block of the equation. The disadvantage of having heat, rather than temperature, in a pivotal role is that it is perhaps a more vague and less intuitive

Equation (1.6) (or variations of it) is typically[13] called "the heat equation".

---

concept than temperature (simply because temperature seems like a notion we are so familiar with, e.g. in view of weather forecasts). However, the notion of heat relates more directly to thermal energy and constitutes an extensive (i.e. additive) property. In this perspective, the heat equation reflects an energy budget in which the heat change in time is equal to the heat produced (source) plus the heat entering through the boundary (flux). Namely, if one is willing to consider heat, instead of temperature, as the building block for the heat equation, the balance equation in (1.1) reads at the level of variation of heat instead of averaged temperature (and note that the constitutive relation in Fourier's law (1.3) is already a relation between the notions of heat and temperature since it assumes the heat flux to be proportional to the gradient of temperature).

The final heat equation in (1.6) would remain essentially unchanged since the change in heat content Q directly relates to the change in temperature via the relation $\partial_t Q = c\rho\partial_t u$, where ρ is mass density and c is the specific heat capacity (a constant that depends on the internal properties of the material, which can be tabulated through experiments).

[13] Often, equations as in (1.6) are referred to as "parabolic". The name comes from the following classification method. A general linear differential operator of second order in the variables $(X_1, \ldots, X_N)$ has the form

$$\sum_{i,j=1}^{N} a_{i,j} \frac{\partial^2}{\partial X_i \partial X_j}.$$

The presence of coefficients a_{ij} (also possibly depending on time, space, or any other structural quantities) is useful in modelization, e.g. to capture the features of inhomogeneous media and complex structures composed of diverse or dissimilar components, see Figures 1.5 and 1.6 for a variety of possible examples (see also [Mar07, Chapter 7] for more beautiful pictures of heterogeneous environments and excellent mathematical explanations).

One classifies the operator, and the corresponding partial differential equation, depending on the sign of the eigenvalues of the coefficient matrix $a_{i,j}$. More specifically, when all the eigenvalues of $a_{i,j}$ have the same sign (either all strictly positive or all strictly negative), the equation is named "elliptic". When one eigenvalue is zero and all the other eigenvalues have the same sign (either all strictly positive or all strictly negative), the equation is referred to as "parabolic". When all the eigenvalues have the same sign except one, which has the opposite sign (i.e. if either there is only one negative eigenvalue and all the remaining are positive, or there is only one positive eigenvalue and all the remaining are negative), the equation is named "hyperbolic". This classification is not exhaustive: the remaining cases not classified here are typically quite hard to study, require specific techniques, and no general theory is available for those.

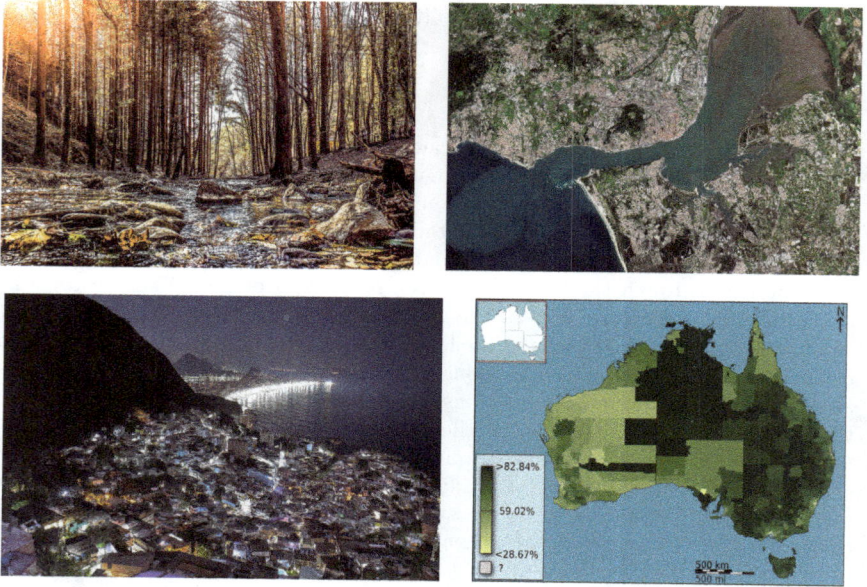

Figure 1.5.   Examples of heterogeneous media which naturally occur in the study of partial differential equations in various contexts. Casentinesi Forests (photo by Francesco-1978, image from Wikipedia, licensed under the Creative Commons Attribution-Share Alike 4.0 International license). Lisbon metropolitan area and the Tagus River (image from Wikipedia by the European Space Agency, contains modified Copernicus Sentinel data 2020). The lights of Vidigal favela in Rio de Janeiro (photo by Wilfredor, image from Wikipedia, licensed under the Creative Commons 1.0 Universal Public Domain Dedication). People whose parents were both born in Australia, as of the 2011 census (diagram by Toby Hudson based on data from the Australian Bureau of Statistics, image from Wikipedia, licensed under the Creative Commons Attribution-Share Alike 3.0 Australia license).

In the case of the heat equation in (1.6), one has $N = n + 1$, $X = (x, t)$ and

$$a_{ij} = \begin{cases} 1 & \text{if } i = j \in \{1, \ldots, n\}, \\ 0 & \text{otherwise.} \end{cases}$$

The corresponding eigenvalues are therefore 0, with multiplicity 1, and 1, with multiplicity n; hence, the equation in (1.6) is parabolic.

Figure 1.6. Examples of heterogeneous media which naturally occur in the study of partial differential equations in various contexts. A Pillar coral at the Florida Keys National Marine Sanctuary (photo by Commander William Harrigan, NOAA Corps, Florida Keys National Marine Sanctuary, National Oceanic and Atmospheric Administration, image from Wikipedia, Public Domain). Structure of a composite material (Public Domain image by PerOX from Wikipedia). Ibex in an alpine habitat (photo by Ibex73, image from Wikipedia, licensed under the Creative Commons Attribution-Share Alike 4.0 International license). Rice terraces in Yuanyang County (photo by Jialiang Gao, www.peace-on-earth.org, Permission GFDL/CC-by-sa-2.5, licensed under the Creative Commons Attribution-Share Alike 3.0 Unported license). Konso Sorghum Market, Ethiopia (photo by Rod Waddington, image from Wikipedia, licensed under the Creative Commons Attribution-Share Alike 2.0 Generic license).

Suppose now that f in (1.6) is independent of time, i.e. $f = f(x, u(x, t))$. In this framework, of particular interest are certainly stationary solutions of (1.6), i.e. solutions $u = u(x)$ which are also independent of time: these solutions are actually equilibria of (1.6) since they provide solutions to (1.6) which remain the same at every instant of time. Of course, when f and u are independent of time, equation (1.6) boils down to[14]

$$\Delta u(x) + f(x, u(x)) = 0. \tag{1.7}$$

In particular, when f is independent of u, equation (1.7) reduces to

$$\Delta u(x) + f(x) = 0, \tag{1.8}$$

which is sometimes called "Poisson's equation".

The case in which f in (1.8) vanishes identically produces

$$\Delta u(x) = 0, \tag{1.9}$$

which is called the "Laplace equation".

Solutions of (1.9) are named "harmonic functions". The study of harmonic functions is essential, one way or another, in virtually every aspect of mathematics (including mathematical analysis, mathematical physics, complex analysis, geometry, probability, finance, statistics, you name it). Also, the theory of harmonic functions is a clear example of adamantine beauty and elegance in which human creativity has reached one of its highest peaks ever.

In the forthcoming pages, we will do our best to present some bits of this theory, but we definitely invite the reader to look to as many other sources as possible to collect the broadest possible amount of information and develop their own point of view on this topic of paramount importance.

---

[14]With respect to the classification presented in footnote 13, equations (1.7), (1.8) and (1.9) are of elliptic type (actually, they provide us with the "paradigmatic" form of elliptic equations). To check their ellipticity, one exploits the setting in footnote 13 with $N = n$, $X = x$ and

$$a_{ij} = \delta_{ij} := \begin{cases} 1 & \text{if } i = j, \\ 0 & \text{if } i \neq j. \end{cases}$$

In this scenario, all the eigenvalues of the coefficient matrix are equal to 1, hence strictly positive.

Further insight on the notion of ellipticity will be given in Sections 2.16 and 4.4 of the companion book [DV23].

The above definition of δ_{ij} will be also used throughout these notes (sometimes this definition of δ_{ij} is referred to with the name "Kronecker notation").

Chapter 2

Population Dynamics, Chemotaxis and Random Walks

Another fascinating situation in which partial differential equations surface quite often occurs in the description of biological populations. This is not only a highly captivating field of study but also one that is truly cross-disciplinary in spirit, bringing together fundamental questions from (at least) mathematics, physics, biology, ethology and social sciences. For the purpose of these pages, we limit ourselves to a simple description of a biological species exhibiting at a point $x \in \mathbb{R}^n$ and at a time $t \in \mathbb{R}$ a population density of the form $u(x,t)$. We assume that a fraction, $\rho \in [0,1]$, of the population has the tendency of moving randomly, while the rest of the population (corresponding to a fraction, $\mu := 1 - \rho$, of the totality of individuals) is driven toward regions of higher concentration of a given attractant (e.g. a chemical signal) distributed according to a function, $w(x,t)$ (the movement of an organism in response to a chemical stimulus, specifically depending on the increasing or decreasing concentration of a particular substance, is called in jargon "chemotaxis"). Additionally, we suppose that the population is subject[1] to a drift (e.g. due to the wind, a stream, or a tide) in a given direction $b(x,t)$. The simpler cases in which there is no chemotactic effect (corresponding to $\mu := 0$), no random motion (corresponding to $\rho := 0$), or no drift (corresponding to $b := 0$) are interesting special situations of the complex phenomena

---

[1] According to many mathematicians, life is a random process with a drift.

for which we now present a mathematical model (the random component of this discussion will also be further generalized in Section 4.4
of the companion book [DV23] to more elaborated environments).
See Figure 2.1 for an illustration of how our ancestors spread across
the world.

To outline, we consider a small spatial scale h and a small time
scale τ. We choose in what follows a suitable relation between the
space and time scales to provide coherent asymptotics in the formal
limit. In this setting, at every unit of time, the variation of the density
of the population is influenced by the previous factors in a rather
explicit way. For the moment, let us focus on the simple case in which
there is no chemotaxis and no drift. In this situation, $\rho := 1$, $\mu := 0$, b
vanishes identically and the only drive for the population comes from
a random walk. To get to the bottom of this case, we consider, at a
given point x and at time $t+\tau$, the new random population $\rho u(x, t+\tau)$
(which actually coincides with the whole population $u(x, t + \tau)$, but
let us keep the fraction ρ for future use, even if for the moment $\rho = 1$).
Then, we have that $\rho u(x, t+\tau)$ is produced by the random population
at the previous time t, which is located at some point $x + h\omega$, for a
given direction $\omega \in \partial B_1$, times the probability that these individuals
move from $x + he$ to x in a unit of time. If all directions are equally
probable, this states that the new random population $\rho u(x, t + \tau)$ is

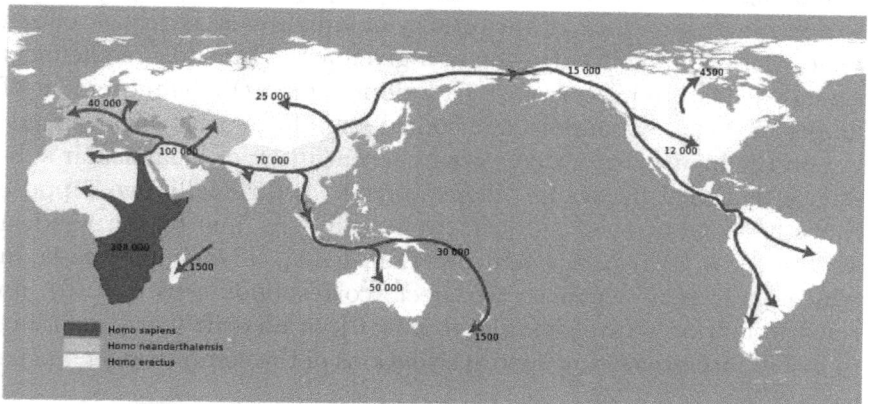

Figure 2.1. Humanity spreading across the world (Public Domain image by
NordNordWest from Wikipedia).

produced by a term of the form $\rho \fint_{\partial B_1} u(x + h\omega, t) \, d\mathcal{H}_\omega^{n-1}$, that is,

$$\rho u(x, t + \tau) = \rho \fint_{\partial B_1} u(x + h\omega, t) \, d\mathcal{H}_\omega^{n-1}. \qquad (2.1)$$

Subtracting $\rho u(x, t)$ from both sides, dividing by τ and choosing

$$h := \sqrt{\tau}, \qquad (2.2)$$

we find that

$$\rho \frac{u(x, t + \tau) - u(x, t)}{\tau} = \frac{1}{\tau} \left[\rho \fint_{\partial B_1} u(x + h\omega, t) \, d\mathcal{H}_\omega^{n-1} - \rho u(x, t) \right]$$

$$= \frac{\rho}{h^2} \fint_{\partial B_1} (u(x + h\omega, t) - u(x, t)) \, d\mathcal{H}_\omega^{n-1}.$$

$$(2.3)$$

Now, we aim to show that, for small h,

$$\fint_{\partial B_1} (u(x + h\omega, t) - u(x, t)) \, d\mathcal{H}_\omega^{n-1} = \frac{ch^2}{2} \Delta u(x, t) + o(h^2), \qquad (2.4)$$

for some $c > 0$ depending on the dimension n (see formula (2.5) as follows). For this, we consider the formal Taylor expansion

$$u(x + h\omega, t) = u(x, t) + h\nabla u(x, t) \cdot \omega + \frac{h^2}{2} D^2 u(x, t)\omega \cdot \omega + o(h^2).$$

We observe, by symmetry, that

$$\fint_{\partial B_1} \nabla u(x, t) \cdot \omega \, d\mathcal{H}_\omega^{n-1} = 0$$

because[2] any (positive or negative) contribution to such a surface integral coming from a point $\omega \in \partial B_1$ is canceled precisely by the

---

[2]Spotting these types of cancellations will often be an essential ingredient of our arguments. After all, one way to obtain significant information about a complex problem is to understand which quantities do not play much of a role since they "average out". At this stage, we discuss these simplifications at an intuitive level. In Section 1.1 of the companion book [DV23], we will give more rigorous arguments to justify them, but for the moment, the main goal is to try to visualize the possibility of these cancellations and, especially, to appreciate their importance.

(negative or positive) contribution coming from $-\omega$. From these observations, we arrive at

$$\fint_{\partial B_1} (u(x+h\omega,t) - u(x,t))\, d\mathcal{H}_\omega^{n-1}$$

$$= \fint_{\partial B_1} \left(h\nabla u(x,t)\cdot\omega + \frac{h^2}{2} D^2 u(x,t)\omega\cdot\omega \right) d\mathcal{H}_\omega^{n-1} + o(h^2)$$

$$= \frac{h^2}{2} \fint_{\partial B_1} D^2 u(x,t)\omega\cdot\omega\, d\mathcal{H}_\omega^{n-1} + o(h^2)$$

$$= \frac{h^2}{2} \sum_{i,j=1}^{n} \fint_{\partial B_1} \partial_{ij} u(x,t)\omega_i\omega_j\, d\mathcal{H}_\omega^{n-1} + o(h^2).$$

We now identify an additional simplification by observing that if $i \neq j$, then

$$\fint_{\partial B_1} \partial_{ij} u(x,t)\omega_i\omega_j\, d\mathcal{H}_\omega^{n-1} = 0$$

because whatever contribution comes from ω_i is canceled precisely by that coming from $-\omega_i$. (Note how helpful it was for all these cancellations that we are integrating over such a symmetric domain as ∂B_1, which remains invariant under all these reflections.)

Therefore,

$$\fint_{\partial B_1} (u(x+h\omega,t) - u(x,t))\, d\mathcal{H}_\omega^{n-1}$$

$$= \frac{h^2}{2} \sum_{i=1}^{n} \fint_{\partial B_1} \partial_{ii} u(x,t)\omega_i^2\, d\mathcal{H}_\omega^{n-1} + o(h^2).$$

Now, once again, we consider the symmetries of the problem under consideration. For this, we note that, for every $i, j \in \{1,\dots,n\}$,

$$\fint_{\partial B_1} \omega_i^2\, d\mathcal{H}_\omega^{n-1} = \fint_{\partial B_1} \omega_j^2\, d\mathcal{H}_\omega^{n-1}$$

because[3] the role played by the ith coordinate in the sphere is precisely the same as the one played by the jth coordinate. Therefore,

---

[3] Once again, we aim here at developing arguments in the most intuitive way possible. This idea with be more rigorously retaken in Section 1.1 of the companion book [DV23].

we can define

$$c := \oint_{\partial B_1} \omega_i^2 \, d\mathcal{H}_\omega^{n-1}, \tag{2.5}$$

and we stress that this quantity does not depend on i.

Consequently,

$$\oint_{\partial B_1} (u(x + h\omega, t) - u(x, t)) \, d\mathcal{H}_\omega^{n-1} = \frac{ch^2}{2} \sum_{i=1}^{n} \partial_{ii} u(x, t) + o(h^2),$$

from which we obtain (2.4), as desired.

Hence, we insert (2.4) into (2.3) to find that

$$\rho \frac{u(x, t + \tau) - u(x, t)}{\tau} = \frac{c\rho}{2} \Delta u(x, t) + o(1). \tag{2.6}$$

By sending $\tau \searrow 0$ (and thus $h \searrow 0$), we thereby find that

$$\rho \partial_t u(x, t) = \frac{c\rho}{2} \Delta u(x, t), \tag{2.7}$$

which coincides, up to a constant, with the heat equation[4] presented in (1.6) (here, with no source term).

This computation is not only instructive from a technical point of view, but it also unveils one of the mainsprings of the mathematical theory of diffusion, revealing the strong conceptual connection of a substance (or a population) moving randomly according to Brownian motion and the way in which heat dissipates; after all, the spread of temperature throughout a region subject to diffusion is, in a sense, simply[5] the tendency of random movements to distribute mass on

---

[4]See e.g. [Law10] for more information about the strong connections between the random walk and the heat equation.

[5]The link between the random motion of molecules and the macroscopic phenomenon of diffusion is indeed a deep feature. While Brownian motion was discovered in 1827 by botanist Robert Brown by looking at dust grains floating in water, Albert Einstein, in his *annus mirabilis* 1905, gave a quantitative model for the motion of floating particles as being moved by individual water molecules [Ein05]. Not only did this article lay the foundation for statistical physics analysis of Brownian motion, but it also provided a way to determine the mass and dimensions of the atoms involved in the process, thus transforming atomic theory from a

Figure 2.2.   Hard-working scientists (Public Domain image from Wikipedia).

---

controversial set of conjectures into an established fact of science. We will come back to the theory proposed by Einstein in Chapter 13.

See Figure 2.2 depicting Albert Einstein together with Niels Bohr.

By the way, Niels Bohr's brother was Harald Bohr, a mathematician who almost pioneered periodic functions, authored, together with Edmund Landau, an elegant result related to the Riemann Conjecture (see [Edw01, Section 9.6]), and was also a soccer player. The Denmark national soccer team in which he was playing took part in the 1908 Summer Olympics, where football was an official event for the first time. Harald Bohr scored two goals in the first match and played in the semifinal (Denmark 17 – France 1, which remains the Olympic record for the most goals scored by one team). After that, the Denmark team lost the final and won the silver medal. In all the subsequent editions of the Olympic Games, the Denmark soccer team has won two more silver medals (in 1912 and 1960) and a bronze (in 1948); hence, Harald Bohr's team also holds the record (*ex aequo*) for the best placement for a Denmark soccer team in the Olympics.

average from regions of high density toward regions of low density. In both cases, the process has a strong tendency to "balance out" differences: as time progresses, hot spots lose their temperature in favor of cold spots, which are heated up by their neighbors, while under a random walk, the regions with low density are occupied by the population coming from the highly populated regions.

This "democratic" tendency of averaging out differences is typical of the Laplace operator, and it will be the underlying feature of all the "regularity theories" for elliptic equations that we will present in the forthcoming pages.

Having well understood the case in which the biological species is dependent solely on a random motion, we now return the interesting case in which a chemotactic agent also comes into play, considering a fraction μ of the population (for the moment, we still suppose that there is no additional drift, hence b vanishes identically). In this situation, the fraction ρ of the population performing the random walk would still be described by (2.6) (which was the reason to include ρ in the previous computation, even if ρ was only equal to 1 earlier), and we focus now on the fraction μ of the population following the attractant w. To appreciate the effect of the chemotactic factor, we suppose that such a fraction μ of the population with density u does not move completely randomly but, at each time step, picks a direction of motion with a probability λ that is proportional to the values of the attractant having a density of w.

More explicitly, given $\omega \in \partial B_1$, we suppose for the chemotactic population that a jump from the point x to the point $x + h\omega$ occurs

---

See Figure 2.3 for a photo of the Danish soccer team in the 1908 Olympic Games. Harald Bohr is the second player from the left in the top row.

Niels Bohr was also a passionate soccer player. He played as goalkeeper in the Copenhagen-based team Akademisk Boldklub (at the time, one of the best clubs in Denmark; actually, the two brothers played several matches together for this club). One can wonder, however, why Niels never made it to the national team. Well, according to *The Guardian* (https://www.theguardian.com/football/2005/jul/27/theknowledge.panathinaikos), in a match against a German team, one of the midfielders of the opposite team launched a very long shot, and Niels, leaning against the post, did not react at all, missing an easy save and letting the German team score. After the game, Niels admitted that on that occasion, he had been distracted by a mathematical problem he was thinking about. Strangely enough, Niels did not play for Akademisk Boldklub after that season.

Figure 2.3. Danish soccer team at the 1908 Olympic Games (Public Domain image from Wikipedia).

with a probability of

$$\lambda(x, \omega, t) := \frac{1}{\mathcal{H}^{n-1}(\partial B_1)} + w(x + h\omega, t) - w(x - h\omega, t). \qquad (2.8)$$

In this setting, we are assuming that the oscillations of w are sufficiently small[6] to make the above quantity positive. In this sense, the probability λ differs from that of the classical random walk (corresponding to w being constant) since it increases in favor of the direction ω in which the concentration of the chemical attractant is

---

[6]More precisely, one could rewrite (2.8) as

$$\lambda(x, \omega, t) := \frac{1}{\mathcal{H}^{n-1}(\partial B_1)} + \varepsilon_0(w(x + h\omega, t) - w(x - h\omega, t)),$$

where $\varepsilon_0 > 0$ is a parameter that takes into account the "sensibility" of the chemotactic population to the chemical attractant. Just to be consistent with the probability scenario, it is convenient to think that $\lambda \geqslant 0$ in view of the smallness of ε_0 with respect to the oscillations of the attractor's density w.

higher. We stress that the above setting provides a normalized probability, since the total probability of jumping to the sphere of radius h
around the given point x is 1 because

$$\int_{\partial B_1} \lambda(x, \omega, t)\, d\mathcal{H}_\omega^{n-1} = 1 + \int_{\partial B_1} (w(x+h\omega, t) - w(x-h\omega, t))\, d\mathcal{H}_\omega^{n-1} = 1,$$

due to odd symmetry cancellations.

We also point out that, in the above setting, the probability of a
jump from a point a to a point b with $|a - b| = h$ is given by

$$\lambda\left(a, \frac{b-a}{h}, t\right) = \frac{1}{\mathcal{H}^{n-1}(\partial B_1)} + w\left(a + h\frac{b-a}{h}, t\right)$$

$$- w\left(a - h\frac{b-a}{h}, t\right)$$

$$= \frac{1}{\mathcal{H}^{n-1}(\partial B_1)} + w(b, t) - w(2a - b, t).$$

In consequence, given $\omega \in \partial B_1$ and taking $a := x + h\omega$ and $b := x$,
we find that the jump from $x + h\omega$ to x occurs with a probability of

$$\lambda(x + h\omega, -\omega, t) = \frac{1}{\mathcal{H}^{n-1}(\partial B_1)} + w(x, t) - w(x + 2h\omega, t).$$

This being so, we can detect the chemotactic counterpart of (2.1).
Indeed, since the chemotactic population $\mu u(x, t + \tau)$ is produced by
a term of the form $\mu \int_{\partial B_1} \lambda(x+h\omega, -\omega, t) u(x+h\omega, t)\, d\mathcal{H}_\omega^{n-1}$, we have
that

$$\mu u(x, t + \tau) = \mu \int_{\partial B_1} \lambda(x + h\omega, -\omega, t) u(x + h\omega, t)\, d\mathcal{H}_\omega^{n-1}$$

$$= \mu \int_{\partial B_1} \left(\frac{1}{\mathcal{H}^{n-1}(\partial B_1)} + w(x, t) - w(x + 2h\omega, t)\right)$$

$$\times u(x + h\omega, t)\, d\mathcal{H}_\omega^{n-1}$$

$$= \mu \fint_{\partial B_1} u(x + h\omega, t)\, d\mathcal{H}_\omega^{n-1}$$

$$+ \mu \int_{\partial B_1} (w(x, t) - w(x + 2h\omega, t)) u(x + h\omega, t)\, d\mathcal{H}_\omega^{n-1}.$$

Thus, we can subtract from both sides $\mu u(x,t)$, divide by $\tau = h^2$ and recall (2.4) and (2.5) to find that

$$\mu \frac{u(x,t+\tau) - u(x,t)}{\tau}$$

$$= \frac{1}{h^2}\left[\mu \fint_{\partial B_1} (u(x+h\omega,t) - u(x,t))\, d\mathcal{H}_\omega^{n-1}\right.$$

$$\left. + \mu \int_{\partial B_1} (w(x,t) - w(x+2h\omega,t))u(x+h\omega,t)\, d\mathcal{H}_\omega^{n-1}\right]$$

$$= \frac{c\mu}{2}\Delta u(x,t) + \frac{\mu}{h^2}\int_{\partial B_1} (w(x,t) - w(x+2h\omega,t))$$

$$\times u(x+h\omega,t)\, d\mathcal{H}_\omega^{n-1} + o(1)$$

$$= \frac{c\mu}{2}\Delta u(x,t) - \frac{\mu}{h^2}\int_{\partial B_1} (2h\nabla w(x,t)\cdot\omega + 2h^2 D^2 w(x,t)\omega\cdot\omega)$$

$$\times (u(x,t) + h\nabla u(x,t)\cdot\omega)\, d\mathcal{H}_\omega^{n-1} + o(1)$$

$$= \frac{c\mu}{2}\Delta u(x,t) - \frac{\mu}{h^2}\int_{\partial B_1} (2h^2\nabla w(x,t)\cdot\omega\,\nabla u(x,t)\cdot\omega$$

$$+ 2h^2 D^2 w(x,t)\omega\cdot\omega\, u(x,t))\, d\mathcal{H}_\omega^{n-1} + o(1)$$

$$= \frac{c\mu}{2}\Delta u(x,t) - 2\mu\int_{\partial B_1} (\nabla w(x,t)\cdot\omega\,\nabla u(x,t)\cdot\omega$$

$$+ D^2 w(x,t)\omega\cdot\omega\, u(x,t))\, d\mathcal{H}_\omega^{n-1} + o(1)$$

$$= \frac{c\mu}{2}\Delta u(x,t) - 2\mu\sum_{i=1}^n \int_{\partial B_1} (\partial_i w(x,t)\partial_i u(x,t)\omega_i^2$$

$$+ \partial_{ii} w(x,t)u(x,t)\omega_i^2)\, d\mathcal{H}_\omega^{n-1} + o(1)$$

$$= \frac{c\mu}{2}\Delta u(x,t) - 2\mu\sum_{i=1}^n \int_{\partial B_1} \partial_i(\partial_i w(x,t)u(x,t))\omega_i^2\, d\mathcal{H}_\omega^{n-1} + o(1)$$

$$= \frac{c\mu}{2}\Delta u(x,t) - \widetilde{c}\mu\sum_{i=1}^n \partial_i(\partial_i w(x,t)u(x,t)) + o(1)$$

$$= \frac{c\mu}{2}\Delta u(x,t) - \widetilde{c}\mu\, \mathrm{div}(u(x,t)\nabla w(x,t)) + o(1)$$

for a suitable $\widetilde{c} > 0$.

Note that in the above we have repeatedly taken advantage of odd symmetry cancellations. Gathering this and (2.6), we thereby find that

$$
\frac{u(x, t+\tau) - u(x, t)}{\tau}
$$

$$
= \rho \frac{u(x, t+\tau) - u(x, t)}{\tau} + \mu \frac{u(x, t+\tau) - u(x, t)}{\tau}
$$

$$
= \frac{c\rho}{2} \Delta u(x, t) + \frac{c\mu}{2} \Delta u(x, t) - \tilde{c}\mu \operatorname{div}(u(x, t)\nabla w(x, t)) + o(1)
$$

$$
= \frac{c}{2} \Delta u(x, t) - \tilde{c}\mu \operatorname{div}(u(x, t)\nabla w(x, t)) + o(1). \tag{2.9}
$$

Sending $\tau \searrow 0$ (and thus $h \searrow 0$), we obtain

$$
\partial_t u(x, t) = \frac{c}{2} \Delta u(x, t) - \tilde{c}\mu \operatorname{div}(u(x, t)\nabla w(x, t)). \tag{2.10}
$$

For constant distributions w of the chemical attractant, equation (2.10) can be seen as a special case (up to normalizing constants) of the heat equation in (1.6).

Stationary solutions of (2.10) satisfy

$$
\frac{c}{2} \Delta u(x, t) - \tilde{c}\mu \operatorname{div}(u(x, t)\nabla w(x, t)) = 0. \tag{2.11}
$$

This is the equation solved by the equilibria corresponding to a biological population in the presence of a chemotactic factor.

Let us now introduce one last complication in the model by accounting for a possible drift (as mentioned on p. 13). In this situation, we may suppose that, in time units, the population is also moved by the velocity vector b by a space displacement $b\tau$. Regarding this, the population density $u(x, t+2\tau)$ has in fact moved to the location $x + b\tau$: correspondingly, to take this drift into account, one can observe that $u(x, t+2\tau) = u(x - b(x, t)\tau, t+\tau)$, and accordingly, recalling (2.9) and using that

$$
u(x - b(x, t)\tau, t+\tau) = u(x, t) - \tau b(x, t) \cdot \nabla u(x, t) + \tau \partial_t u(x, t) + o(\tau),
$$

we have that

$$\frac{u(x,t+2\tau)-u(x,t)}{\tau}$$

$$= \frac{u(x,t+2\tau)-u(x,t+\tau)}{\tau} + \frac{u(x,t+\tau)-u(x,t)}{\tau}$$

$$= \frac{u(x-b(x,t)\tau,t+\tau)-u(x,t+\tau)}{\tau}$$

$$+ \frac{c}{2}\Delta u(x,t) - \tilde{c}\mu \operatorname{div}(u(x,t)\nabla w(x,t)) + o(1)$$

$$= \frac{\begin{array}{c}u(x,t)-\tau b(x,t)\cdot\nabla u(x,t)+\tau\partial_t u(x,t)\\ -u(x,t)-\tau\partial_t u(x,t)+o(\tau)\end{array}}{\tau}$$

$$+ \frac{c}{2}\Delta u(x,t) - \tilde{c}\mu \operatorname{div}(u(x,t)\nabla w(x,t)) + o(1)$$

$$= -b(x,t)\cdot\nabla u(x,t) + \frac{c}{2}\Delta u(x,t) - \tilde{c}\mu \operatorname{div}(u(x,t)\nabla w(x,t)) + o(1)$$

Thus, we can pass to the limit and obtain

$$\partial_t u(x,t) = \frac{c}{2}\Delta u(x,t) - \tilde{c}\mu \operatorname{div}(u(x,t)\nabla w(x,t)) - b(x,t)\cdot\nabla u(x,t).$$
$$(2.12)$$

Stationary solutions of (2.12) give rise to the equation

$$\Delta u(x,t) - \operatorname{div}(u(x,t)\nabla w(x,t)) - b(x,t)\cdot\nabla u(x,t) = 0, \qquad (2.13)$$

up to normalizing constants, which we have omitted for the sake of simplicity. In this spirit, (2.13) models[7] the equilibria reached by a

---

[7]The equation in (2.13), together with the diffusion of the attractant, reveals the complexity of the model under consideration, which, depending on different parameter thresholds, exhibits a variety of different stable geometrical patterns, colony formation, blowup mechanisms, flocculation and aggregation phenomena.

All these traits play a fundamental role in life as we know it since these configurations can arise from equations such as the ones presented here and possess quite a precise counterpart in nature, when the corresponding behaviors occur, for example, in response to external stress or to a lack of resources (sometimes the chemical attractant is produced by the biological species itself, e.g. by amoebae in the case of food scarcity) or in response to predation (autoaggregation biological species possibly subject to chemotaxis and to an external drift.

Note that (2.13) reduces to (2.11) in the absence of drift.

We have written (2.13) in a form that emphasizes its "divergence structure": if instead one wants to highlight the presence of a second-order differential operator, it suffices to note that $\operatorname{div}(u\nabla w) = u\Delta w + \nabla w \cdot \nabla u$ for recasting (2.13) into

$$\Delta u(x,t) - (b(x,t) + \nabla w(x,t)) \cdot \nabla u(x,t) - u(x,t)\Delta w(x,t) = 0, \quad (2.14)$$

which is also a telling expression since it reveals that the effect of the chemotaxis is to alter the diffusion with an additional drift term ∇w that favors movements toward regions with higher concentrations of the attractant.

We refer to [KS70, Mur02, Mur03, HP09] and the references therein for further information about chemotaxis and its biological implications. See also e.g. [Isa65, Fri71, BV16, Section 2.2] and [BR19] for applications of the theory of random walks to game theory.

---

provides protection and defense against predators). These aggregation processes often produce slimy yet durable coatings called biofilms. Even blowup phenomena can be exploited by bacteria to enhance their probability of survival in hostile environments (e.g. the individual at the top of the tower formed by the blowup has a greater chance of being picked up, say, by the wind or by an external factor, and deposited possibly in a less hostile environment), so all in all the variety of patterns exhibited by the solutions of a relatively simple differential equation is, in some sense, the mathematical counterpart of the variety of ways in which "life finds a way" (as Ian Malcolm utters in *Jurassic Park*).

Chapter 3

Pattern Formation, or How the Leopard Gets Its Spots

We know well that the natural world offers a variety of amazing regular patterns, such as symmetries, tessellations, spirals and spots, see e.g. Figures 3.1–3.4. These patterns visible in nature are determinant factors in the processes of natural and sexual selection, providing organisms with structures recognizable by their conspecifics (thus favoring social interactions and reproduction) and often realizing optimal configurations for mobility, hunt or camouflage (see e.g. [HT96] and the references therein).

The biological process leading to the development of the specific shape of an organism is dubbed "morphogenesis", and a scientific investigation of the formation of patterns in nature is relatively recent. One of the pioneer scientists interested in the mathematical analysis of the growth, form and evolution of plants and animals was Sir D'Arcy Wentworth Thompson; his 1116-page book [Tho42] (793 pages in the first edition of 1917) combined classical natural philosophy, biology and mathematics to give insights on a number of biological shapes and analyzed the differences in the forms in nature in the light of mathematical transformations.

A modern mathematical treatment of morphogenesis was initiated by Alan Mathison Turing[1] in 1952, see [Tur52]. Turing's brilliant idea

---

[1]A curious coincidence is that Alan Turing's second name was Mathison. Which can be interpreted as Math-is-on. Not only the father of mathematical

Figure 3.1. A painting by Todd Schaffer in the "Tingatinga" style (image from Wikipedia, available under the Creative Commons Attribution-Share Alike 3.0 Unported license).

---

morphogenesis, computer science and artificial intelligence, Alan Turing was also an exceptional long-distance runner, capable of world-class marathon standards. See https://kottke.org/18/04/alan-turing-was-an-excellent-runner for a nice photo of Turing during one of his running performances.

During World War II, Turing worked as a codebreaker for British Intelligence, devising an electromechanical machine named "Bombe", to decipher the encrypted messages created by the German cipher device "Enigma". The headquarters of the British codebreakers was located in a mansion named Bletchley Park, which today hosts a statue of Turing made with slates (by Stephen Kettle). In this statue, Turing is depicted seated and looking at a German Enigma machine, see Figure 3.5.

Turing was 39 years old when he was charged with "gross indecency" on the basis of a homosexual relationship under Section 11 of the Criminal Law Amendment Act 1885. The case, Regina v. Turing and Murray, was brought to trial on March 31, 1952, and Turing pleaded guilty, insisting that he saw nothing wrong with his actions.

Figure 3.2. Leopard mating dance (image from Wikipedia, licensed under the Creative Commons Attribution 2.0 Generic license; for the complete series of shots, see Steve Jurvetson's website: https://www.flickr.com/photos/jurvetson/5913330010/).

is that simple physical laws were sufficient to justify the shaping of complex patterns, such as animal markings and the arrangement of leaves and florets in plants.

In a nutshell, Turing proposed that the process of morphogenesis is regulated by the interaction of chemical substances, called morphogens, which diffuse through a tissue: these substances could be hormones, genes or any other essence which can act and react in the presence of another one. In practice (see e.g. [GM73]), one of the

---

He was convicted and given a choice between imprisonment and chemical castration via synthetic estrogens. As a result of choosing the second option, Turing suffered severe mutations of his body, was barred from his work for Government Communication Headquarters and was denied entry into the United States. On June 8, 1954, Turing was found dead of cyanide poisoning with a half-eaten apple next to his bed. The inquest into his death recorded a verdict of suicide, but the apple was not tested for cyanide.

Turing's favorite fairy tale was Snow White and the Seven Dwarfs.

Figure 3.3. Examples of seven-spotted ladybugs ((a) photo by Dominik Stodulski; image from Wikipedia, licensed under the Creative Commons Attribution-Share Alike 3.0 Unported license; (b) photo by André Karwath; image from Wikipedia, licensed under the Creative Commons Attribution-Share Alike 2.5 Generic license).

Figure 3.4. (a) Two young female tigers in a playful mood (photo by Vedang Vadalkar; image from Wikipedia, licensed under the Creative Commons Attribution-Share Alike 4.0 International license). (b) Lithographic plate by John Gould, representing two examples of thylacines, the extinct (or maybe not?) carnivorous marsupials, a.k.a. Tasmanian tigers (Public Domain image from Wikipedia).

morphogens may act as an "activator", which is self-sustaining and introduces positive feedback, while the other may play the role of an "inhibitor", which tends to suppress the self-amplification of the activator. In this interplay, the pattern may be created by the different speeds of diffusion of the two substances: namely, the faster diffusion of the inhibitor can catch up with the activator's self-replication (that is, in rough terms, on the one side the activator's capacity of self-replication could be strong enough to produce local patches, but the predominant speed of the inhibitor could prevent these patches

Figure 3.5.   Statue of Alan Turing (photo by Jon Callas; image from Wikipedia, licensed under the Creative Commons Attribution 2.0 Generic license).

from growing incessantly). All in all, the whole process can thereby be considered a "diffusion-driven instability". See [Mur03, p. 76] for a very clear explanation of the roles of activators and inhibitors via an example with sweating grasshoppers.

The mathematical formulation of Turing's idea combines the notion of diffusion, as modeled for instance by the heat equation (2.7), and that of reaction, taking into account that each morphogen can chemically react with the others and the effect of this interaction can depend on the concentration of the diffusing substances in the tissue. The combination of reaction and diffusion in this type of system of partial differential equations justified the name of "reaction-diffusion equations".

For instance, one can consider the case of two morphogens with densities u and v and a system of reaction-diffusion equations of the form

$$\begin{cases} \partial_t u = \mu \Delta u + f(u,v), \\ \partial_t v = \nu \Delta v + g(u,v), \end{cases} \tag{3.1}$$

where μ and ν are positive coefficients which model the speed of diffusion of the chemical substances with densities u and v, respectively.

Now, the formation of patterns as an outcome of (3.1) is, in a sense, not completely intuitive: on the contrary, given the "democratic" tendency of the Laplace operator (as discussed on p. 24), one may imagine that the solution $(u(t), v(t))$, as $t \to +\infty$, will evolve spontaneously toward some constant values (u_0, v_0), which are only the common zeros of f and g (that is, such that $f(u_0, v_0) = g(u_0, v_0) = 0$). And this is indeed one of the possible destinations for the solutions of (3.1). However, there is also a more intriguing possibility: the constant state (u_0, v_0) may well exist (and it may also be "stable" for small perturbations when $\mu := 0$ and $\nu := 0$), but it may be triggered off by random disturbances. In this situation, when $\mu > 0$ and $\nu > 0$, the solutions may end up drifting away from the constant (u_0, v_0). Thus (unless for some reason the solution diverges), several interesting patterns may arise, such as oscillations between equilibria, stationary waves and moving fronts.

Though a full understanding of Turing's theory of morphogenesis goes well beyond the scope of this set of notes, following [Tur52], one can at least grasp some of the ideas involved. For instance, one can consider a simple subcase of (3.1) in which f and g are linear functions (this simplification can also be inspiring to treat the general case since, in the vicinity of the constant equilibrium (u_0, v_0), one can try to "linearize the equation" to obtain information about its dynamics). That is, let us consider the system of equations

$$\begin{cases} \partial_t u = \mu \Delta u + a(u - u_0) + b(v - v_0), \\ \partial_t v = \nu \Delta v + c(u - u_0) + d(v - v_0), \end{cases} \qquad (3.2)$$

for some a, b, c, $d \in \mathbb{R}$. For simplicity, let us also suppose that the problem is set on a circle, namely $x \in \mathbb{R}$ and u and v are periodic functions of period 2π. One can thus look for solutions of (3.2) in Fourier series of the form

$$u(x, t) = u_0 + \sum_{j \in \mathbb{Z}} U_j(t) \, e^{ijx} \quad \text{and} \quad v(x, t) = v_0 + \sum_{j \in \mathbb{Z}} V_j(t) \, e^{ijx}, \quad (3.3)$$

with U_j and V_j to be determined.

Substituting (3.3) into (3.2), we have that

$$\sum_{j\in\mathbb{Z}} \dot{U}_j(t)\, e^{ijx} = \partial_t u = \mu\Delta u + a(u - u_0) + b(v - v_0)$$

$$= -\mu\sum_{j\in\mathbb{Z}} j^2 U_j(t)\, e^{ijx} + a\sum_{j\in\mathbb{Z}} U_j(t)\, e^{ijx} + b\sum_{j\in\mathbb{Z}} V_j(t)\, e^{ijx}$$

and, similarly,

$$\sum_{j\in\mathbb{Z}} \dot{V}_j(t)\, e^{ijx} = -\nu\sum_{j\in\mathbb{Z}} j^2 V_j(t)\, e^{ijx} + c\sum_{j\in\mathbb{Z}} U_j(t)\, e^{ijx} + d\sum_{j\in\mathbb{Z}} V_j(t)\, e^{ijx}.$$

From these equations, we arrive at

$$\begin{pmatrix} \dot{U}_j(t) \\ \dot{V}_j(t) \end{pmatrix} = \begin{pmatrix} a - j^2\mu & b \\ c & d - j^2\nu \end{pmatrix} \begin{pmatrix} U_j(t) \\ V_j(t) \end{pmatrix}. \tag{3.4}$$

This is a first-order ordinary differential equation with constant coefficients. Hence, we suppose for simplicity that the (possibly complex) eigenvalues of the matrix $\begin{pmatrix} a - j^2\mu & b \\ c & d - j^2\nu \end{pmatrix}$, which we denote by λ_j and Λ_j, are distinct. We also denote by w_j and $W_j \in \mathbb{C}^2$ the corresponding eigenvectors. With this notation, we find (see e.g. [Kap15, Theorem 3.6]) that the solutions of (3.4) are of the form

$$\begin{pmatrix} U_j(t) \\ V_j(t) \end{pmatrix} = \xi_j\, e^{\lambda_j t} w_j + \Xi_j\, e^{\Lambda_j t} W_j \tag{3.5}$$

for some $\xi_j, \Xi_j \in \mathbb{C}$.

It is convenient to use the vector notation

$$w_j = \begin{pmatrix} w_{j1} \\ w_{j2} \end{pmatrix} \quad\text{and}\quad W_j = \begin{pmatrix} W_{j1} \\ W_{j2} \end{pmatrix}$$

and let

$$A_j := \xi_j w_{j1}, \quad B_j := \Xi_j W_{j1}, \quad C_j := \xi_j w_{j2} \quad\text{and}\quad D_j := \Xi_j W_{j2}.$$

In this way, (3.5) yields that

$$U_j(t) = A_j e^{\lambda_j t} + B_j e^{\Lambda_j t} \quad\text{and}\quad V_j(t) = C_j e^{\lambda_j t} + D_j e^{\Lambda_j t}. \tag{3.6}$$

Moreover,

$$\begin{pmatrix} a - j^2\mu & b \\ c & d - j^2\nu \end{pmatrix} \begin{pmatrix} A_j \\ C_j \end{pmatrix} = \begin{pmatrix} a - j^2\mu & b \\ c & d - j^2\nu \end{pmatrix} \begin{pmatrix} \xi_j w_{j1} \\ \xi_j w_{j2} \end{pmatrix}$$

$$= \xi_j \begin{pmatrix} a - j^2\mu & b \\ c & d - j^2\nu \end{pmatrix} w_j = \xi_j \lambda_j w_j = \lambda_j \begin{pmatrix} A_j \\ C_j \end{pmatrix},$$

leading to

$$(a - j^2\mu - \lambda_j)A_j + bC_j = 0. \tag{3.7}$$

Similarly,

$$(a - j^2\nu - \Lambda_j)B_j + bD_j = 0. \tag{3.8}$$

In jargon, (3.7) and (3.8) are sometimes dubbed "dispersion relations": their interest lies in the fact that they relate the speed of oscillation in the time variable (quantified in (3.6) by the eigenvalues λ_j and Λ_j) with the spatial periodicity of the medium (characterized by the eigenvalues $-j^2$ of the one-dimensional Laplacian and modulated by the speeds of diffusion μ and ν).

To recap briefly, from the system of reaction-diffusion equations in (3.2), one arrives at the solutions introduced in (3.3), with U_j and V_j as in (3.6), where

$$\lambda_j \text{ and } \Lambda_j \text{ are (distinct) complex}$$

$$\text{eigenvalues of the matrix } \begin{pmatrix} a - j^2\mu & b \\ c & d - j^2\nu \end{pmatrix}, \tag{3.9}$$

and with the parameters satisfying (3.7) and (3.8).

As detailed in [Tur52, Sections 7 and 8] and [Mur03, Chapters 2 and 3], this explicit mathematical construction has a number of important biological consequences and presents a sufficiently rich structure to account for many patterns visible in nature. To see these features, one may focus on the case in which one of the eigenvalues has the largest real part (in rough terms, one expects that the other modes are dominated by this one). Also, it is convenient to distinguish between the case in which the dominant eigenvalue is real from the one in which it is complex and with a nonzero imaginary part:

indeed, real eigenvalues will be related to stationary states and complex eigenvalues to oscillatory cases.

More specifically, suppose that

$$\Lambda_{j_0} \text{ is the eigenvalue with the largest real part.} \qquad (3.10)$$

We note that $\Lambda_{-j_0} = \Lambda_{j_0}$ as well since

$$\text{the matrix } \begin{pmatrix} a - j^2\mu & b \\ c & d - j^2\nu \end{pmatrix} \text{ remains the same}$$

$$\text{if we exchange } j \text{ with } -j. \qquad (3.11)$$

Hence, dropping the higher-order terms, we can assume that the dynamics of the solutions in (3.3) is governed by the following long-time asymptotics:

$$u(x,t) \simeq u_0 + U_{j_0}(t)\, e^{ij_0 x} + U_{-j_0}(t)\, e^{-ij_0 x}$$

$$\simeq u_0 + e^{\Lambda_{j_0} t}\left(B_{j_0} e^{ij_0 x} + B_{-j_0} e^{-ij_0 x}\right) \qquad (3.12)$$

$$\text{and} \quad v(x,t) \simeq v_0 + e^{\Lambda_{j_0} t}\left(D_{j_0} e^{ij_0 x} + D_{-j_0} e^{-ij_0 x}\right).$$

The invariance in (3.11) also suggests that if A_j, B_j, C_j and D_j are solutions of (3.7) and (3.8), then so are A_{-j}, B_{-j}, C_{-j} and D_{-j}; for this reason, we can suppose that $B_{-j_0} = B_{j_0}$ and $D_{-j_0} = D_{j_0}$ in (3.12), obtaining that

$$u(x,t) \simeq u_0 + B_{j_0} e^{\Lambda_{j_0} t}\left(e^{ij_0 x} + e^{-ij_0 x}\right)$$

$$= u_0 + 2B_{j_0} e^{\Lambda_{j_0} t} \cos(j_0 x) \qquad (3.13)$$

$$\text{and} \quad v(x,t) \simeq v_0 + 2D_{j_0} e^{\Lambda_{j_0} t} \cos(j_0 x).$$

Without aiming at exhausting all the possible patterns included in (3.13), let us now show a concrete case of interest.

For instance, let $\vartheta \in \mathbb{N}$,

$$\gamma := \frac{33\,\vartheta^2}{13\sqrt{3} - 9} \tag{3.14}$$

and also

$$u_0 := 0, \qquad v_0 := 0,$$
$$a := \gamma, \qquad b := -2\gamma, \qquad c := 2\gamma \quad \text{and} \quad d := -2\gamma. \tag{3.15}$$

In this case, the system in (3.2) describes an activator with density u, which is an autocatalytic activator; that is, such a substance stimulates the production of itself (since $a > 0$) and also activates the production of the substance with density v (since $c > 0$ as well). Also, the substance with density v corresponds to a self-degrading inhibitor: indeed, higher concentrations of this reactant are noxious to itself (since $d < 0$) and for the activator with density u (since $b < 0$).

Interestingly, the origin, which corresponds to the equilibrium (u_0, v_0), is a stable sink for the system of ordinary differential equations corresponding to (3.2) when $\mu := 0$ and $\nu := 0$. See Figure 3.6 for a sketch of the trajectories of

$$\begin{cases} \partial_t u = \gamma u - 2\gamma v, \\ \partial_t v = 2\gamma u - 2\gamma v. \end{cases} \tag{3.16}$$

Quite remarkably, as discovered by Turing, the stability of (3.16) can be destroyed by random fluctuations arising from the diffusivity of the chemical reactant. To appreciate this, given u_0, v_0, a, b, c and d as in (3.15), we take $\mu := 1$ and $\nu := 12$ in (3.2). Note that this corresponds to a situation in which the diffusion of the inhibitor is faster than that of the activator. This scenario gives that

$$\begin{pmatrix} a - j^2\mu & b \\ c & d - j^2\nu \end{pmatrix} = \begin{pmatrix} \gamma - j^2 & -2\gamma \\ 2\gamma & -2\gamma - 12j^2 \end{pmatrix},$$

which possesses eigenvalues of the form

$$-\frac{1}{2}\left(\gamma + 13j^2 \pm \sqrt{(11j^2 + 7\gamma)(11j^2 - \gamma)}\right). \tag{3.17}$$

By (3.10), we aim at detecting the greatest possible real part in (3.17). To this end, note that when $(11j^2 + 7\gamma)(11j^2 - \gamma) \leqslant 0$,

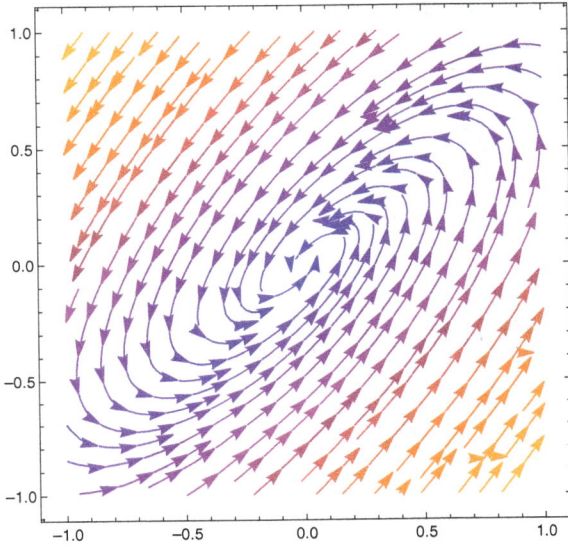

Figure 3.6.   Stream plot of the system of ordinary differential equations in (3.16).

the real part in (3.17) is equal to $-\frac{1}{2}(\gamma + 13j^2) < 0$. Instead, if $(11j^2 + 7\gamma)(11j^2 - \gamma) > 0$, then the largest possible real part in (3.17) is equal to

$$
\sup_{j \in \mathbb{Z}} \frac{1}{2} \left(\sqrt{(11j^2 + 7\gamma)(11j^2 - \gamma)} - \gamma - 13j^2 \right)
$$

$$
= \frac{\gamma}{2} \sup_{j \in \mathbb{Z}} \left(\sqrt{\left(\frac{11j^2}{\gamma} + 7 \right) \left(\frac{11j^2}{\gamma} - 1 \right)} - 1 - \frac{13j^2}{\gamma} \right) \qquad (3.18)
$$

$$
= \frac{\gamma}{2} \sup_{j \in \mathbb{Z}} \Phi \left(\frac{j^2}{\gamma} \right),
$$

where

$$
\Phi(\tau) := \sqrt{(11\tau + 7)(11\tau - 1)} - 1 - 13\tau.
$$

Using elementary calculus, one checks that

$$
\max_{\tau \geqslant 0} \Phi(\tau) = \frac{4}{11} \left(7 - 4\sqrt{3} \right) = \Phi \left(\frac{13\sqrt{3} - 9}{33} \right) = \Phi \left(\frac{\vartheta^2}{\gamma} \right),
$$

thanks to (3.14).

This observation and (3.18) give that the eigenvalues with the largest possible real part in (3.17) correspond to the choice $j := \vartheta$ and are of the form

$$\frac{2\gamma}{11}(7 - 4\sqrt{3}) = \frac{66(7 - 4\sqrt{3}), \vartheta^2}{11(13\sqrt{3} - 9)}.$$

Hence, up to constants, the corresponding setting in (3.13) takes the form

$$u(x,t) \simeq \exp\left(\frac{66\left(7 - 4\sqrt{3}\right)\vartheta^2}{11\left(13\sqrt{3} - 9\right)}t\right)\cos(\vartheta x)$$

$$\text{and} \qquad v(x,t) \simeq \exp\left(\frac{66\left(7 - 4\sqrt{3}\right)\vartheta^2}{11\left(13\sqrt{3} - 9\right)}t\right)\cos(\vartheta x).$$

(3.19)

See Figure 3.7 for an example with $\vartheta := 2$.

Of course, the asymptotics in (3.19) are divergent as $t \to +\infty$, which would correspond to the chemical substances to reach infinite density. This is certainly unfeasible in practice; hence, the meaning

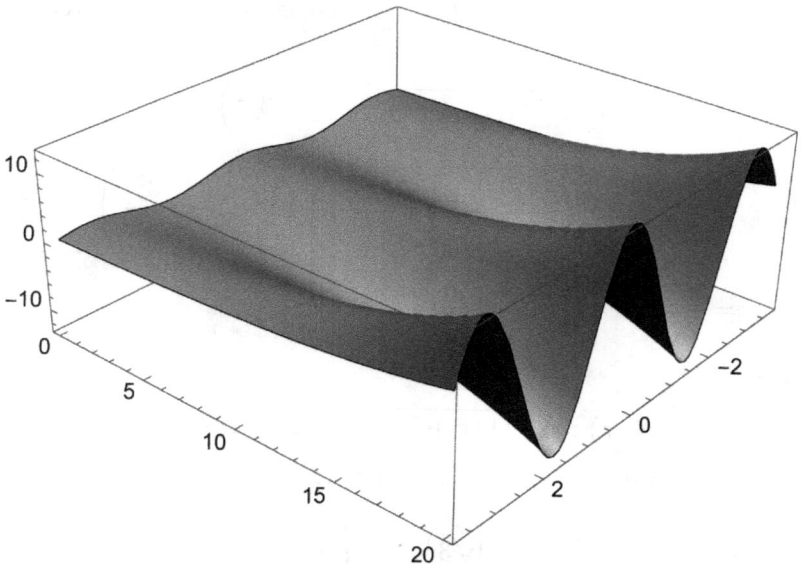

Figure 3.7. Plot of the function $(x,t) \mapsto \exp\left(\frac{264\left(7 - 4\sqrt{3}\right)}{11\left(13\sqrt{3} - 9\right)}t\right)\cos(2x)$.

of (3.19) has to be understood only in the vicinity of the equilibrium (u_0, v_0), which was set to be the origin for simplicity. Indeed, in practice, the linear system in (3.2) must be considered as an efficient linearization only in the vicinity of the equilibrium (u_0, v_0), while the general situation is more accurately described by a nonlinear system as in (3.1). For practical purposes, the nonlinear sources f and g would force a bound on the densities u and g and possibly favor the convergence of the solution for large times to steady solutions $u = u(x)$ and $v = v(x)$ of

$$\begin{cases} \mu\Delta u(x) + f(u(x), v(x)) = 0, \\ \nu\Delta v(x) + g(u(x), v(x)) = 0. \end{cases} \tag{3.20}$$

That is, for small times, the linear mechanism identified in (3.19) is helpful to detect how diffusion can drive instability and place the kinetics of the system out of the "trivial" state (u_0, v_0); then, at a longer time scale, the nonlinear structure of (3.1) becomes instrumental to confine the solution and lead it toward spatially inhomogeneous patterns, as described by the steady solutions of (3.20) (as another option, the evolution of the equation may be stopped after a certain time in case the release of the chemical substances stops; this could be the case in which the pattern is formed at an embryonic stage for the animal due to chemical substances that are released only during specific periods of the early-stage development of an organism).

We also remark that, for the sake of simplicity, here we confined ourselves to the case in which the spatial domain is a circle (i.e. the real line with periodic assumptions in x): in general, if one considers more complicated domains (say, closer to biological situations of specific interest), then it is convenient to replace e^{ijx} in (3.3) with the eigenfunctions of the Laplacian in the domain of interest (with the corresponding boundary conditions). In a similar manner, the dispersion relations in (3.7) and (3.8) must take into account the corresponding eigenvalues in the place of $-j^2$. These eigenvalues replace $-j^2$ in the matrix in (3.9) too, and the diffusion eigenvalues λ_j and Λ_j must be modified accordingly. The diffusion eigenvalues corresponding to diffusive instability still correspond to the ones with a positive real part, and one can focus for concreteness on the diffusive eigenvalue with the largest real part in (3.10). The structural difference in the general case is, however, that the excited modes, i.e. the

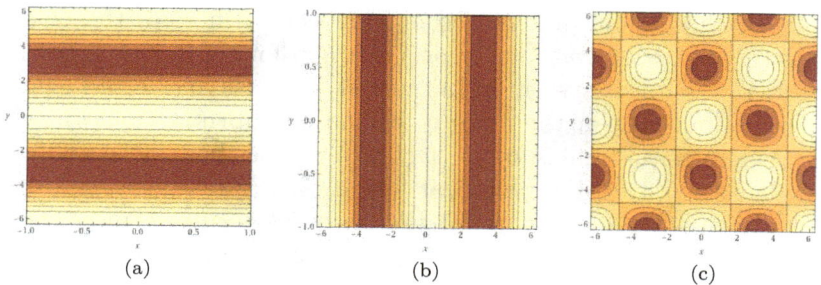

Figure 3.8. Level sets of (a) $(x_1, x_2) \mapsto 2\cos x_2$, (b) $(x_1, x_2) \mapsto 2\cos x_1$ and (c) $(x_1, x_2) \mapsto 2\cos x_1 \cos x_2$.

diffusive eigenvalues with positive real parts, depend on the domain. Since their regions of positive and negative values create a visible pattern in the dynamics of the solution (compare with the positive and negative values of the solution depicted in Figure 3.7), it is conceivable that these regions have a connection with the visible patterns in nature. For instance, two-dimensional rectangular domains present eigenfunctions of the type $\cos(\frac{2\pi j_1 x_1}{\ell_1})\cos(\frac{2\pi j_2 x_2}{\ell_2})$, where ℓ_1 and ℓ_2 account for the lengths of the sides of the rectangle and $j_1, j_2 \in \mathbb{N}$. In this scenario, the excited states corresponding to $j_1 = 0$ give rise to horizontal stripes, the ones corresponding to $j_2 = 0$ to vertical stripes, and the ones with $j_1 \neq 0$ and $j_2 \neq 0$ to maculate patterns, see e.g. Figure 3.8.

With respect to this, notice also that different excite modes correspond to different sizes of the pattern (such as width of the stripes or possible elongation of the spots). Of course, the analysis of simple rectangular regions is insufficient to capture the whole complexity of animal patterns; yet, it is suggestive to "approximate" an animal's coat with rectangular regions and observe how, for instance, the orientation of stripes "locally" follows the proportion of the approximating rectangles, see e.g. Figures 3.9 and 3.10, which clearly show a change between vertical and horizontal stripe patterns in zebras at the junctions between the body and the legs and between the body and the tail.

Animal patterns may also bifurcate from stripes to spots in different regions[2] of the organism, possibly in response of a variation

---

[2]It is indeed quite tempting to divide the whole fur of zebras and cheetahs into "rectangular" regions to approximately reproduce the eigenfunction patterns

Figure 3.9. Mutually grooming zebras (photo by Duvignau Alain from Wikipedia, available under the Creative Commons CC0 1.0 Universal Public Domain Dedication).

of the "local geometry", compare e.g. with the tail markings of the cheetah, in which the typical spots of the animal become stripes at the tip of the tail, see Figure 3.11.

---

depicted in Figure 3.8. Of course, such a reduction is a bit simplistic since these rectangles do overlap a bit in regions which cannot be accurately described by the simple eigenfunction patterns (equivalently, to understand the nodal lines of eigenfunctions, it is too crude to approximate a complicated domain with non-overlapping rectangles, not only because these rectangles may not shade faithfully the precise contour of the domain but also, and more importantly, because a rigid rectangular subdivision imposes new artificial boundary conditions): in any case, an attempt to perform such a rectangular reduction is sketched in Figure 3.12. See also Figures 3.2, 3.6 and 3.8 in [Mur03] for accurate simulations accounting for realistic body surfaces.

On a different note, a common curiosity question is: are zebras white animals with black stripes or black animals with white stripes? Well, the most reasonable answer is that, arguably, zebras are black with white stripes. Indeed, a zebra's fur grows from follicles that contain melanocyte cells, which generate the pigment that gives color to the hair. In this context, the zebra's white fur represents an absence of melanin. That is, white is not really the pigment of the stripe, and white stripes only exist because the pigment is denied.

Figure 3.10. A painting of East African wildlife in the "Tingatinga" style. Painted by Rubuni Rashid Said and scanned by Ingo Koll (image from Wikipedia, available under the Creative Commons Attribution-Share Alike 4.0 International license).

Figure 3.11. Caress of the Sphinx by Fernand Khnopff (Public Domain image from Wikipedia).

After [Tur52], Turing's ideas about morphogenesis have become a cornerstone in mathematical biology and led to a number of fantastic accomplishments; see e.g. the mathematical reconstruction of highly sophisticated animal patterns presented in [Mur88, Mur03]. Also, the name "Turing pattern" is nowadays commonly used to denote

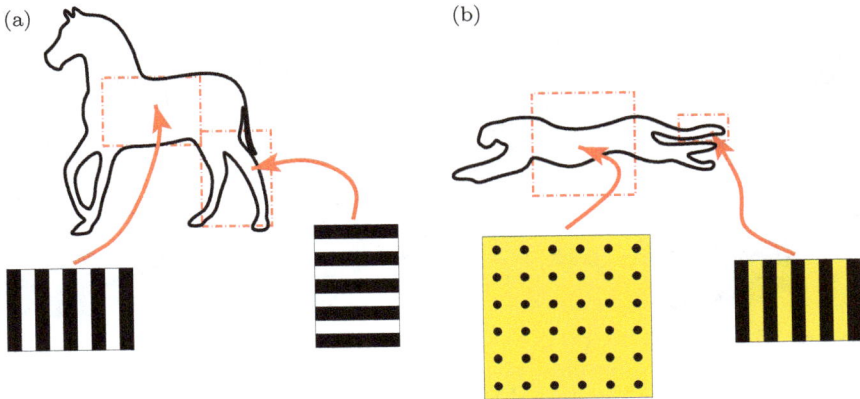

Figure 3.12. Change in visual markers in different areas of (a) zebras and (b) cheetahs.

biological structures, specifically those arising from the imbalances between the diffusion rates of different chemical agents, which make a stable system sensitive to perturbations.

See [Mur03, Chapters 2 and 3], [JPS10, Section 12.5], [Per15] and the references therein for a thorough analysis of diffusion-driven pattern formation, with several remarkable examples inspired by concrete natural phenomena. See also [Smo83] and the reference therein for more information about reaction-diffusion equations.

Let us mention anyway that reaction-diffusion equations may not be the end of the story for Turing patterns, since often other mechanisms come into play. For instance, a validation of the model purely based on reaction-diffusion equations relies on the enlargement of the patterned region obtained by increasing the number of cells in which the phenomenon occurs. In this situation, a model based only on eigenfunctions of partial differential equations would probably predict that the stripes or spots do not enlarge when the available area becomes bigger; rather, the intrinsic size of the pattern would be preserved by the insertion of additional stripes or spots. This phenomenon has been verified by numerical simulations in [Bal99] and in real-life biological experiments in [KA95] on the angelfish *Pomacanthus imperator*, thus contributing to the consolidation of reaction-diffusion-based models in this case.

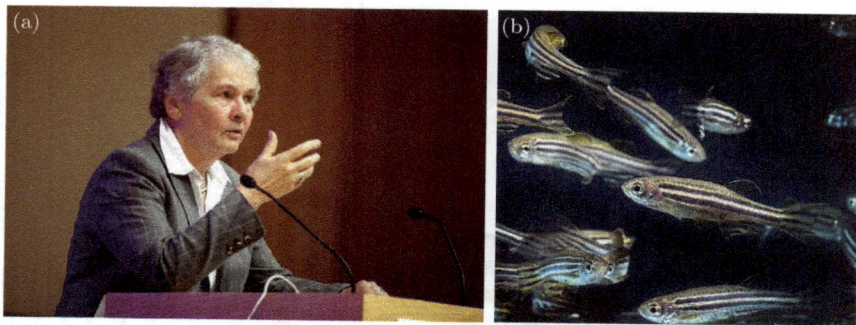

Figure 3.13.   (a) Nobel Prize Laureate Christiane Nüsslein-Volhard (photo by Rama, image from Wikipedia, licensed under the Creative Commons Attribution-Share Alike 2.0 France license). (b) A school of Zebrafish at Oregon State University (photo by Lynn Ketchum, image from Wikipedia, licensed under the Creative Commons Attribution-Share Alike 2.0 Generic license).

In other biological circumstances, however, the situation appears to be different. One interesting case is that of the zebrafish (see e.g. [KFW+14] and Figure 3.13), in which the stripes of the fish do not appear to exhibit the above dynamic rearrangement during the growth of the fish.

The case of the zebrafish is indeed intriguing. On the one hand, if some of the stripes are partially ablated in the zebrafish, other stripes fill the empty space by sliding, and this situation is perfectly coherent with the numerics produced by reaction-diffusion models (see [YYK07]). On the other hand, cell–cell interactions also seem to play a major role in the pattern formation of the zebrafish due to mutual repulsion and signal transmission occurring in cell protrusions (see [IYK12]). In this spirit, long-distance signaling and cell interactions have been used to develop alternative and complementary models to molecular diffusion, extending the theory of animal markings beyond the original realm of reaction-diffusion equations. In particular, classical partial differential equations have often been substituted or flanked by convolution and integral[3] equations of the nonlocal type, producing the so-called kernel-based Turing models (or KT models), see [Kon17]. See also [KWM21] for a review on these topics.

---

[3]We will meet other nonlocal models related to cell interactions in Chapter 37.

Chapter 4

Space Invaders

A topical argument in mathematical biology consists of the study of invasive species and their territorial colonization ability, see e.g. Figure 4.1. To introduce ourselves to this subject, it is opportune to first understand the so-called logistic equation. This model was introduced by Pierre François Verhulst (see Figure 4.2 for his portrait) and describes the evolution in time of a biological population. The number of individuals $N(t)$ is supposed to grow at an intrinsic growth rate, parameterized by a given $\rho > 0$, which somewhat accounts for the ideal birth rate of the population (N individuals would give rise to ρN newborns in a unit of time). Additionally, the population undergoes intraspecific competition, which causes the death of some individuals due to possible overcrowding (assuming, for instance, that the environmental resources only allow a maximum number of individuals K). The combination of these effects lead to the logistic ordinary differential equation

$$\frac{dN}{dt}(t) = \rho N(t) \left(1 - \frac{N(t)}{K}\right). \tag{4.1}$$

This equation considers the entire population to be located basically at the same place, but for many practical purposes, it is also convenient to describe biological individuals in a spatial environment, in which case the function N depends on the time variable t as well as on the space variable x. If the population performs some kind of random walk, as discussed in Chapter 2 (see in particular equation (2.7)), by combining (4.1) with the random diffusive tendency

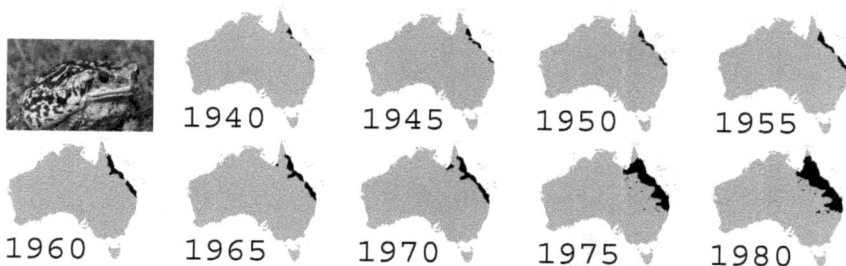

Figure 4.1. A cane toad (*bufo marinus*) and the spread of cane toads in Australia from 1940 to 1980 in five-year intervals (photo by Bill Waller and animated map by Froggydarb; images from Wikipedia, licensed under the Creative Commons Attribution-Share Alike 3.0 Unported license).

Figure 4.2. Pierre Verhulst (Public Domain image from Wikipedia).

of the population, we arrive at the partial differential equation

$$\partial_t N(x,t) = c\Delta N(x,t) + \rho N(x,t)\left(1 - \frac{N(x,t)}{K}\right), \qquad (4.2)$$

for some diffusion coefficient $c > 0$.

We can actually simplify (4.2) via a simple rescaling; namely, by defining

$$u(x,t) := \frac{1}{K}N\left(\sqrt{\frac{c}{\rho}}x, \frac{t}{\rho}\right), \qquad (4.3)$$

equation (4.2) reduces to

$$\partial_t u = \Delta u + u(1 - u), \qquad (4.4)$$

which is dubbed[1] in jargon the Fisher–Kolmogorov–Petrovsky–Piskunov equation (or Fisher–KPP equation for short). Interestingly,

---

[1]Equation (4.4) is named after Sir Ronald Aylmer Fisher [Fis37a], Andrey Nikolaevich Kolmogorov, Ivan Georgievich Petrovsky and Nikolai Semenovich Piskunov [Kol91, pp. 248–270].

Though biological invasions and the spread of genetic traits were the prime motivations for the study of (4.4), nowadays this equation is also intensively studied for its applications in combustion and flame propagation – that's the unifying power of good mathematics!

Besides introducing equation (4.4) and some questionable studies in eugenics, Fisher is considered one of the founders of population genetics and modern statistical science. His name is also linked to the so-called Fisherian runaway sexual selection mechanism, which aims to explain certain exaggerated, costly and apparently maladaptive male ornamentation in nature (seemingly in conflict with natural selection), as dictated by a persistent female choice (possibly triggered initially by the ornament signaling greater potential fitness, hence the likelihood of leaving more descendants). A classical example of Fisherian runaway is the elaborate peacock plumage, see Figure 4.3.

A classical story about Fisher's contribution to the standardization of statistical experiments is that of a "lady tasting tea" [Fis37b]. In a nutshell, once Fisher offered a cup of tea to phycologist Blanche Muriel Bristol. She politely declined it, saying that she preferred the flavor when the milk was poured into the cup before (not after) the tea. So, we have an intriguing dilemma: can the order of pouring milk affect the flavor of the tea? And, specifically, can Bristol notice the difference?

To test this, Fisher provided Bristol with eight randomly ordered cups of tea: four of which were prepared by first pouring the tea and then adding milk, and four by first pouring the milk and then adding the tea. Bristol had to identify the different cups. Fisher interpreted the data to distinguish the cases in which Bristol could actually tell the difference in the pouring order and those in which she could identify correctly some cups but simply by chance.

For this, with a bit of combinatorics, Fisher computed that there are $\binom{8}{4} = \frac{8!}{4!(8-4)!} = 70$ possible combinations of cups. Since Bristol was aware that there were four cups of each type, her answer would have included four of each.

There is, of course, precisely only one possibility of making the correct identification; hence, achieving a complete identification purely by chance occurs with a probability of $\frac{1}{70} = 0.01428571428\ldots$.. There are instead 16 possibilities of making exactly one error (indeed, one error comes from a single swap of two cups of different types, say the ith cup of the first type with the jth cup of the second type, for $i, j \in \{1, 2, 3, 4\}$). Hence, there are $1 + 16 = 17$ possibilities of making this equation is a reaction-diffusion equation in the setting introduced on p. 32.

The question of determining the invasivity of biological species is strongly related to (4.4). In this setting, in light of (4.3), the level $u = 0$ corresponds to the absence of the population, while the level $u = 1$ corresponds to the maximal density permitted by environmental resources. Remarkably, for any given $v \in [2, +\infty)$:

equation (4.4) admits a solution u describing the full

environmental invasion of the biological species with velocity v.

$$(4.5)$$

More precisely, given any $v \in [2, +\infty)$ and any $\omega \in \partial B_1$, there exists a smooth function $U : \mathbb{R} \to (0, 1)$, with

$$\lim_{\tau \to +\infty} U(\tau) = 0 \quad \text{and} \quad \lim_{\tau \to -\infty} U(\tau) = 1, \qquad (4.6)$$

and a solution u for (4.4) of the form

$$u(x, t) := U(\omega \cdot x - vt). \qquad (4.7)$$

The solutions[2] of this form are often called "solitary waves". See Figure 4.4 for a representation of a solitary wave.

---

at most one mistake; therefore, if we adopt a conventional probability criterion of success to be below the 5% threshold, Bristol's ability to correctly categorize the cups of tea would have been confirmed if and only if she was able to correctly identify all of them, without making any mistakes.

This story is relevant since it became an example of a randomized experiment for checking a "null hypothesis" in statistics, and it contributed to a scientific establishment of randomization analysis of experimental data.

For the record, Muriel Bristol allegedly succeeded in classifying all eight cups correctly.

See [Gri59, Sal01] for further reading on the "lady tasting tea" experiment.

See also Figure 24.2 for a picture of Kolmogorov and footnote 1 on p. 253 for more information about the prominent figure of Kolmogorov.

[2]The concrete interest in the very special solutions of (4.7) arises from the fact that, under suitable initial conditions, "reasonable" solutions of (4.4) evolve into a solitary wave, see [Kol91, AW75, AW78]. In rough terms, wells in the initial conditions are quickly reabsorbed by the evolving dynamics since regions with u close to 1 tend to grow and eat up adjacent regions with u close to 0, producing, for large times, a monotonic traveling wave with a constant speed. This phenomenon makes solitary waves an essential building block for understanding the dynamics of the Fisher–Kolmogorov–Petrovsky–Piskunov equation.

To establish (4.5), we consider[3] the ordinary differential equation

$$\frac{d^2U}{d\tau^2}(\tau) + v\frac{dU}{d\tau}(\tau) + U(\tau)(1 - U(\tau)) = 0. \qquad (4.8)$$

By setting $V(\tau) := \frac{dU}{d\tau}(\tau)$, we obtain the system of equations

$$\begin{pmatrix} \dfrac{dU}{d\tau} \\ \dfrac{dV}{d\tau} \end{pmatrix} = \begin{pmatrix} V \\ -vV - U(1 - U) \end{pmatrix}. \qquad (4.9)$$

The equilibria of (4.9) are $E_0 := (0,0)$ and $E_1 := (1,0)$. The linearized dynamics in the vicinity of E_1 is described by the matrix $M_1 := \begin{pmatrix} 0 & 1 \\ 1 & -v \end{pmatrix}$, which possesses the eigenvalues $\frac{1}{2}\left(-v \pm \sqrt{v^2 + 4}\right)$ with the eigenvectors $\left(\frac{1}{2}\left(v \pm \sqrt{v^2 + 4}\right), 1\right)$.

In particular, E_1 is a hyperbolic saddle for all values of $v \in \mathbb{R}$. So, we consider the unstable manifold (see e.g. [Per01, Section 2.7]) of E_1 in the direction $-\left(\frac{1}{2}\left(v + \sqrt{v^2 + 4}\right), 1\right)$. This provides us with an orbit $(U(\tau), V(\tau))$ such that

$$\lim_{\tau \to -\infty} (U(\tau), V(\tau)) = E_1 \qquad (4.10)$$

and

$$\lim_{\tau \to -\infty} \frac{1}{\lambda e^{\lambda t}}\left(\frac{dU}{d\tau}(\tau), \frac{dV}{d\tau}(\tau)\right) = -\left(\frac{1}{2}\left(v + \sqrt{v^2 + 4}\right), 1\right),$$

$$\text{with } \lambda := \frac{1}{\sqrt{v^2 + 4} - v}, \qquad (4.11)$$

see Figure 4.5.

---

[3] As remarked in [AZ79], when $v := \frac{5}{\sqrt{6}} = 2.04124145232\ldots$, one can find explicit solutions for (4.8) in the form of

$$U(\tau) = \frac{1}{\left(1 + C\exp\left(\frac{\tau}{\sqrt{6}}\right)\right)^2},$$

for all $C > 0$.

Figure 4.4 was obtained using this function with $C := 1$, and the frames of the second image in the figure were obtained at times $t \in \{0, 2, 4, 6, 8\}$.

See also [Mur02, Section 13.4] for more information about this explicit solution. We are not aware of other solutions of (4.8) which can be written explicitly in a nice and closed form.

Figure 4.3.   A blue peacock presenting its feathers (image from Wikipedia, available under the Creative Commons Attribution-Share Alike 3.0 Unported).

Given $v \in [2, +\infty)$, we now consider the planar region

$$\mathcal{R} := \left\{ (U, V) \in \mathbb{R}^2 \text{ s.t. } U < 1, \ V \in \left(-\frac{1}{v}, 0 \right) \quad \text{and} \right.$$

$$\left. \frac{v + \sqrt{v^2 - 4}}{2} U + V > 0 \right\}.$$

Note that, in light of (4.10) and (4.11), there exists $\tau_\star \in \mathbb{R}$ such that $(U(\tau), V(\tau)) \in \mathcal{R}$ for all $\tau \in (-\infty, \tau_\star)$.

We claim that

$$(U(\tau), V(\tau)) \in \mathcal{R} \quad \text{for all} \ \tau \in (-\infty, +\infty). \tag{4.12}$$

Indeed, suppose not. Then, there exists $\tau_0 \geqslant \tau_\star$ such that $(U(\tau), V(\tau)) \in \mathcal{R}$ for all $\tau \in (-\infty, \tau_0)$, but either $V(\tau_0) = -1/v$, $V(\tau_0) = 0$, $U(\tau_0) = 1$, or $\frac{v + \sqrt{v^2 - 4}}{2} U(\tau_0) + V(\tau_0) = 0$. We exclude all these cases to prove (4.12).

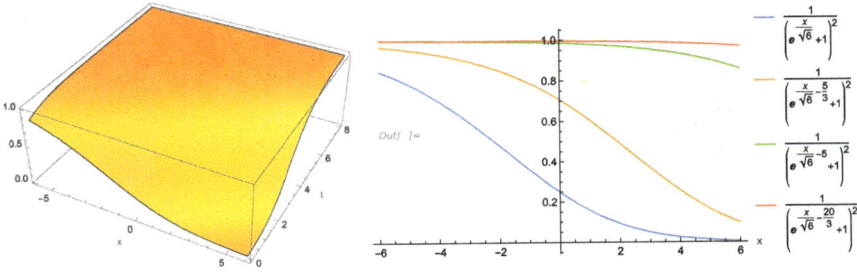

Figure 4.4. A traveling wave of a species invading the spatial domain from left to right.

First, we show that $V(\tau_0) = -1/v$ cannot hold. This is because otherwise $\frac{dV}{d\tau}(\tau_0) \leqslant 0$, but we know from (4.9) that

$$\frac{dV}{d\tau}(\tau_0) = -vV(\tau_0) - U(\tau_0)(1 - U(\tau_0)) \geqslant -vV(\tau_0) - \frac{1}{4} = 1 - \frac{1}{4} > 0.$$

Similarly, $V(\tau_0) = 0$ cannot occur, as otherwise

$$0 \leqslant \frac{dV}{d\tau}(\tau_0) = -vV(\tau_0) - U(\tau_0)(1 - U(\tau_0)) \leqslant -vV(\tau_0) = 0.$$

This entails that either $U(\tau_0) = 0$ or $U(\tau_0) = 1$, whence either $(U(\tau_0), V(\tau_0)) = E_0$ or $(U(\tau_0), V(\tau_0)) = E_1$, which is impossible, as both E_0 and E_1 are equilibria of the system.

Also, $U(\tau_0) = 1$ cannot hold true, as otherwise, by (4.9),

$$0 \leqslant \frac{dU}{d\tau}(\tau_0) = V(\tau_0) \leqslant 0,$$

which gives that $(U(\tau_0), V(\tau_0)) = E_1$, which is impossible.

Finally, the case of $\frac{v+\sqrt{v^2-4}}{2} U(\tau_0) + V(\tau_0) = 0$ cannot occur either; otherwise, we argue as follows. Let

$$F(\tau) := \frac{v + \sqrt{v^2 - 4}}{2} U(\tau) + V(\tau).$$

We note that $F(\tau) > 0$ for $\tau \in (-\infty, \tau_0)$ and $F(\tau_0) = 0$. As a result, from (4.9), we arrive at

$$
\begin{aligned}
0 \geqslant \frac{dF}{d\tau}(\tau_0) \\
&= \frac{v + \sqrt{v^2 - 4}}{2} \frac{dU}{d\tau}(\tau_0) + \frac{dV}{d\tau}(\tau_0) \\
&= \frac{v + \sqrt{v^2 - 4}}{2} V(\tau_0) - vV(\tau_0) - U(\tau_0) + U^2(\tau_0) \\
&= \frac{\sqrt{v^2 - 4} - v}{2} V(\tau_0) - U(\tau_0) + U^2(\tau_0) \\
&= -\frac{\left(\sqrt{v^2 - 4} - v\right)\left(v + \sqrt{v^2 - 4}\right)}{4} U(\tau_0) - U(\tau_0) + U^2(\tau_0) \\
&= -\frac{(v^2 - 4) - v^2}{4} U(\tau_0) - U(\tau_0) + U^2(\tau_0) \\
&= U^2(\tau_0).
\end{aligned}
$$

This gives that $U(\tau_0) = 0$, whence $V(\tau_0) = 0$, but this contradicts the fact that E_0 is an equilibrium of the system.

The proof of (4.12) is thereby complete.

As a byproduct of (4.12), we also have that, for all $t \in \mathbb{R}$,

$$
1 > U(\tau) > -\frac{2}{v + \sqrt{v^2 - 4}} V > 0. \tag{4.13}
$$

Now, we claim that

$$
\lim_{\tau \to +\infty} (U(\tau), V(\tau)) = E_0. \tag{4.14}
$$

Indeed, by (4.12), we know that $V(\tau) < 0$ for all $\tau \in \mathbb{R}$; therefore, by (4.9), $U(\tau)$ is decreasing. Since also by (4.12) we have that $U(\tau) \in (0, 1)$ for all $\tau \in \mathbb{R}$, we infer that there exists $U_\infty \in [0, 1)$ such that $U(\tau) \to U_\infty$ as $\tau \to +\infty$.

Now, let $\delta > 0$, which is to be taken as small as we wish in the following. Using dots to denote derivatives, recalling (4.9) and (4.12), we observe that

$$
\begin{aligned}
|V(\tau)| &= \frac{1}{\delta}\left|\int_{\tau}^{\tau+\delta}(V(\tau)-V(\theta)+V(\theta))\,d\theta\right| \\
&\leq \frac{1}{\delta}\left|\int_{\tau}^{\tau+\delta}(V(\tau)-V(\theta))\,d\theta\right| + \frac{1}{\delta}\left|\int_{\tau}^{\tau+\delta}V(\theta)\,d\theta\right| \\
&= \frac{1}{\delta}\left|\int_{\tau}^{\tau+\delta}\left(\int_{\theta}^{\tau}\dot V(\zeta)\,d\zeta\right)d\theta\right| + \frac{1}{\delta}\left|\int_{\tau}^{\tau+\delta}\dot U(\theta)\,d\theta\right| \\
&= \frac{1}{\delta}\left|\int_{\tau}^{\tau+\delta}\left(\int_{\tau}^{\theta}vV(\zeta)+U(\zeta)(1-U(\zeta))\,d\zeta\right)d\theta\right| \\
&\quad + \frac{1}{\delta}|U(\tau+\delta)-U(\tau)| \\
&\leq \frac{2}{\delta}\int_{\tau}^{\tau+\delta}\left(\int_{\tau}^{\theta}d\zeta\right)d\theta + \frac{1}{\delta}|U(\tau+\delta)-U(\tau)| \\
&= \delta + \frac{1}{\delta}|U(\tau+\delta)-U(\tau)|.
\end{aligned}
\tag{4.15}
$$

From this, we arrive at

$$
\limsup_{\tau\to+\infty}|V(\tau)| \leq \limsup_{\tau\to+\infty}\delta + \frac{1}{\delta}|U(\tau+\delta)-U(\tau)|
$$
$$
= \delta + \frac{1}{\delta}|U_\infty - U_\infty| = \delta,
$$

and accordingly, by choosing δ to be arbitrarily small,

$$
\lim_{\tau+\infty}V(\tau)=0.
$$

As a result, we have that $(U(\tau),V(\tau))\to(U_\infty,0)\in[0,1)\times\{0\}$. Since the only equilibrium of (4.9) in $[0,1)\times\{0\}$ is E_0, the proof of (4.14) is complete.

Hence, in view of (4.9), (4.10), (4.13) and (4.14), the trajectory $U(\tau)$ provides a solution for (4.8) that satisfies the asymptotics

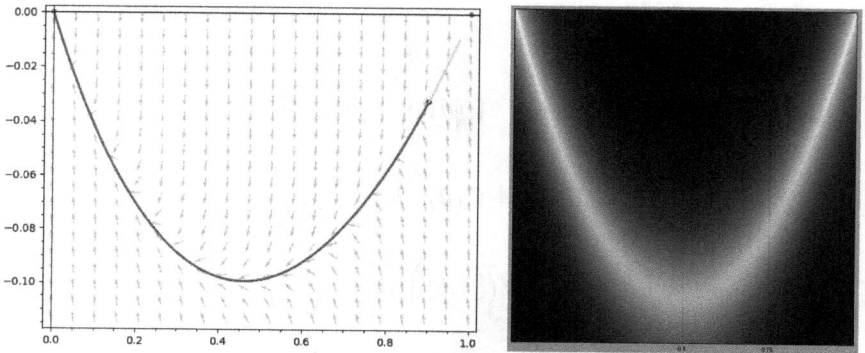

Figure 4.5.   Phase portrait and heteroclinic connection for the dynamical system in (4.9) with $v := 5/2$.

in (4.6). Then, by (4.7),

$$-\partial_t u(x,t) + \Delta u(x,t) + u(x,t)(1 - u(x,t))$$
$$= v\frac{dU}{d\tau}(\omega \cdot x - vt) + \frac{d^2U}{d\tau^2}U(\omega \cdot x - vt)$$
$$+ U(\omega \cdot x - vt)(1 - U(\omega \cdot x - vt)) = 0.$$

This shows that u is the desired solitary wave solution of the Fisher–Kolmogorov–Petrovsky–Piskunov equation in (4.4).

For completeness, we also show that biologically interesting solitary waves of the Fisher–Kolmogorov–Petrovsky–Piskunov equation do not exist when $v \in [0,2)$ since

when $v \in [0,2)$, any solution $U : \mathbb{R} \to (-\infty, 1]$ of (4.6) and (4.8)

necessarily attains negative values.   (4.16)

We recall indeed that biologically interesting values of the solutions are those modeling a population density and are hence confined to the interval $[0,1]$ due to (4.3).

To check the claim for (4.16), suppose by contradiction that a solution $U : \mathbb{R} \to (0,1]$ to (4.6) and (4.8) exists for some $v \in [0,2)$, and let $V := \frac{dU}{d\tau}$ to find the system of equations in (4.9).

We claim that, for all $\tau \geqslant 0$,

$$|V(\tau)| < |V(0)| + \frac{1}{v}. \tag{4.17}$$

Indeed, suppose not. Then, there exists $\underline{\tau} > 0$ such that $|V(\tau)| < |V(0)| + \frac{1}{v}$ for all $\tau \in [0, \underline{\tau})$ and $|V(\underline{\tau})| = |V(0)| + \frac{1}{v}$. Two cases can occur: either $V(\underline{\tau}) \geqslant 0$ or $V(\underline{\tau}) < 0$. In the first case, we have that

$$0 \leqslant \frac{dV}{d\tau}(\underline{\tau}) = -vV(\underline{\tau}) - U(\underline{\tau})(1 - U(\underline{\tau}))$$

$$= -v\left(|V(0)| + \frac{1}{v}\right) - U(\underline{\tau})(1 - U(\underline{\tau})) \leqslant -1,$$

which is a contradiction.

If instead $V(\underline{\tau}) < 0$, then

$$0 \geqslant \frac{dV}{d\tau}(\underline{\tau}) = -vV(\underline{\tau}) - U(\underline{\tau})(1 - U(\underline{\tau}))$$

$$= v\left(|V(0)| + \frac{1}{v}\right) - U(\underline{\tau})(1 - U(\underline{\tau})) \geqslant 1 - \frac{1}{4} > 0,$$

which is a contradiction too, and hence (4.17) is proved.

Using this and computing as in (4.15), for all $\tau \geqslant 0$ and $\delta > 0$, we arrive at

$$|V(\tau)| \leqslant \frac{1}{\delta} \left| \int_{\tau}^{\tau+\delta} \left(\int_{\tau}^{\theta} vV(\zeta) + U(\zeta)(1 - U(\zeta)) \, d\zeta \right) d\theta \right|$$

$$+ \frac{1}{\delta} |U(\tau + \delta) - U(\tau)|$$

$$\leqslant C\delta + \frac{1}{\delta} |U(\tau + \delta) - U(\tau)|,$$

with $C := v\left(|V(0)| + \frac{1}{v}\right) + 1$. Therefore,

$$\limsup_{\tau \to +\infty} |V(\tau)| \leqslant C\delta + \frac{1}{\delta} |0 - 0| = C\delta,$$

and accordingly, taking δ as small as we wish, we find that $V(\tau) \to 0$ as $\tau \to +\infty$.

In this way, we have established that $(U(\tau), V(\tau)) \to (0,0)$ as $\tau \to +\infty$. Hence, we pick $\eta > 0$, to be chosen conveniently small in what follows, and we find $\tau_\eta \in \mathbb{R}$ such that $|(U(\tau), V(\tau))| \leqslant \eta$, for all $\tau \geqslant \tau_\eta$. Up to a time shift, i.e. up to replacing $(U(\tau), V(\tau))$ with $(U(\tau - \tau_\eta), V(\tau - \tau_\eta))$, we can assume hereinafter that $\tau_\eta = 0$ and write

$$|(U(\tau), V(\tau))| \leqslant \eta \quad \text{for all } \tau \geqslant 0. \tag{4.18}$$

Now, we develop an argument related to the nonlinear stability of equilibria in planar dynamical systems. Namely, we take $\varrho(\tau) > 0$ and $\varphi(\tau) \in \mathbb{R}$ such that

$$U(\tau) = \varrho(\tau) \cos(\varphi(\tau)) \quad \text{and} \quad V(\tau) = \varrho(\tau) \sin(\varphi(\tau)).$$

Actually, since $U(\tau) > 0$, we have that

$$\text{we can pick } \varphi(\tau) \text{ in the interval } \left(-\frac{\pi}{2}, \frac{\pi}{2}\right). \tag{4.19}$$

Thus, denoting by a dot the derivative with respect to τ,

$$\dot{U} = \dot{\varrho} \cos \varphi - \varrho \dot{\varphi} \sin \varphi \quad \text{and} \quad \dot{V} = \dot{\varrho} \sin \varphi + \varrho \dot{\varphi} \cos \varphi.$$

From this, we arrive at

$$\dot{U} \sin \varphi - \dot{V} \cos \varphi$$
$$= (\dot{\varrho} \cos \varphi - \varrho \dot{\varphi} \sin \varphi) \sin \varphi - (\dot{\varrho} \sin \varphi + \varrho \dot{\varphi} \cos \varphi) \cos \varphi$$
$$= \dot{\varrho} \sin \varphi \cos \varphi - \varrho \dot{\varphi} \sin^2 \varphi - \dot{\varrho} \sin \varphi \cos \varphi - \varrho \dot{\varphi} \cos^2 \varphi$$
$$= -\varrho \dot{\varphi}.$$

Since, by (4.9),

$$\dot{U} \sin \varphi - \dot{V} \cos \varphi = V \sin \varphi + (vV + U - U^2) \cos \varphi$$
$$= \varrho \sin^2 \varphi + v\varrho \sin \varphi \cos \varphi + \varrho \cos^2 \varphi - \varrho^2 \cos^3 \varphi$$
$$= \varrho + \frac{v\varrho}{2} \sin(2\varphi) - \varrho^2 \cos^3 \varphi,$$

we conclude that

$$-\varrho \dot{\varphi} = \varrho + \frac{v\varrho}{2} \sin(2\varphi) - \varrho^2 \cos^3 \varphi$$

and accordingly

$$-\dot\varphi = 1 + \frac{v}{2}\sin(2\varphi) - \varrho\cos^3\varphi.$$

In particular, using (4.18) and the assumption that here $v \in [0,2)$,

$$-\dot\varphi \geqslant 1 - \frac{v}{2} - \varrho \geqslant 1 - \frac{v}{2} - \eta \geqslant \frac{1 - \frac{v}{2}}{2} > 0,$$

as long as η is small enough. Hence, there exists $\tau_\star > 0$ such that $\varphi(\tau_\star) < -\frac{\pi}{2}$, in contradiction to (4.19). This completes[4] the proof of (4.16).

---

[4]Here is an alternative argument to prove (4.16), relying on the additional simplifying assumptions that

$$\lim_{\tau\to+\infty}\dot U(\tau) = 0$$

and the following limit exists

$$\ell := \lim_{\tau\to+\infty}\frac{\ddot U(\tau)}{\dot U(\tau)}.$$

From this, if $U(\tau) \in (0,1)$ for all $\tau \in \mathbb{R}$ and $U(\tau) \to 0$ as $\tau \to +\infty$, we can apply L'Hôpital's rule and deduce that

$$\lim_{\tau\to+\infty}\frac{\dot U(\tau)}{U(\tau)} = \lim_{\tau\to+\infty}\frac{\ddot U(\tau)}{\dot U(\tau)} = \ell.$$

As a result, dividing (4.8) by $U(\tau)$ and taking the limit, we find that

$$0 = \lim_{\tau\to+\infty}\frac{\ddot U(\tau)}{U(\tau)} + v\frac{\dot U(\tau)}{U(\tau)} + 1 - U(\tau)$$

$$= \lim_{\tau\to+\infty}\frac{\ddot U(\tau)}{\dot U(\tau)}\frac{\dot U(\tau)}{U(\tau)} + v\frac{\dot U(\tau)}{U(\tau)} + 1$$

$$= \ell^2 + v\ell + 1$$

$$= \left(\ell + \frac{v}{2}\right)^2 + 1 - \frac{v^2}{4}$$

$$\geqslant 1 - \frac{v^2}{4},$$

which gives $v^2 \geqslant 4$ and hence (in the convention that $v \geqslant 0$) necessarily $v \geqslant 2$, allowing invading fronts.

These computations are also interesting since, for more general equations such as $\ddot U + v\dot U + f(U) = 0$ with f smooth, $f(r) > 0$ for all $r \in (0,1)$, $f(0) = f(1) = 0$

For additional information on the Fisher–Kolmogorov–Petrovsky–Piskunov equation, see [Mur02, GK04, Per15, Chapter 13] and the references therein.

---

and $f'(1) < 0$, a calculation as above leads to

$$0 = \ell^2 + v\ell + \lim_{\tau \to +\infty} \frac{f(U(\tau))}{U(\tau)} = \ell^2 + v\ell + f'(0)$$

$$= \left(\ell + \frac{v}{2}\right)^2 + f'(0) - \frac{v^2}{4} \geqslant f'(0) - \frac{v^2}{4},$$

which results in the necessary condition for invasion: $v \geqslant 2\sqrt{f'(0)}$.

This is quite interesting since it also points out that the speed of the invasive front is dictated by its low-density fringes, not by the high-density regions (namely, the velocity of invasion is completely determined by the forcing term f, specifically by the values of f in the vicinity of 0 and not in the vicinity of 1).

A function f with the above structure is often called a bistable nonlinearity.

For further details on bistable nonlinearities and plane wave solutions, see [AW78, Section 4].

Chapter 5

Equations from Hydrodynamics

A classical arena for partial differential equations is provided by fluid mechanics and hydrodynamics. Let us present some situations that naturally occur in this framework. Let us consider a fluid with density $\rho(x,t)$, vectorial velocity $v(x,t)$ and pressure $p(x,t)$. The position of a parcel of fluid at time t is given by $x(t)$; therefore, such a parcel travels according to the velocity prescription

$$\dot{x}(t) = v(x(t), t). \tag{5.1}$$

Also, the total mass of a fluid element occupying a region Ω of the space at a given time t_0 is given by the quantity $\int_\Omega \rho(x, t_0)\, dx$. This amount of fluid will move around in a small time τ, possibly occupying a different region, which we name Ω_τ, that collects all the evolution trajectories $x(t_0+\tau)$ of the fluid parcels such that $x(t_0) \in \Omega$. Since we are assuming that matter is neither created nor disappear in this process, we can deduce a conservation law, given by the equation

$$\int_\Omega \rho(x, t_0)\, dx = \int_{\Omega_\tau} \rho(x, t_0 + \tau)\, dx$$

or, equivalently,

$$\frac{d}{d\tau} \int_{\Omega_\tau} \rho(x, t_0 + \tau)\, dx = 0. \tag{5.2}$$

Note that we can change the variable in the previous integral simply by reversing the direction of fluid flow. Specifically, given a point $q \in \mathbb{R}^n$, we denote by $\Phi^t(q)$ the evolution according to the fluid vector field v with the initial position at time t_0 equal to q, or, more formally, we define $\Phi^t(q)$ as the solution of the following Cauchy problem for ordinary differential equations:

$$
\begin{cases}
\dfrac{d}{dt}\Phi^t(q) = v(\Phi^t(q), t), \\[2mm]
\Phi^{t_0}(q) = q.
\end{cases}
$$

This is nothing but a reformulation of (5.1), but here we explicitly keep track of the initial position. With this notation, we have that $\Omega_\tau = \{\Phi^\tau(y),\ y \in \Omega\}$. As a result, one can perform the change of variable

$$
x = \Phi^\tau(y) = y + \tau v(y, t_0) + o(\tau),
$$

which produces

$$
|\det \partial_y x| = 1 + \tau \operatorname{div} v(y, t_0) + o(\tau)
$$

and therefore

$$
\int_{\Omega_\tau} \rho(x, t_0 + \tau)\, dx
$$

$$
= \int_\Omega \rho(y + \tau v(y, t_0), t_0 + \tau)\,(1 + \tau \operatorname{div} v(y, t_0))\, dy + o(\tau)
$$

$$
= \int_\Omega (\rho(y, t_0) + \tau \nabla \rho(y, t_0) \cdot v(y, t_0) + \tau \partial_t \rho(y, t_0))
$$
$$
\times (1 + \tau \operatorname{div} v(y, t_0))\, dy + o(\tau)
$$

$$
= \int_\Omega (\rho(y, t_0) + \tau \nabla \rho(y, t_0) \cdot v(y, t_0) + \tau \partial_t \rho(y, t_0)
$$
$$
+ \tau \rho(y, t_0) \operatorname{div} v(y, t_0))\, dy + o(\tau)
$$

$$
= \int_\Omega (\rho(y, t_0) + \tau \partial_t \rho(y, t_0) + \tau \operatorname{div}(\rho(y, t_0)\, v(y, t_0)))\, dy + o(\tau).
$$

This and (5.2) formally lead to

$$
0 = \frac{d}{d\tau} \int_{\Omega_\tau} \rho(x, t_0 + \tau)\, dx \Big|_{\tau = 0}
$$

$$
= \int_\Omega (\partial_t \rho(y, t_0) + \operatorname{div}(\rho(y, t_0)\, v(y, t_0)))\, dy. \tag{5.3}
$$

Since this holds true for all domains Ω and all times t_0, we can obtain the equation

$$\partial_t \rho + \text{div}(\rho v) = 0. \tag{5.4}$$

This is called[1] in jargon the "mass transport equation" (or simply "transport equation") or "continuity equation".

Of course, mass transport, corresponding to the conservation of matter, is not the only ingredient to accurately describe the dynamics of a fluid. Hence, one has to complement (5.4) with additional information. In particular, one can take into account Newton's law regarding momentum conservation. In this setting, a change in fluid momentum must correspond to the forces acting on the fluid (for simplicity, we assume here that the fluid is subject only to gravity and its own pressure force, with no other external forces acting on it). Note that the momentum corresponding to the mass of the fluid in a region of space Ω is given by the quantity $\int_\Omega \rho(x,t)v(x,t)\,dx$. The corresponding gravity force is provided by $-g e_n \int_\Omega \rho(x,t)\,dx$, where g is the gravitational acceleration constant (and $e_n = (0,\ldots,0,1)$, supposing that gravity is acting downward in the vertical direction). As for the pressure force, it arises from p in the normal direction along the surface $\partial\Omega$, with a conventional minus sign (so that a resistance to increasing velocity occurs when moving toward regions of high pressures, while low-pressure regions produce a sucking effect). These considerations transform Newton's law of momentum conservation into the formula

$$\frac{\partial}{\partial t} \int_\Omega \rho(x,t)v(x,t)\,dx = -g e_n \int_\Omega \rho(x,t)\,dx - \int_{\partial\Omega} p(x,t)\nu(x)\,d\mathcal{H}_x^{n-1},$$

$$\tag{5.5}$$

where, as usual, $\nu(x)$ is the unit external normal at $x \in \partial\Omega$.

---

[1] It is interesting to observe that in the framework of equation (5.4), one can reinterpret the divergence term in (2.11) as a transport term with velocity ∇w (that is, the effect of chemotaxis is to produce a transport of the biological population with a velocity proportional to the gradient of the chemical attractant).

Moreover, in chemistry, one often models the spread of some substrate with concentration ρ in a fluid with velocity v: in this context. Equation (5.4) is often called the "convection equation".

Furthermore, in light of (5.3) (applied here to ρv_j instead of ρ), we have that, for all $j \in \{1, \ldots, n\}$,

$$\frac{\partial}{\partial t} \int_\Omega \rho(x, t) v_j(x, t)\, dx$$

$$= \int_\Omega \left(\partial_t(\rho v_j)(x, t) + \text{div}((\rho v_j)(x, t)\, v(x, t)) \right) dx$$

$$= \int_\Omega \left(\rho(x, t) \partial_t v_j(x, t) + \rho(x, t) v(x, t) \cdot \nabla v_j(x, t) \right) dx,$$

where (5.4) has been exploited in the latter step.

Thus, using the notation $v \cdot \nabla$ to denote the operator $\sum_{i=1} v_i \partial_i$, possibly applied to a vector-valued function component-wise, we can write that

$$\frac{\partial}{\partial t} \int_\Omega \rho(x, t) v(x, t)\, dx$$

$$= \int_\Omega \left(\rho(x, t) \partial_t v(x, t) + \rho(x, t)(v(x, t) \cdot \nabla) v(x, t) \right) dx.$$

From this and (5.5), we obtain that

$$\int_\Omega \left(\rho(x, t) \partial_t v(x, t) + \rho(x, t)(v(x, t) \cdot \nabla) v(x, t) \right) dx$$

$$= -ge_n \int_\Omega \rho(x, t)\, dx - \int_{\partial\Omega} p(x, t) \nu(x)\, d\mathcal{H}_x^{n-1}. \qquad (5.6)$$

Also, from the divergence theorem, for all $j \in \{1, \ldots, n\}$,

$$\int_\Omega \partial_j p(x, t)\, dx = \int_\Omega \text{div}(p(x, t) e_j)\, dx = \int_{\partial\Omega} p(x, t) e_j \cdot \nu(x)\, d\mathcal{H}_x^{n-1}$$

and, in consequence,

$$\int_\Omega \nabla p(x, t)\, dx = \int_{\partial\Omega} p(x, t) \nu(x)\, d\mathcal{H}_x^{n-1}.$$

From these considerations and (5.6), we arrive at

$$\int_\Omega \left(\rho(x, t) \partial_t v(x, t) + \rho(x, t)(v(x, t) \cdot \nabla) v(x, t) \right) dx$$

$$= -ge_n \int_\Omega \rho(x, t)\, dx - \int_\Omega \nabla p(x, t)\, dx$$

and, accordingly,

$$\rho\partial_t v + \rho(v \cdot \nabla)v = -ge_n\rho - \nabla p. \tag{5.7}$$

The system consisting of the mass conservation equation in (5.4) and the momentum balance equation in (5.7) (plus possibly a constitutive relation linking pressure and density) constitutes what in jargon[2] are called the "Euler fluid flow equations".

In a sense, these equations are obtained for a rather ideal situation, and several modifications of the previous setting can be performed to account for more complex models. Among these modifications, one of the most popular consists of accounting for friction between fluid molecules, which creates a "viscosity" that resists the change in fluid velocity. More specifically, let us quantify[3] the difference between the velocity of a fluid parcel at x and that of the parcels nearby by the difference between $v(x, t)$ and its average in a

---

[2]The fountainhead of these equations is indeed the work of Leonhard Euler, see Figure 5.1.

See also Figure 5.2 for Euler's iconic portrait (by Jakob Emanuel Handmann). Many of us wonder why Euler is depicted in pajamas with a towel on his head. Arguably, it was not a pajama but a fashionable banyan and not a towel but a silk cloth. It was possibly believed at the time that loose and informal dress (such as the banyan, without a wig) contributed to the exercise of the faculties of the mind (and indeed, we also find it much more comfortable to do mathematics wearing informal clothes).

Note that Euler is portrayed as facing left, possibly to draw less attention to his right eyelid ptosis and a divergent strabismus. At age 31, Euler became almost blind in his right eye [AA13] (which might be a possible explanation why Euler was nicknamed "the cyclops" by Frederick II, the King of Prussia).

This partial loss of vision did not discourage Euler from producing the finest possible mathematics (actually, he allegedly stated, "Now I will have fewer distractions"). At age 59, a surgical restoration for a cataract in his left eye rendered Euler almost completely blind. Yet, Euler's passion and talent for mathematics remained undefeated, and he kept producing a vast number of impressive works which changed the course of mathematics until he died at age 76. In his eulogy, Marquis de Condorcet wrote *il cessa de calculer et de vivre* (French for "he ceased to calculate and to live").

[3]This idea of confronting the pointwise value of a function with the local average will be extensively used in Chapter 2 of the companion book [DV23], and it will be one of the leitmotifs of the study of harmonic functions.

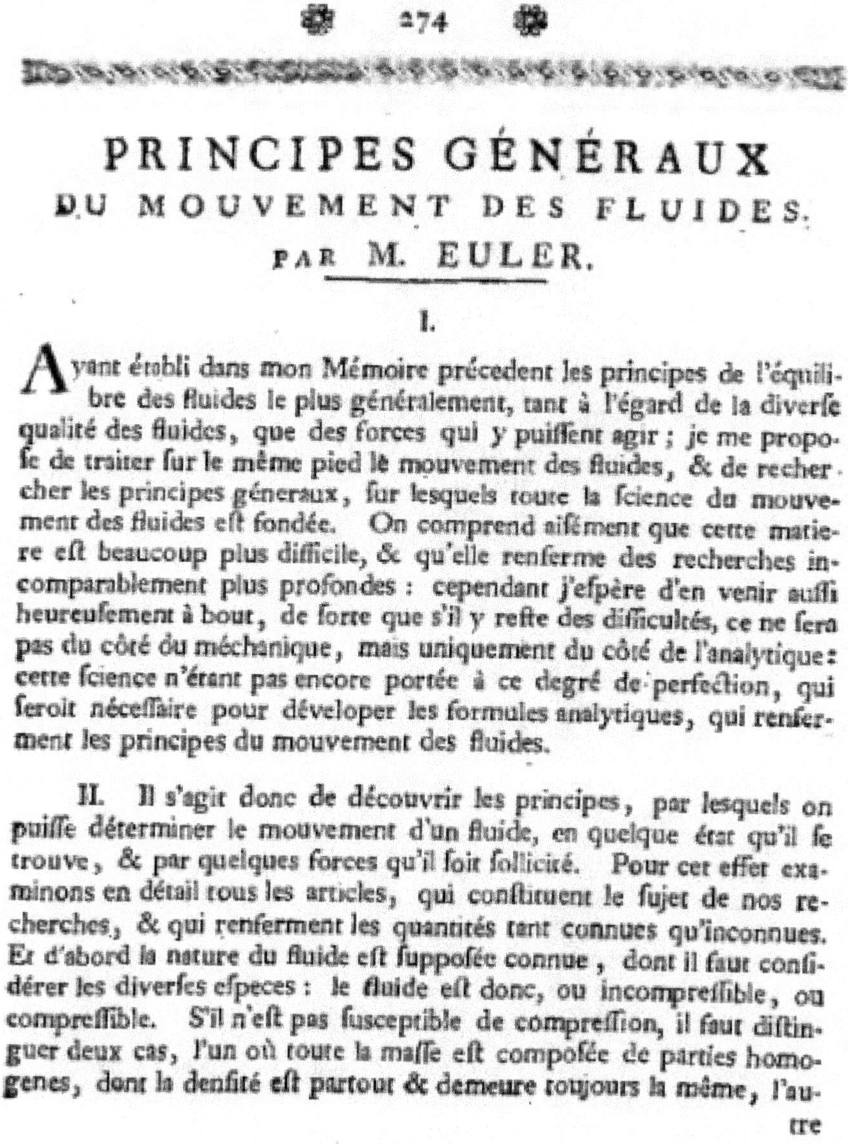

⚘   ᵃ74   ⚘

PRINCIPES GÉNÉRAUX
DU MOUVEMENT DES FLUIDES.
par M. EULER.

I.

Ayant établi dans mon Mémoire précedent les principes de l'équili-
bre des fluides le plus généralement, tant à l'égard de la diverse
qualité des fluides, que des forces qui y puiffent agir ; je me propo-
fe de traiter fur le même pied le mouvement des fluides, & de recher-
cher les principes généraux, fur lesquels toute la fcience du mouvement des fluides eft fondée. On comprend aifément que cette matie-
re eft beaucoup plus difficile, & qu'elle renferme des recherches in-
comparablement plus profondes : cependant j'efpère d'en venir auffi
heureufement à bout, de forte que s'il y refte des difficultés, ce ne fera
pas du côté du méchanique, mais uniquement du côté de l'analytique:
cette fcience n'étant pas encore portée à ce degré de perfection, qui
feroit néceffaire pour déveloper les formules analytiques, qui renferment les principes du mouvement des fluides.

II. Il s'agit donc de découvrir les principes, par lesquels on
puiffe déterminer le mouvement d'un fluide, en quelque état qu'il fe
trouve, & par quelques forces qu'il foit follicité. Pour cet effet exa-
minons en détail tous les articles, qui conftituent le fujet de nos recherches, & qui renferment les quantités tant connues qu'inconnues.
Et d'abord la nature du fluide eft fuppofée connue , dont il faut confi-
dérer les diverfes efpeces : le fluide eft donc, ou incompreffible, ou
compreffible. S'il n'eft pas fusceptible de compreffion, il faut diftin-
guer deux cas, l'un où toute la maffe eft compofée de parties homo-
genes, dont la denfité eft partout & demeure toujours la même, l'au-
tre

Figure 5.1. The title page to Euler's original article "Principes généraux du
mouvement des fluides", published in Mémoires de l'Académie des Sciences de
Berlin in 1757 (Public Domain source from Internet Archive).

Figure 5.2.   Portrait of Leonhard Euler (Public Domain image from Wikipedia).

small ball, say of radius h centered at x, namely

$$\mathcal{D}(x,t) := v(x,t) - \fint_{B_h(x)} v(y,t)\, dy.$$

By exploiting (2.4) and the polar coordinates, we find that

$$\mathcal{D}(x,t) = \fint_{B_h(x)} (v(x,t) - v(y,t))\, dy$$

$$= -\frac{1}{|B_h|} \int_0^h \left(\int_{\partial B_1} r^{n-1}(v(x+r\omega,t) - v(x,t))\, d\mathcal{H}_\omega^{n-1} \right) dr$$

$$= -\frac{|B_1|}{|B_h|} \int_0^h r^{n-1} \left(\frac{cr^2}{2} \Delta v(x,t) + o(h^2) \right) dr$$

$$= -\mu h^2 \Delta v(x,t) + o(h^2),$$

for some $\mu > 0$.

Then, we suppose that the velocity of the fluid is reduced when $\mathcal{D}$ is positive (since the parcel at x is faster than the ones in its vicinity; therefore, it is dragged back by them) and is enhanced when $\mathcal{D}$ is negative (since the parcel at x is slower than the ones in its vicinity; therefore, it is pulled forward by them). For simplicity, we can therefore assume that such an additional acceleration term is proportional to $\mathcal{D}$ (say, to make it a finite quantity in the limit, proportional to $\mathcal{D}/h^2$). By taking $h \searrow 0$, we thus obtain an additional acceleration (or deceleration) given by a term of the form $\mu \Delta v$. By incorporating this correction into (5.7), we thus find the equation

$$\rho \partial_t v + \rho(v \cdot \nabla)v = \mu \Delta v - g e_n \rho - \nabla p, \tag{5.8}$$

which is called in jargon[4] the Navier–Stokes equation.

We stress that (5.8) is a vectorial equation (or, equivalently, a system of equations in each component of v). There are also interesting situations in which (5.8) reduces to a scalar equation, such as when the fluid is constrained within an infinite pipe with a given direction. For instance, to make things as simple as possible, let us suppose that the fluid is incompressible; hence, we have that the density of the fluid remains constant during the flow (namely, a change in density over time would imply that the fluid had either compressed or expanded). That is, for incompressible fluids, the quantity $\rho(x(t), t)$ must be constant in time, leading to

$$0 = \frac{d}{dt}\rho(x(t), t) = \nabla \rho(x(t), t) \cdot \dot{x}(t) + \partial_t \rho(x(t), t)$$
$$= \nabla \rho(x(t), t) \cdot v(x(t), t) + \partial_t \rho(x(t), t).$$

---

[4]Equation (5.8) (possibly complemented with other structural equations of the fluid) was given by Claude Louis Marie Henri Navier and Sir George Gabriel Stokes, 1st Baronet. Our experience with the Laplacian as a "democratic" operator (recall the discussion on p. 24) may suggest that the Laplacian in (5.8) provides a smoothing effect. Although this may be the case, our current understanding of the solutions of (5.8) is unsatisfactory and highly incomplete. In particular, it is not yet known whether smooth (say, globally defined and meeting certain natural conditions) solutions always exist. If the reader finds an answer to this question, they will not only earn immortal fame among the circle of PDE enthusiasts but also obtain a substantial amount of money since the problem is listed as one of the Millennium Prize Problems, and a prize of US\$1 million will be awarded to the first person who provides a solution (the case of incompressible fluids in $\mathbb{R}^3$ is hard enough). See e.g. [Fef06, CJW06] for further details about this rather difficult way to become rich.

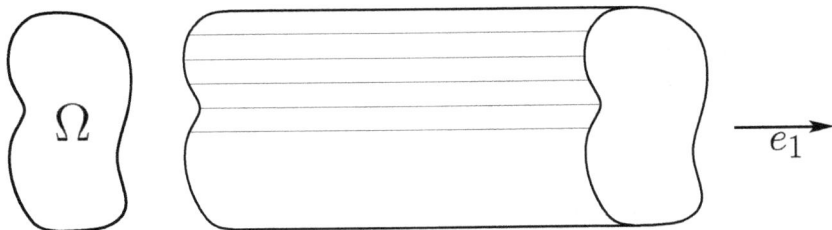

Figure 5.3. A horizontal pipe in the direction e_1 with cross-section Ω.

Combining this with (5.4),

$$\rho \operatorname{div} v = \operatorname{div}(\rho v) - \nabla\rho \cdot v = \operatorname{div}(\rho v) + \partial_t \rho(x(t), t) = 0,$$

whence

$$\operatorname{div} v = 0, \tag{5.9}$$

that is, the velocity is a divergence-free vector field.

For this reason, recalling (5.8), the stationary solutions of incompressible viscous fluids are the solutions of

$$\begin{cases} \mu \Delta v - \rho(v \cdot \nabla)v - ge_n\rho - \nabla p = 0, \\ \operatorname{div} v = 0. \end{cases} \tag{5.10}$$

Suppose now that the above fluid moves in parallel streamlines through a horizontal straight pipe of a given cross-sectional shape. Up to rotations which leave invariant the vertical directions, we can assume that the pipe has the form $\mathbb{R} \times \Omega$ for some $\Omega \subset \mathbb{R}^{n-1}$ and $v = (u, 0, \dots, 0)$ for some scalar function u (that is, the pipe directing the fluid is oriented toward the e_1-direction, see Figure 5.3). In this situation, we have that $\Delta v = \Delta u\, e_1$, $(v \cdot \nabla)v = u\partial_1 u\, e_1$ and $\operatorname{div} v = \partial_1 u$. Consistent with these observations, (5.10) reduces to

$$\begin{cases} \mu \Delta u\, e_1 - \rho u\partial_1 u\, e_1 - ge_n\rho - \nabla p = 0, \\ \partial_1 u = 0. \end{cases} \tag{5.11}$$

Taking the components $\{2, \dots, n-1\}$ of the first equation in (5.11), we see that $\partial_i p = 0$, for each $i \in \{2, \dots, n-1\}$, implying that the fluid pressure is constant in these directions. Taking the last component instead, we find $\partial_n p = -g\rho$.

The most interesting information, however, comes from the first component of the first equation in (5.11) since this provides an elliptic equation for $u : \Omega \to \mathbb{R}$. Indeed, focusing on this aspect, we infer from (5.11) that u is a solution of

$$\begin{cases} \mu \Delta u - \rho u \partial_1 u - \partial_1 p = 0, \\ \partial_1 u = 0. \end{cases}$$

Hence, plugging the second equation into the first one,

$$\mu \Delta u = \partial_1 p \quad \text{in } \Omega. \tag{5.12}$$

A special case of interest, which constitutes one of the main motivations for the theory that will be also presented in Sections 7.3 and 8.2 of the companion book [DV23], is when

the pressure rate is constant,

the velocity on the boundary of the pipe is constant $\qquad$ (5.13)

and the tangential stress on the pipe is constant.

The problem, originally posed by James Serrin [Ser71], is precisely to identify the possible shapes of the pipe that allow such special configurations (that is, to determine all the cross-sections Ω that give rise to solutions of (5.12) which are compatible with the prescriptions in (5.13)): this is certainly a problem of great practical importance since, ideally, one would like to design pipes that do not produce excessive tangential stress, to reduce material wear and the resulting formation of defects, holes and licking (for this, maintaining a constant stress along the pipe could be a nice way to balance the wearing somewhat uniformly along the whole surface of the pipe).

To efficiently address a question of this type, we first need to translate the physical conditions described in (5.13) into a mathematical framework. In order to achieve this, we point out that the prescription that the pressure rate is constant simply means that $\partial_1 p = c_1$, for some $c_1 \in \mathbb{R}$.

Also, the assumption that the velocity on the boundary of the pipe is constant means that $u = c_2$ along $\partial \Omega$, for some $c_2 \in \mathbb{R}$. As a matter of fact, since equation (5.12) remains invariant if we replace u

with $u - c_2$, we can simply assume[5] that $c_2 = 0$ and hence $u = 0$ along $\partial\Omega$.

As for the stress on the pipe, we can suppose that this is due to the fact that fluid parcels near the pipe may move faster or slower than the ones directly adhering to the pipe (that is, in the normalization above, the fluid parcels in the vicinity of the pipe could possess strictly positive or strictly negative velocities). We can additionally suppose that the wearing of the pipe is directly influenced by this change in velocity (i.e. some particles of the pipe are removed by fluid parcels moving either forward or backward). In this spirit, one can model the stress along the pipe to be proportional to the normal derivative of the fluid velocity. With this, the prescription of having a constant stress along the pipe is translated into $\partial_\nu u = c_3$ along $\partial\Omega$, for some $c_3 \in \mathbb{R}$.

In consideration of this, as well as normalizing constants for simplicity, we write (5.12) and (5.13) as

$$\begin{cases} \Delta u = 1 & \text{in } \Omega, \\ u = 0 & \text{on } \partial\Omega, \\ \partial_\nu u = c & \text{on } \partial\Omega. \end{cases} \tag{5.14}$$

This equation will indeed[6] be the starting point of the topics presented in Sections 7.3 and 8.2 of the companion book [DV23]. A different model leading to the same mathematical problem will also be presented in Chapter 16.

---

[5]Equivalently, one can argue that the invariance in the form of the description of physical problems among mutually translating reference frames allows us to choose an inertial frame of reference moving at a constant speed c_2. In this frame, the previous prescription reduces to that the velocity on the boundary of the pipe is equal to zero.

[6]We observe that (5.14) does possess a solution when Ω is a ball. Specifically, given $x_0 \in \mathbb{R}^n$ and $r > 0$, the function $u(x) := \frac{|x-x_0|^2 - r^2}{2n}$ is a solution of (5.14) in the ball $B_r(x_0)$ (and note that this solution corresponds to our physical intuition of the problem since it parallels a fluid reaching its maximal speed at the center of a circular pipe, with the velocity of the fluid being minimal at the boundary of the pipe due to the viscous effects).

We will discuss in Sections 7.3 and 8.2 of the companion book [DV23] whether this is the only type of possible solution or if there exist domains other than a ball that allow different kinds of solutions.

We end this section by recalling a simple, but helpful, use of the "principle of inertia" in the fluid dynamics setting. Namely, a case of special interest is provided by a rigid body (say, with a given shape described by a set Ω) moving in a fluid with a constant speed. In this situation, the equations presented in this section hold outside the moving domain described, for instance, by $\Omega(t) := \{x \in \mathbb{R}^3 \text{ s.t. } x + v_0 t \in \Omega\}$, for a given constant vector $v_0 \in \mathbb{R}^3$. Since fluid dynamics is already quite complex in a given domain, it is usually more convenient to change the inertial frame to describe the motion by following the moving body (and we expect by Newton's first law that this new choice of coordinates does not alter the physical description of the phenomena).

More explicitly, it comes in handy to define

$$\widetilde{v}(x,t) := v(x - v_0 t, t) + v_0, \quad \widetilde{p}(x,t) := p(x - v_0 t, t)$$
$$\text{and} \quad \widetilde{\rho}(x,t) := \rho(x - v_0 t, t) \tag{5.15}$$

and observe that equations (5.4), (5.7)–(5.9) are all preserved by this transformation since $\operatorname{div} \widetilde{v} = \operatorname{div} v$,

$$\partial_t \widetilde{\rho} + \operatorname{div}(\widetilde{\rho v}) = \partial_t \rho - v_0 \cdot \nabla \rho + \operatorname{div}(\rho(v + v_0))$$
$$= \partial_t \rho + \operatorname{div}(\rho v)$$

and

$$\widetilde{\rho}\partial_t \widetilde{v} + \widetilde{\rho}(\widetilde{v} \cdot \nabla)\widetilde{v} - \mu \Delta \widetilde{v} + g e_n \widetilde{\rho} + \nabla \widetilde{p}$$
$$= \rho(\partial_t v - v_0 \cdot \nabla v) + \rho((v + v_0) \cdot \nabla)v - \mu \Delta v + g e_n \rho + \nabla p$$
$$= \rho \partial_t v + \rho v \cdot \nabla v - \mu \Delta v + g e_n \rho + \nabla p.$$

Also, the new domain of reference is now simply the complement of Ω.

Note that the transformation in (5.15) involves simply replacing the situation of a rigid body moving at a constant speed in a fluid at rest at infinity with that of a still body in a fluid flowing at a constant speed at infinity. This is precisely the idea that led to the construction of wind tunnels: to replicate the aerodynamic interactions between a moving object and the surrounding air, it is common to construct large tubes with air blowing through them against a static model of the object.

In the forthcoming Chapters 8 and 9, which consider rigid bodies moving at a constant speed, we will tacitly assume that the transformation in (5.15) has been performed, allowing us to narrow our focus on the case of still objects and domains that do not vary with time.

Chapter 6

Irrotational Fluids

Among all possible fluids, a class deserving special attention is that consisting of fluids which have "no vortexes". In light of Stokes' theorem, this notion is made precise by the mathematical concept[1] of curl, that is, we say that a fluid is irrotational if the curl of its velocity field vanishes.

A natural question in this setting is whether vortexes can be produced out of nothing; for instance, if the initial conditions of a fluid present no vortexes, is it possible that vortexes arise at a later stage?

To increase our familiarity with multivariate calculus and partial differential equations, we show that this is impossible, at least for inviscid, incompressible and barotropic flows (in this setting, a flow is said to be barotropic if its density is a function of pressure alone, say $\rho = g(p)$ for some positive function g). More precisely, we consider a barotropic solution in the absence of gravity for the momentum balance equation in (5.7), satisfying the incompressibility condition in (5.9), that is,

$$
\begin{cases}
\rho \partial_t v + \rho(v \cdot \nabla)v = -\nabla p, \\
\operatorname{div} v = 0, \\
\rho = g(p),
\end{cases}
\tag{6.1}
$$

---

[1]While most of the arguments in other sections are valid in $\mathbb{R}^n$, when dealing with the curl, we restrict our attention to the case of $\mathbb{R}^3$. The case of $\mathbb{R}^2$ is also included simply by identifying $\mathbb{R}^2$ with $\mathbb{R}^2 \times \{0\} \subset \mathbb{R}^3$. The curl of the velocity field is often called "vorticity".

and we show that

if $\operatorname{curl} v = 0$ at time $t = 0$, then $\operatorname{curl} v = 0$ at every time t.   (6.2)

To achieve this goal, we first recall a vector calculus identity, which is valid for all smooth vector fields $V = (V_1, V_2, V_3) : \mathbb{R}^3 \to \mathbb{R}^3$:

$$(V \cdot \nabla)V = \frac{1}{2}\nabla|V|^2 - V \times (\operatorname{curl} V), \qquad (6.3)$$

where $\times$ denotes the vector product operation. To prove (6.3), up to exchanging the order of the coordinates, we can concentrate our attention on the first coordinate: hence, we aim at proving that

$$\sum_{j=1}^{3} V_j \partial_j V_1 = \frac{1}{2}\partial_1|V|^2 - (V \times (\operatorname{curl} V))_1. \qquad (6.4)$$

To check this, we calculate

$$(V \times (\operatorname{curl} V))_1 - \frac{1}{2}\partial_1|V|^2$$
$$= V_2(\operatorname{curl} V)_3 - V_3(\operatorname{curl} V)_2 - V \cdot \partial_1 V$$
$$= V_2(\partial_1 V_2 - \partial_2 V_1) - V_3(\partial_3 V_1 - \partial_1 V_3)$$
$$\quad - V_1\partial_1 V_1 - V_2\partial_1 V_2 - V_3\partial_1 V_3$$
$$= -V_2\partial_2 V_1 - V_3\partial_3 V_1 - V_1\partial_1 V_1,$$

which proves (6.4) and thus (6.3).

We also recall another vector calculus identity, which is valid for all smooth vector fields $V = (V_1, V_2, V_3)$, $W = (W_1, W_2, W_3)$: $\mathbb{R}^3 \to \mathbb{R}^3$:

$$\operatorname{curl}(V \times W) = \operatorname{div} W\, V - \operatorname{div} V\, W + (W \cdot \nabla)\, V - (V \cdot \nabla)\, W. \quad (6.5)$$

To prove this, we can write $V = V_1 e_1 + V_2 e_2 + V_3 e_3$ and note that, since (6.5) is linear in V, it suffices to prove it when V reduces to each of the components $V_j e_j$.

Thus, up to reordering the coordinates, we can focus on the case in which V is actually $V_1 e_1$. Hence, we calculate

$$\mathrm{curl}((V_1 e_1) \times W) + \mathrm{div}(V_1 e_1)\, W - (W \cdot \nabla)\, V_1 e_1 + ((V_1 e_1) \cdot \nabla)\, W$$

$$= \mathrm{curl}(0, -V_1 W_3, V_1 W_2) + \partial_1 V_1\, W - \sum_{j=1}^{3} W_j \partial_j V_1 e_1 + V_1 \partial_1 W$$

$$= (\partial_2(V_1 W_2) + \partial_3(V_1 W_3), -\partial_1(V_1 W_2), -\partial_1(V_1 W_3))$$

$$+ \partial_1 V_1\, W - \sum_{j=1}^{3} W_j \partial_j V_1 e_1 + V_1 \partial_1 W$$

$$= (\partial_2 V_1\, W_2 + V_1 \partial_2 W_2 + \partial_3 V_1\, W_3 + V_1 \partial_3 W_3 + \partial_1 V_1\, W_1$$

$$+ V_1 \partial_1 W_1, 0, 0)$$

$$= -\sum_{j=1}^{3} W_j \partial_j V_1 e_1$$

$$= (V_1 \partial_2 W_2 + V_1 \partial_3 W_3 + V_1 \partial_1 W_1, 0, 0)$$

$$= \mathrm{div}\, W\, V_1 e_1,$$

and this completes the proof of (6.5).

It is also useful to recall the vector calculus identity

$$\mathrm{div}(\mathrm{curl}\, V) = 0. \tag{6.6}$$

For this, we take any ball $B \subset \mathbb{R}^3$ and consider the spherical surface $S := \partial B$. We split S into two hemispheres, S^+ and S^-, with a common equator η, which is the intersection between the boundaries of the surfaces S^+ and S^-. Looking at η as a curve, the natural direction of travel along η when considered as the boundary of S^+ is opposite to the one obtained when considering it as the boundary of S^-, see Figure 6.1 (here, we are endowing S^+ and S^- with the external unit vector field of ∂B). As a result, the circulation of a vector field V along η considered as the boundary of S^+ (which we denote by $\mathcal{C}^+(V)$) is opposite to the circulation of V along η considered as

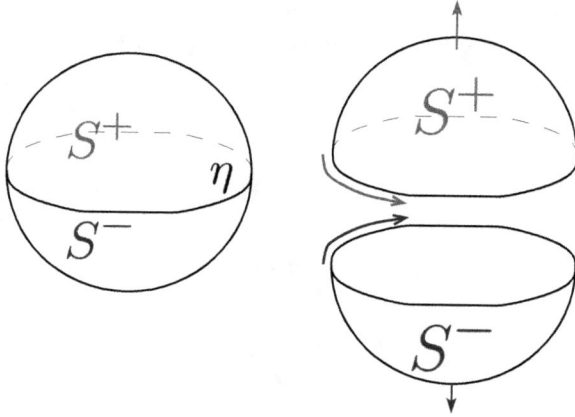

Figure 6.1.   The geometric argument used to prove (6.6).

the boundary of S^- (which we denote by $\mathcal{C}^-(V)$), that is,

$$\mathcal{C}^+(V) = -\mathcal{C}^-(V).$$

In addition, by Stokes' theorem, we know that $\mathcal{C}^\pm(V)$ agrees with the flux of $\operatorname{curl} V$ through the surface S^+.

Using these bits of information and the divergence theorem, we conclude that

$$0 = \mathcal{C}^+(V) + \mathcal{C}^-(V) = \int_{S^+} \operatorname{curl} V \cdot \nu + \int_{S^-} \operatorname{curl} V \cdot \nu$$
$$= \int_{\partial B} \operatorname{curl} V \cdot \nu = \int_B \operatorname{div}(\operatorname{curl} V).$$

From the arbitrariness of the ball B, we obtain (6.6), as desired.

We also point out that, for any smooth function ψ,

$$\operatorname{curl} \nabla \psi = 0. \qquad\qquad (6.7)$$

This can be checked by employing Stokes' theorem. Indeed, if Σ is any smooth oriented element of surface in $\mathbb{R}^3$ with boundary $\partial \Sigma$, the flux of the $\operatorname{curl} \nabla \psi$ through the surface Σ is equal to the circulation of $\nabla \psi$ over the loop $\partial \Sigma$. The latter object, if we describe $\partial \Sigma$ by a closed curve $\gamma : [0, L] \to \mathbb{R}^3$ with arc length parameterization, can be written as

$$\int_0^L \nabla\psi(\gamma(\tau))\cdot\dot\gamma(\tau)\,d\tau = \int_0^L \frac{d}{d\tau}\psi(\gamma(\tau))\,d\tau = \psi(\gamma(L))-\psi(\gamma(0)) = 0.$$

Since this is valid for an arbitrary surface Σ, the proof of (6.7) is complete.

Now, to prove (6.2), we let $\omega := \operatorname{curl} v$, and we use the momentum balance equation and the barotropicity in (6.1) in combination with (6.3). Then, we see that

$$\begin{aligned}
0 &= \operatorname{curl}\left(\partial_t v + (v\cdot\nabla)v + \frac{\nabla p}{\rho}\right) \\
&= \operatorname{curl}\left(\partial_t v + \frac{1}{2}\nabla|v|^2 - v\times(\operatorname{curl} v) + \frac{\nabla p}{g(p)}\right) \qquad (6.8)\\
&= \partial_t\omega + \operatorname{curl}\left(\frac{1}{2}\nabla|v|^2 - v\times\omega + \frac{\nabla p}{g(p)}\right).
\end{aligned}$$

Now, we observe that

$$\begin{aligned}
\operatorname{curl}\left(\frac{\nabla p}{g(p)}\right) &= \left(\partial_2\left(\frac{\partial_3 p}{g(p)}\right) - \partial_3\left(\frac{\partial_2 p}{g(p)}\right), \partial_3\left(\frac{\partial_1 p}{g(p)}\right) - \partial_1\left(\frac{\partial_3 p}{g(p)}\right),\right.\\
&\qquad \left.\partial_1\left(\frac{\partial_2 p}{g(p)}\right) - \partial_2\left(\frac{\partial_1 p}{g(p)}\right)\right)\\
&= \left(\frac{\partial_{23} p}{g(p)} - \frac{\partial_3 p\, g'(p)\,\partial_2 p}{g^2(p)} - \frac{\partial_{23} p}{g(p)} + \frac{\partial_2 p\, g'(p)\,\partial_3 p}{g^2(p)},\right.\\
&\qquad \frac{\partial_{13} p}{g(p)} - \frac{\partial_1 p\, g'(p)\,\partial_3 p}{g^2(p)} - \frac{\partial_{13} p}{g(p)} + \frac{\partial_3 p\, g'(p)\,\partial_1 p}{g^2(p)},\\
&\qquad \left.\frac{\partial_{12} p}{g(p)} - \frac{\partial_2 p\, g'(p)\,\partial_1 p}{g^2(p)} - \frac{\partial_{12} p}{g(p)} + \frac{\partial_1 p\, g'(p)\,\partial_2 p}{g^2(p)}\right)\\
&= (0,0,0).
\end{aligned}$$

From this, (6.7) and (6.8), it follows that

$$0 = \partial_t\omega - \operatorname{curl}(v\times\omega).$$

This and (6.5) lead to

$$\partial_t\omega = \operatorname{curl}(v\times\omega) = \operatorname{div}\omega\, v - \operatorname{div} v\,\omega + (\omega\cdot\nabla)v - (v\cdot\nabla)\omega.$$

Thus, combining the incompressibility condition in (6.1) and (6.6),

$$\partial_t \omega = (\omega \cdot \nabla) v - (v \cdot \nabla) \omega;$$

therefore following the fluid parcel $x(t)$,

$$\frac{d}{dt}\omega(x(t), t) = \partial_t \omega(x(t), t) + (\dot{x}(t) \cdot \nabla)\omega(x(t), t)$$

$$= \partial_t \omega(x(t), t) + (v(x(t), t) \cdot \nabla)\omega(x(t), t)$$

$$= (\omega(x(t), t) \cdot \nabla) v(x(t), t).$$

As a result, if $Z(t) = (Z_1(t), Z_2(t), Z_3(t)) := \omega(x(t), t)$, we have that Z is a solution of the Cauchy problem for the ordinary differential equations

$$\begin{cases} \dot{Z}(t) = \sum_{k=1}^{3} Z_k(t)\, \partial_k v(x(t), t), \\ Z(0) = 0. \end{cases}$$

By the uniqueness result for solutions of ordinary differential equations, we thereby infer that $Z(t) = 0$ for all $t \in \mathbb{R}$, and this completes[2] the proof of (6.2).

---

[2]As a technical detail, we point out the following: we are freely assuming that if, for a given function f, we know that $f(x(t), t) = 0$ for all times t, then we can infer $f(x, t) = 0$ for all spatial positions x and times t. This is because we are implicitly assuming that the flow of the fluid exists for all times; hence, given any position x, we can consider "flow x backward for a time t", that is, we consider the fluid parcel evolution starting at some point x_0 at time 0 arriving at x at time t (which gives us the possibility of choosing $x(t) = x$).

Chapter 7

Propagation of Sound Waves

Now, we reconsider the set of Euler fluid flow equations for the velocity v of an inviscid fluid; namely, we look at the mass conservation equation in (5.4), the momentum balance equation in (5.7) and a constitutive relation linking the pressure p and the density ρ (neglecting the gravity effects):

$$\begin{cases} \partial_t \rho + \mathrm{div}(\rho v) = 0, \\ \rho \partial_t v + \rho(v \cdot \nabla)v = -\nabla p, \\ p = f(\rho), \end{cases} \qquad (7.1)$$

for some function f. In practice, it is useful to take f to be strictly increasing (the higher the density of the fluid, the higher the pressure produced; this would also give that the fluid is barotropic in the setting used in Chapter 6). We can think of (7.1) as a very simple model for a gas, and our aim here is to understand the propagation of sound waves in such a medium.

For this objective, we first point out that a solution to (7.1) is provided by $(v, p, \rho) = (0, p_0, \rho_0)$, for every $\rho_0 \in (0, +\infty)$ and $p_0 := f(\rho_0)$. This configuration corresponds to the physical situation of a gas at rest, with constant density and pressure.

Suppose now that we perturb this configuration by creating a small variation in the gas pressure, for instance by singing or by playing[1] a guitar. The idea is thus to look for (at least approximate)

---

[1] To play a guitar, see Chapter 14.

solutions of the form

$$v(x,t) = \varepsilon v_1(x,t), \quad p(x,t) = p_0 + \varepsilon p_1(x,t) \quad \text{and}$$
$$\rho(x,t) = \rho_0 + \varepsilon \rho_1(x,t).$$

In this setting, ε is a small parameter, and our goal is to determine the functions (v_1, p_1, ρ_1) in order to satisfy the set of equations in (7.1), formally up to negligible errors in ε. To this end, we observe that

$$p_0 + \varepsilon p_1 = p = f(\rho) = f(\rho_0 + \varepsilon \rho_1)$$
$$= f(\rho_0) + \varepsilon f'(\rho_0)\rho_1 + o(\varepsilon) = p_0 + \varepsilon f'(\rho_0)\rho_1 + o(\varepsilon),$$

leading to the choice of

$$p_1 = f'(\rho_0)\rho_1,$$

up to higher orders in ε, which we here sloppily disregard.

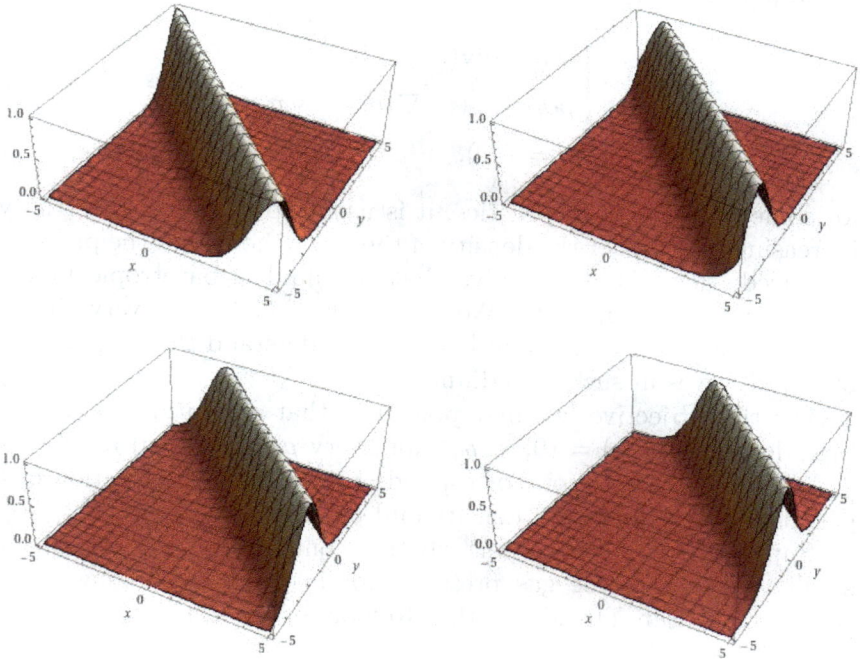

Figure 7.1. A plane wave.

Moreover,

$$0 = \partial_t \rho + \operatorname{div}(\rho v) = \varepsilon \partial_t \rho_1 + \varepsilon \rho_0 \operatorname{div} v_1 + o(\varepsilon)$$

and

$$0 = \rho \partial_t v + \rho(v \cdot \nabla)v + \nabla p = \varepsilon \rho_0 \partial_t v_1 + \varepsilon \nabla p_1 + o(\varepsilon)$$
$$= \varepsilon \rho_0 \partial_t v_1 + \varepsilon f'(\rho_0)\nabla \rho_1 + o(\varepsilon).$$

These observations give

$$\begin{cases} \partial_t \rho_1 + \rho_0 \operatorname{div} v_1 = 0, \\ \rho_0 \partial_t v_1 + f'(\rho_0)\nabla \rho_1 = 0. \end{cases}$$

As a consequence,

$$\partial_{tt}\rho_1 = -\partial_t(\rho_0 \operatorname{div} v_1) = -\operatorname{div}(\rho_0 \partial_t v_1) = \operatorname{div}(f'(\rho_0)\nabla \rho_1)$$
$$= f'(\rho_0)\Delta \rho_1.$$

That is, the perturbed density ρ_1 satisfies the equation

$$\partial_{tt} u = c^2 \Delta u, \tag{7.2}$$

with $c := \sqrt{f'(\rho_0)} > 0$. Equation (7.2) is called[2] in jargon the "wave equation".

To get a feel for the "propagation of waves" encoded in equation (7.2), we can consider a smooth function $u_\star : \mathbb{R} \to \mathbb{R}$ and a direction $\omega \in \partial B_1$, thus defining $u_\omega(x,t) := u_\star(\omega \cdot x - ct)$. Note that u_ω is indeed a solution of (7.2), physically corresponding to a traveling plane wave. Indeed, its evolution in time corresponds to a translation of $u_\star$ along the direction ω at speed c. Also, for $\kappa \in \mathbb{R}$, the parallel hyperplanes $\{\omega \cdot x - ct = \kappa\}$ (again, traveling with a constant

---

[2]In terms of the classification mentioned in footnote 13 on p. 9, equation (7.2) is hyperbolic. Indeed, we can take here $N = n+1$, $X = (x,t)$ and

$$a_{ij} = \begin{cases} c^2 & \text{if } i = j \in \{1,\ldots,n\}, \\ -1 & \text{if } i = j = n+1, \\ 0 & \text{otherwise}, \end{cases}$$

thus producing n strictly positive and one strictly negative eigenvalues.

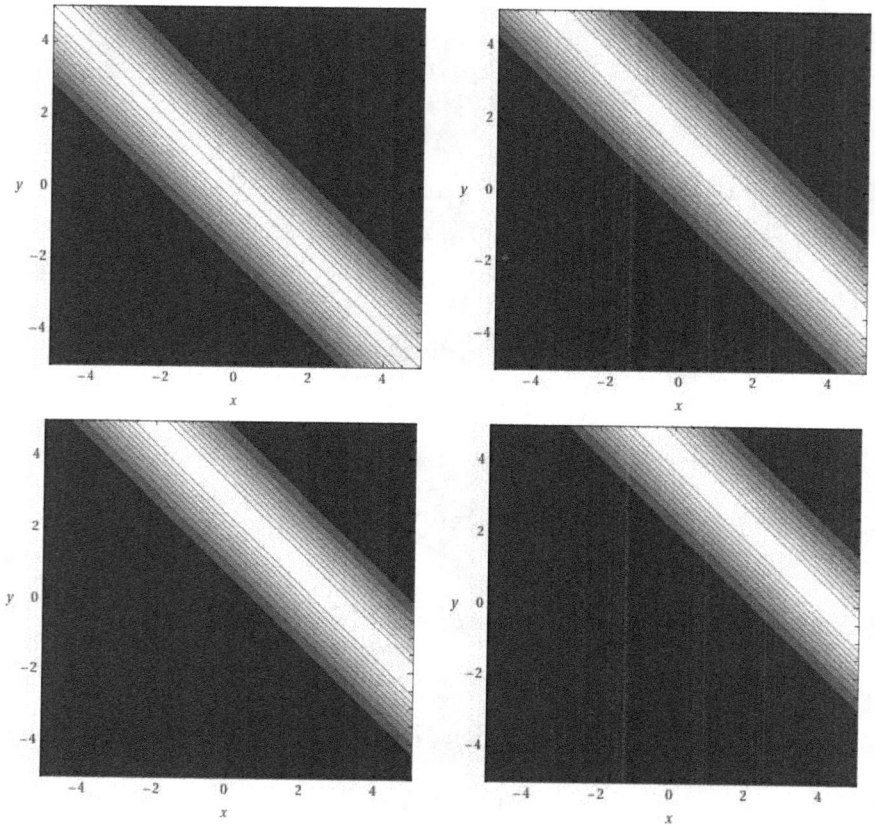

Figure 7.2. Wavefronts of the plane wave in Figure 7.1.

speed c in the direction ω) correspond to the level sets $\{u_\omega = \kappa_\star\}$, where $\kappa_\star := u_\star(\kappa)$ (the level sets of a wave are often called "wavefronts", and they hold special importance since they correspond to the surfaces on which, at a given moment of time, all particles of the medium undergo the same motion).

See Figure 7.1 for a series of frames of a traveling plane wave (this image is obtained by choosing $u_\star(r) := e^{-r^2}$, $\omega := \left(\frac{\sqrt{2}}{2}, \frac{\sqrt{2}}{2}\right)$, $c = 1$ and instants of time $t \in \{0, 1, 2, 3\}$). See also Figure 7.2 for the images of the corresponding level sets of the wave (i.e. the corresponding wavefronts).

Chapter 8

Hydrodynamics Doesn't Always Work Right

Now that we have built some confidence in the equations of hydrodynamics, it is time to question and challenge our own knowledge and to appreciate its power as well as its limitation. After all, it's not so important that one knows, but rather that one knows what one does not know.

In particular, the great Euler ("Master of us all", according to Laplace) allowing, we recall here a rather striking, and perhaps surprising, flaw in the hydrodynamic equations presented in Chapter 5 when describing aerodynamic forces.

We focus here on the notion of "drag", which is the force acting in the opposite direction to an object moving in a fluid (say, air resistance, for example). This is not merely a mathematical curiosity; for instance, according to Wikipedia: https://en.wikipedia.org/wiki/Drag_(physics), "induced drag tends to be the most important component for airplanes during take-off or landing flight" and, more importantly, "in the physics of sports, the drag force is necessary to explain the motion of balls, javelins, arrows and frisbees and the performance of runners and swimmers".

The issue of drag that we discuss here was discovered by Jean le Rond d'Alembert in 1752. Often, this remarkable discovery is referred to as "d'Alembert's paradox"; see Figure 8.1 for a portrait of d'Alembert's intense glance (by Maurice Quentin de La Tour).

Figure 8.1.   Portrait of Jean Le Rond d'Alembert (Public Domain image from Wikipedia).

In a nutshell, we will prove that

   if a body is moving with constant velocity

   in an inviscid, irrotational and steady flow with constant density,

   then the corresponding drag force is zero.   (8.1)

Of course, we may wonder how such a statement can be coherent with our own everyday experience, in which the effect of air resistance is usually apparent and quite decisive (at least when we catch a flight or kick a ball). Though a full understanding of hydrodynamics is likely well beyond the present possibilities of science, one common explanation for the discrepancy between the claim in (8.1) and real-life situations relies on the fact that the description of a fluid in (8.1) is too idealistic. In particular, the occurrence of the paradox is usually attributed to the neglected effects of viscosity[1]

---

[1]The importance of viscosity actually reflects the importance of the Laplace operator in the model: compare equations (5.7) and (5.8).

(in rough terms, one of the effects of viscosity [Pra05] is to produce thin boundary layers near the surface of the moving object, which may entail friction, flow separation and a low-pressure wake, leading to pressure drag). However, the official resolution of the question is possibly debatable in its full generality; for instance, in relation to the paradoxical features of (8.1), Garrett Birkhoff [Bir50] states[2] that, "I think that to attribute them all to the neglect of viscosity is an unwarranted oversimplification. The root lies deeper, in lack of precisely that deductive rigor whose importance is so commonly minimized by physicists and engineers".

Now, we work out the mathematical details needed to prove the claim in (8.1). To achieve this goal, we recall the Euler equation in (5.7), and we complement it with the mass transport equation in (5.4), the assumption that the density is constant and the irrotationality condition. We also assume that the flow is "steady", a term that usually refers to a situation in which the fluid properties at a point in the system do not change over time (and they only depend on the point); concretely, to avoid ambiguities, we suppose here that

---

Though we do not explicitly use this here, for completeness, we point out that an irrotational fluid with constant density is automatically incompressible (because in this case, $\operatorname{div} v = \frac{1}{\rho} \operatorname{div}(\rho v) = \frac{1}{\rho}(\partial_t \rho + \operatorname{div}(\rho v)) = 0$ by the continuity equation in (5.4)), and an incompressible and irrotational fluid is automatically inviscid (hence, the lack of viscosity is also a byproduct of two assumptions, namely incompressibility and the lack of vortexes, which, together, entail very restrictive byproducts); indeed, using the vector calculus identity

$$\nabla(\operatorname{div} V) - \operatorname{curl}(\operatorname{curl} V) = \Delta V$$

(see e.g. (19.2) for a proof), we have that for incompressible and irrotational fluid, the following holds:

$$\Delta v = \nabla(\operatorname{div} v) - \operatorname{curl}(\operatorname{curl} v) = 0 - 0 = 0.$$

Hence, the Navier–Stokes equation in (5.8) boils down to the Euler equation in (5.7).

[2]Garrett Birkhoff's statements were in turn criticized by James J. Stoker [Sto51] (who was concerned that readers were "very likely to get wrong ideas about some of the important and useful achievements in hydrodynamics", being misled by the "negative aspects of the theory"). Some of the original statements in [Bir50] were revised in the second edition [Bir60].

All in all, much more knowledge has to be developed before we can reach a complete understanding of fluid mechanics and hydrodynamics.

the relevant quantities of the fluid, such as velocity and pressure, follow the body moving at a constant velocity v_0 (that is, they are not functions of x and t separately but of $x - v_0 t$). Equivalently, according to the principle of inertia discussed on p. 65, if we consider a system of reference moving together with the body at the constant speed v_0, then the relevant quantities of the fluid would be constant in time (in particular, $\partial_t v = 0$).

Hence, in this system, by (5.4) and (5.7), the velocity v of a steady and irrotational fluid, subject to no gravitational effects and with a constant density (say $\rho := 1$) can be described by the following set of equations (valid outside the body Ω, which is supposed to remain within this reference system):

$$\begin{cases} (v \cdot \nabla)v = -\nabla p, \\ \operatorname{div} v = 0, \\ \operatorname{curl} v = 0. \end{cases} \tag{8.2}$$

Now, combining (6.3) with the irrotationality condition in (8.2), we deduce that

$$(v \cdot \nabla)v = \frac{1}{2}\nabla|v|^2. \tag{8.3}$$

We also consider the differential form $\omega := v_1 \, dx_1 + v_2 \, dx_2 + v_3 \, dx_3$, and we observe that

$$\begin{aligned} d\omega &= -\partial_2 v_1 \, dx_1 \wedge dx_2 - \partial_3 v_1 \, dx_1 \wedge dx_3 + \partial_1 v_2 \, dx_1 \wedge dx_2 \\ &\quad - \partial_3 v_2 \, dx_2 \wedge dx_3 + \partial_1 v_3 \, dx_1 \wedge dx_3 + \partial_2 v_3 \, dx_2 \wedge dx_3 \\ &= (\operatorname{curl} v)_3 \, dx_1 \wedge dx_2 - (\operatorname{curl} v)_2 \, dx_1 \wedge dx_3 + (\operatorname{curl} v)_1 \, dx_1 \wedge dx_2 \\ &= 0, \end{aligned}$$

thanks again to the irrotationality condition in (8.2).

Accordingly, the differential form ω is closed and, therefore, exact thanks to the Poincaré lemma (see e.g. [Con93, Theorem 8.3.8]). This implies there exists a velocity potential function φ satisfying $\omega = d\varphi$, that is,

$$v = \nabla\varphi. \tag{8.4}$$

We stress that φ is harmonic outside Ω since $0 = \operatorname{div} v = \operatorname{div} \nabla\varphi = \Delta\varphi$. Moreover, the first equation in (8.2)–(8.4) entail[3] that

$$0 = (v \cdot \nabla)v + \nabla p = \frac{1}{2}\nabla|v|^2 + \nabla p = \frac{1}{2}\nabla|\nabla\varphi|^2 + \nabla p$$

$$= \nabla\left(\frac{1}{2}|\nabla\varphi|^2\right) + \nabla p. \tag{8.5}$$

With this, we can now compute the drag exerted by air on the moving body. For this, we recall that the force F of the air is produced by the pressure, acting normally on the surface of the body (see Equation (5.5)). Hence, denoting the moving object as Ω and ν as its outer normal, we have that

$$F = -\int_{\partial\Omega} p\nu. \tag{8.6}$$

As for the minus sign here, we recall the convention that positive pressures are assumed to act in the opposite direction with respect to the exterior normal. Since the drag D is the force acting in the direction of motion of the body, using the divergence theorem, we thus conclude that

$$D = F \cdot \frac{v_0}{|v_0|} = -\int_{\partial\Omega} p\frac{v_0 \cdot \nu}{|v_0|} = \int_{\mathbb{R}^3 \setminus \Omega} \operatorname{div}\left(p\frac{v_0}{|v_0|}\right) = \int_{\mathbb{R}^3 \setminus \Omega} \frac{\nabla p \cdot v_0}{|v_0|}. \tag{8.7}$$

Note that to apply the divergence theorem here to the infinite region $\mathbb{R}^3 \setminus \Omega$ (which corresponds to the region occupied by the fluid), we implicitly suppose that the pressure decays sufficiently fast at infinity (that is, the pressure disturbance is essentially localized in the vicinity of the moving object, which is a reasonable[4] physical assumption, though quantifying it precisely and checking it rigorously would require some technical skills).

---

[3]In Chapter 9, we will have a closer look at expressions such as the one in (8.5), and we will interpret them as a form of Bernoulli's principle.

[4]Once again, in this chapter, if an argument sounds convincing, we buy it! We will be much more skeptical about heuristic reasoning from the following chapter onward. For instance, when we need to integrate on exterior domains, as described

Figure 8.2. Portrait of Nikolai Yegorovich Joukowski, Museum of Moscow Aviation Institute (image by Just from Wikipedia, licensed under the Creative Commons Attribution-Share Alike 4.0 International license).

---

in Section 6.1 of the companion book [DV23], the reader will appreciate the care taken in explicitly checking the appropriate decay properties of the functions involved (which is perhaps a delicate and annoying, albeit absolutely necessary, detail to be taken care of to avoid nonsensical computations).

In any case, let us mention that utilizing the divergence theorem in an exterior domain in (8.7) is quite a delicate matter (see e.g. a thoughtful argument in [Bat99, formulas (6.4.2) and (6.4.3)]), and it is a byproduct of the decay at infinity of the fluid velocity and pressure, which is modeled on that of harmonic functions.

Arguably, in the setting of these notes, alternative arguments to the ones in [Bat99, formulas (6.4.2) and (6.4.3)] could be provided by using Green's representation formula in equation (2.7.11) in the companion book [DV23] (say, in the domain $B_R \setminus \overline{\Omega}$, sending $R \to +\infty$); for this, one would need to assume that the velocity potential has a limit at infinity (thus, from Cauchy's estimates in Theorem 2.17.1 of the companion book [DV23], one can also bound the derivative of the potential along ∂B_R).

In any case, the decay estimates of the velocity fields rely on the fact that the fluid cannot penetrate into the object (as formalized in (8.9)) since this

We also point out that, for all $k \in \{1, 2, 3\}$,

$$\partial_k \left(\frac{1}{2} |\nabla \varphi|^2 \right) = \sum_{j=1}^{3} \partial_j \varphi \, \partial_{jk} \varphi = \sum_{j=1}^{3} \partial_j (\partial_j \varphi \, \partial_k \varphi) - \Delta \varphi \partial_k \varphi$$

$$= \operatorname{div}(\partial_k \varphi \nabla \varphi) - 0.$$

Consequently,

$$v_0 \cdot \nabla \left(\frac{1}{2} |\nabla \varphi|^2 \right) = \sum_{k=1}^{3} v_{0,k} \partial_k \left(\frac{1}{2} |\nabla \varphi|^2 \right)$$

$$= \sum_{k=1}^{3} v_{0,k} \operatorname{div}(\partial_k \varphi \nabla \varphi) = \operatorname{div}((v_0 \cdot \nabla \varphi) \nabla \varphi).$$

This entails that

$$\int_{\mathbb{R}^3 \setminus \Omega} \frac{v_0}{|v_0|} \cdot \nabla \left(\frac{1}{2} |\nabla \varphi|^2 \right) = \frac{1}{|v_0|} \int_{\mathbb{R}^3 \setminus \Omega} \operatorname{div}((v_0 \cdot \nabla \varphi) \nabla \varphi)$$

$$= -\frac{1}{|v_0|} \int_{\partial \Omega} (v_0 \cdot \nabla \varphi)(\nabla \varphi \cdot \nu) = -\frac{1}{|v_0|} \int_{\partial \Omega} (v_0 \cdot v)(v \cdot \nu).$$

Hence, recalling (8.5) and (8.7),

$$D = \int_{\mathbb{R}^3 \setminus \Omega} \frac{\nabla p \cdot v_0}{|v_0|} = -\int_{\mathbb{R}^3 \setminus \Omega} \frac{v_0}{|v_0|} \cdot \nabla \left(\frac{1}{2} |\nabla \varphi|^2 \right)$$

$$= \frac{1}{|v_0|} \int_{\partial \Omega} (v_0 \cdot v)(v \cdot \nu). \tag{8.8}$$

---

information allows one to get rid of one order of magnitude in the corresponding estimates. This is an interesting physical feature since it reveals the "global" character of fluid dynamics, where some features in the proximity of the moving body have a significant influence on the velocity field at infinity, and vice versa.

In general, the decay analysis for solutions of fluid mechanics equations is a delicate matter, see e.g. [Kat72] and the references therein.

We now stress that since the air cannot penetrate into the object, necessarily,

the normal component of the fluid velocity vanishes along $\partial\Omega$,

$$(8.9)$$

namely $v \cdot \nu = 0$ on $\partial\Omega$. This and (8.8) yield that $D = 0$, which completes the proof of d'Alembert's paradox in (8.1).

See e.g. [Ste81, CM93, Bat99, RM18] and the references therein for additional information on d'Alembert's paradox.

Chapter 9

Lift of an Airfoil

To safely recover from the shock received in Chapter 8 by d'Alembert's paradox, we present now a positive result in terms of the aerodynamic lift of an airfoil. This result was established independently by Martin Kutta and Nikolai Yegorovich Joukowski in the early twentieth century and, in its simplest formulation, can be stated[1] as follows:

an airfoil in relative motion with a constant velocity $-v_0$

to an ambient inviscid homogeneous, irrotational fluid

has a lift force (that is the component of the force

perpendicular to v_0) of magnitude $\rho |v_0| \Gamma$, $\qquad$ (9.1)

where Γ denotes a circulation of v_0 along the cross-section

of the airfoil.

It is interesting to observe that (9.1) explains, for instance, the generation of a lift on a wing as a result of the contributions from the circulation Γ of the velocity field around the wing. Remarkably, the fact that (9.1) takes into account the component of the force

---

[1]Though the theory that we present in these pages is far too crude to capture aerodynamics in its full complexity, the importance of a statement like (9.1) in the theory of flight appears to be paramount, as confirmed by the high recognition bestowed to the scientists involved. For instance, the Russian Air Force Academy and one of the airports in Moscow were named after Joukowski. See also Figure 8.2.

perpendicular to v_0 clearly shows the importance of such a result for the theory of flight (that is, this lift force is what allows, in principle at least, an airplane to take off, provided they manage to create a sufficiently large circulation Γ).

To model an airfoil, i.e. a "long wing" (actually, infinitely long for simplicity), we consider a (nice, contractible) planar domain $\Omega \subset \mathbb{R}^2$ and the wing given by $\Omega_\star := \Omega \times \mathbb{R}$, see Figure 9.1. We reconsider the Euler equation in (5.7) (in the absence of gravity) and the incompressibility condition in (5.9) outside $\Omega_\star$, assuming that, by symmetry, the fluid parameters v and p are actually independent of x_3 as well as time (leading to a completely steady solution of the problem). In this way, we may consider the following set of equations for $v = v(x,y)$ and $p = p(x,y)$, with $X = (x,y) \in \mathbb{R}^2 \setminus \Omega$, for a given constant $\rho \in (0, +\infty)$ (the constancy of the density being the mathematical translation of the fact that the fluid is homogeneous):

$$\begin{cases} \rho(v \cdot \nabla)v = -\nabla p, \\ \operatorname{div} v = 0, \\ \operatorname{curl} v = 0. \end{cases} \qquad (9.2)$$

In terms of real-world applications, a common belief is indeed that the setting in (9.1) is too idealized to detect the intricate patterns produced by real fluids in the vicinity of a traveling object; however, it is also believed that the circulation detected in (9.1) does reflect significant physical information since, in many concrete situations, the flow around a thin airfoil is composed of a narrow viscous region near the body (a sort of "boundary layer") outside which the idealized and inviscid description of the flow in (9.1) turns out to be sufficiently realistic.

The gist is then to apply the setting in (9.1), not quite to the traveling body, but rather to the aggregate of the body and its own boundary layer; in particular, the loop to compute the circulation in (9.1) must be chosen outside this boundary layer.

With this, the boundary layer gives any traveling object an "effective shape" that may be different from its physical shape by accounting for the region in which the velocity changes from zero at the surface to the stream value away from the surface. The effectiveness of (9.1) thus lies in incorporating, as much as possible, the effects of turbulence and skin friction into the effective shape described by this boundary layer, while maintaining a more mathematically treatable description away from it.

In this sense, the success of the Kutta–Joukowski theory lies in providing a simple, but not trivial, approach to aerodynamics which incorporates some aspects of viscous effects while neglecting others.

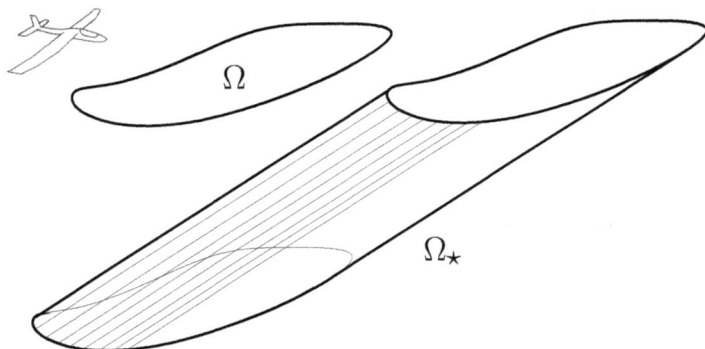

Figure 9.1.   Two-dimensional model of an airfoil, as inspired by a sailplane.

Note that in this setting (considering the vector field $(v_1(x,y),$ $v_2(x,y),0)$ to be three-dimensional but with a trivial last entry to compute the curl), we have that

$$0 = \operatorname{curl} v = (\partial_x v_2 - \partial_y v_1)e_3. \tag{9.3}$$

Moreover, since, according to the principle of inertia discussed on p. 65, we take an inertial frame in which the body appears to be still:

$$\lim_{|(x,y)| \to +\infty} v(x,y) = v_0. \tag{9.4}$$

Also, the Euler equation in (9.2) and the assumption that the density is constant give that

$$v \cdot \nabla \left(\frac{|v|^2}{2} + \frac{p}{\rho} \right) = v \cdot \left(\sum_{k=1}^{2} v_k \nabla v_k + \frac{\nabla p}{\rho} \right)$$

$$= v \cdot \left(\sum_{k=1}^{2} v_k \nabla v_k - (v \cdot \nabla)v \right)$$

$$= \sum_{j,k=1}^{2} v_j v_k \partial_j v_k - v \cdot \left(\sum_{j=1}^{2} v_j \partial_j v \right)$$

$$= \sum_{j,k=1}^{2} v_j v_k \partial_j v_k - \sum_{j,k=1}^{2} v_k v_j \partial_j v_k = 0.$$

As a consequence, if $\beta(X) := \frac{|v(X)|^2}{2} + \frac{p(X)}{\rho}$,

$$v \cdot \nabla \beta = 0,$$

and thus

$$\frac{d}{dt}\beta(X(t)) = \nabla\beta(X(t)) \cdot \dot{X}(t) = \nabla\beta(X(t)) \cdot v(X(t)) = 0.$$

As a result, $\beta(X(t)) = \beta(X(0))$ for every time t, that is,

$$\frac{|v(X(t))|^2}{2} + \frac{p(X(t))}{\rho} = \frac{|v(X(0))|^2}{2} + \frac{p(X(0))}{\rho}. \qquad (9.5)$$

Equations such as (9.5) are often referred to by the name Bernoulli's principle.

Now, we rely on the two-dimensional structure of the problem combined with the divergence-free condition in (9.2) to see that the differential form $\zeta := v_2\,dx - v_1\,dy$ is closed since

$$d\zeta = -(\partial_y v_2 + \partial_x v_1)\,dx \wedge dy = -\operatorname{div} v\,dx \wedge dy = 0. \qquad (9.6)$$

As a matter of fact, the exterior of Ω is not simply connected in $\mathbb{R}^2$; however, we can consider two regions $\mathcal{R}_1$ and $\mathcal{R}_2$ which are simply connected, and their union is the exterior of Ω, see Figure 9.2. In this setting, we can apply the Poincaré lemma (see e.g. [Con93, Theorem 8.3.8]) to each of the regions $\mathcal{R}_1$ and $\mathcal{R}_2$, thereby finding two

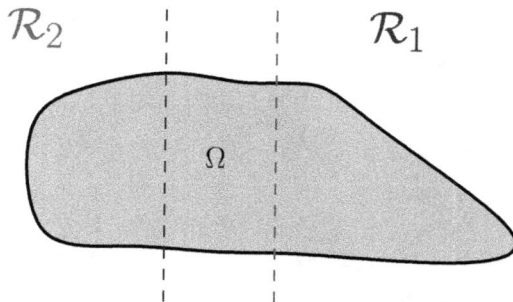

Figure 9.2.   A topological argument used to account for the fact that $\mathbb{R}^2 \setminus \Omega$ has a hole.

potentials $\psi_j : \mathcal{R}_j \to \mathbb{R}$ such that $d\psi_j = \zeta$ in $\mathcal{R}_j$, with $j \in \{1, 2\}$. In particular,

$$\partial_x \psi_j = v_2 \quad \text{and} \quad \partial_y \psi_j = -v_1. \tag{9.7}$$

Note that $\mathcal{R}_1 \cap \mathcal{R}_2$ consists of two connected components which reach infinity, say $\mathcal{S}_1$ and $\mathcal{S}_2$. We observe that

$$\nabla(\psi_1 - \psi_2) = \nabla\psi_1 - \nabla\psi_2 = (v_2, -v_1) - (v_2, -v_1) = 0,$$

in $\mathcal{S}_1$; hence, there exists $c_1 \in \mathbb{R}$ such that $\psi_1 - \psi_2 = c_1$ in $\mathcal{S}_1$.

Similarly, there exists $c_2 \in \mathbb{R}$ such that $\psi_1 - \psi_2 = c_2$ in $\mathcal{S}_2$. Up to incorporating the constant into ψ_2 (which leaves (9.7) invariant), we can suppose that $c_2 = 0$.

We claim that

$$c_1 = 0. \tag{9.8}$$

For this, let $A \in (\partial\Omega) \cap \overline{\mathcal{S}_1}$ and $B \in (\partial\Omega) \cap \overline{\mathcal{S}_2}$, and let $\vartheta : [0, \ell] \to (\partial\Omega) \cap \overline{\mathcal{R}_1}$ be a curve, parameterized by its arc length, lying in $(\partial\Omega) \cap \overline{\mathcal{R}_1}$ and joining A to B, say with $\vartheta(0) = A$ and $\vartheta(\ell) = B$. It follows from the impenetrability condition in (8.9) that the velocity field is tangential to $\partial\Omega$ and, consequently,

$$v(\vartheta(\tau)) = \alpha(\tau)\,\dot{\vartheta}(\tau) \quad \text{for all } \tau \in [0, \ell], \tag{9.9}$$

for some scalar function $\alpha(\tau)$.

Thus, since ϑ lies in $\overline{\mathcal{R}_1}$, we can utilize (9.7), with $j := 1$, and write that

$$(-\partial_y \psi_1(\vartheta(\tau)), \partial_x \psi_1(\vartheta(\tau))) = \alpha(\tau)\,\dot{\vartheta}(\tau)$$

and, therefore,

$$\nabla\psi_1(\vartheta(\tau)) = \alpha(\tau)\,(\dot{\vartheta}_2(\tau), -\dot{\vartheta}_1(\tau)).$$

From this, we obtain that

$$\frac{d}{d\tau}\psi_1(\vartheta(\tau)) = \nabla\psi_1(\vartheta(\tau)) \cdot \dot{\vartheta}(\tau)$$

$$= \alpha(\tau)\,(\dot{\vartheta}_2(\tau), -\dot{\vartheta}_1(\tau)) \cdot (\dot{\vartheta}_1(\tau), \dot{\vartheta}_2(\tau)) = 0$$

and, consequently,

$$\psi_1(B) = \psi_1(\vartheta(\ell)) = \psi_1(\vartheta(0)) = \psi_1(A).$$

Similarly, by considering a curve lying in $(\partial\Omega)\cap\overline{\mathcal{R}_2}$, one obtains that

$$\psi_2(B) = \psi_2(A).$$

As a result,

$$0 = c_2 = \psi_1(B) - \psi_2(B) = \psi_1(A) - \psi_2(A) = c_1,$$

and the proof of (9.8) is complete.

Hence, we can now define, in the exterior of Ω,

$$\psi := \begin{cases} \psi_1 & \text{in } \mathcal{R}_1, \\ \psi_2 & \text{in } \mathcal{R}_2, \end{cases}$$

and we stress that this is a fair definition since $\psi_1 = \psi_2$ in $\mathcal{R}_1 \cap \mathcal{R}_2$, owing to (9.8).

Furthermore, by (9.7), in the exterior of Ω,

$$\partial_x\psi = v_2 \quad \text{and} \quad \partial_y\psi = -v_1. \tag{9.10}$$

Note in particular that $|\nabla\psi| = v$. Moreover, ψ is harmonic, thanks to the irrotationality situation pointed out in (9.3).

The function ψ is often referred to as the "stream function". This name comes from a physical intuition. Indeed, the streamlines, i.e. the trajectories of fluid particles, are described by the level sets of the stream function because

$$\frac{d}{dt}\psi(X(t)) = \nabla\psi(X(t)) \cdot \dot{X}(t) = \nabla\psi(X(t)) \cdot v(X(t)) = 0, \tag{9.11}$$

thanks to (9.10).

Let us now reassess Bernoulli's principle in (9.5). In our setting, we need to refine (9.5) along the boundary of our moving object, namely we claim that, for every $x \in \partial\Omega$,

$$|v(x)|^2 + \frac{2p(x)}{\rho} = p_0, \tag{9.12}$$

for some constant $p_0 \in \mathbb{R}$.

To prove this, we take the simplifying assumption that the velocity field possesses at most finitely many zeros along $\partial\Omega$. We denote by $Z_1,\ldots,Z_N$ these zeros (if any). We pick $X_0 \in (\partial\Omega)\setminus\{Z_1,\ldots,Z_N\}$. Up to rigid motion, we assume that $X_0 = 0$ and that the tangent vector of $\partial\Omega$ at 0 is horizontal. Say, we describe Ω near 0 as the subgraph of a smooth function $f : \mathbb{R} \to \mathbb{R}$, with $f(0) = 0$ and $f'(0) = 0$. We can also replace ψ with $\psi - \psi(0)$ and thus assume additionally that $\psi(0) = 0$. In this setting, from the impenetrability condition in (8.9), we have that the fluid vector field at the origin is horizontal and thus, by (9.10), we have that $\nabla\psi(0)$ is vertical and different from zero. Let us suppose that

$$\partial_y\psi(0) > 0, \tag{9.13}$$

with the case $\partial_y\psi(0) < 0$ being similar, except that the level sets of ψ correspond to negative, instead of positive, values.

In this framework, near 0, the level sets of ψ can be written as a graph of smooth functions; more precisely, near the origin, for small $\delta > 0$, the level set $\{\psi = \delta\}$ can be identified with the graph $\{y = f_\delta(x)\}$ for a suitable f_δ. It follows from (9.13) that $f_{\delta'} \geqslant f_\delta$ if $\delta' \geqslant \delta$. Also, since, by (9.11), $\{\psi = \delta\}$ describes the streamlines of the fluid, we have that $\{\psi = \delta\}$ is contained in the complement of Ω for all $\delta > 0$, and consequently, near the origin, $\partial\Omega$ lies below the graph of f_δ for all $\delta > 0$.

Thus, given the monotonicity of f_δ, near the origin, we can define

$$f_0(x) := \lim_{\delta\searrow 0} f_\delta(x),$$

and we have that $f \leqslant f_0$. Moreover, since

$$\psi(0, f_0(0)) = \lim_{\delta\searrow 0} \psi(0, f_\delta(0)) = \lim_{\delta\searrow 0} \delta = 0 = \psi(0,0) = \psi(0, f(0)),$$

we deduce from (9.13) that $f_0(0) = f(0)$. Accordingly, $f_0 = f$ (otherwise, different fluid trajectories would meet at the origin with the same velocity, in contradiction with the uniqueness results for ordinary differential equations), see Figure 9.3.

This shows that the streamline emanating from the origin (that is, from every point of $\partial\Omega$ with a nonzero velocity field) remains on $\partial\Omega$.

Thus, if the velocity field does not vanish on $\partial\Omega$, then $\partial\Omega$ consists of a streamline, and subsequently (9.12) follows from (9.5).

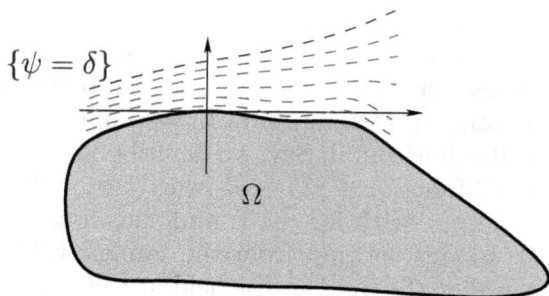

Figure 9.3.   Foliation of streamlines in the vicinity of an airfoil.

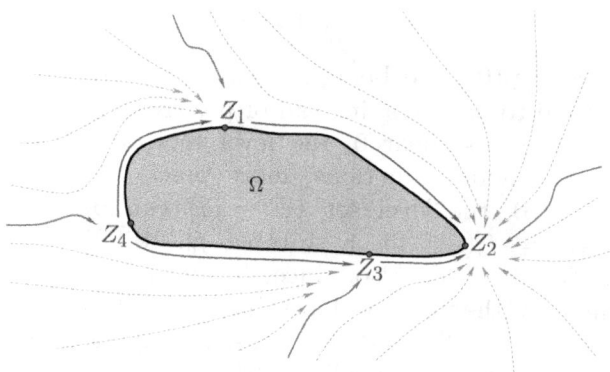

Figure 9.4.   Partitioning $\partial\Omega$ as in (9.14) (not necessarily a realistic picture).

If instead the velocity field vanishes at $Z_1, \ldots, Z_N \in \partial\Omega$, we can decompose $\partial\Omega$ into open arcs, $\Lambda_1, \ldots, \Lambda_N$, such that

$$\partial\Omega = \Lambda_1 \cup \cdots \cup \Lambda_N \cup \{Z_1, \ldots, Z_N\}. \qquad (9.14)$$

Consequently, the velocity field does not vanish on each Λ_j, and Λ_j is a streamline connecting Z_j to Z_{j+1} in infinite time (with $Z_{N+1} := Z_1$), see Figure 9.4. In this setting, we can consider a trajectory $X(t)$ starting at a given point of Λ_j and deduce from (9.5) that, for every $x \in \Lambda_j$,

$$|v(x)|^2 + \frac{2p(x)}{\rho} = p_j,$$

for some $p_j \in \mathbb{R}$, and

$$p_j = \lim_{t \to \pm\infty} |v(X(t))|^2 + \frac{2p(X(t))}{\rho},$$

which leads to

$$p_j = |v(Z_j)|^2 + \frac{2p(Z_j)}{\rho} = |v(Z_{j+1})|^2 + \frac{2p(Z_{j+1})}{\rho}.$$

This actually states that $p_1 = \cdots = p_N$, and thus (9.12) plainly follows, as desired.

Now, we describe $\partial\Omega$ as a smooth and closed curve, traversed counterclockwise and parameterized by its arc length, say $\gamma : [0, L] \to \mathbb{C}$, for some $L > 0$ (in particular, note that $\gamma(0) = \gamma(L)$). We identify the points of $\mathbb{R}^2$ with complex numbers. In this way, given $\tau \in [0, L]$, a unit tangent vector to $\partial\Omega$ at $\gamma(\tau)$ is given by $\dot\gamma(\tau)$, and the unit normal vector $\nu(\gamma(\tau))$ pointing outward is given by a clockwise rotation of $\dot\gamma(\tau)$ by an angle $\frac{\pi}{2}$, corresponding to the complex multiplication by $e^{-i\pi/2} = -i$ (that is, the outward unit normal can be written as $-i\dot\gamma(\tau)$).

For this reason, in light of (8.6), we have that the force acting on the airfoil (after identifying $\mathbb{R}^2$ with $\mathbb{C}$) takes the form

$$F = -\int_{\partial\Omega} p\nu = -\int_0^L p(\gamma(\tau))\,\nu(\gamma(\tau))\,d\tau = i\int_0^L p(\gamma(\tau))\,\dot\gamma(\tau)\,d\tau.$$

From this and (9.12), we arrive at

$$\frac{2F}{\rho} = i\int_0^L \frac{2p(\gamma(\tau))}{\rho}\,\dot\gamma(\tau)\,d\tau = i\int_0^L \left(p_0 - |\nabla\psi(\gamma(\tau))|^2\right)\dot\gamma(\tau)\,d\tau. \tag{9.15}$$

Since

$$\int_0^L \dot\gamma(\tau)\,d\tau = \gamma(L) - \gamma(0) = 0,$$

we deduce from (9.15) that

$$\frac{2F}{\rho} = -i\int_0^L |\nabla\psi(\gamma(\tau))|^2\,\dot\gamma(\tau)\,d\tau. \tag{9.16}$$

We also recall that, in light of the impenetrability condition in (8.9), combined with (9.10), we have that $\nabla\psi$ is orthogonal to the unit tangent vector $\dot{\gamma}$. Thus, we write that

$$\nabla\psi(\gamma(\tau)) = \pm i |\nabla\psi(\gamma(\tau))| \, \dot{\gamma}(\tau). \tag{9.17}$$

Also, in the complex variable setting, i.e. using the notation $z = x+iy$ and identifying $x + iy$ with (x, y), we set

$$w(z) = w(x + iy) := \partial_x\psi(x, y) - i\partial_y\psi(x, y), \tag{9.18}$$

and it follows that w is the complex conjugate of $\nabla\psi$ and $|w| = |\nabla\psi|$. Hence, taking the complex conjugation in (9.17),

$$w(\gamma(\tau)) = \pm i \overline{|w(\gamma(\tau))| \, \dot{\gamma}(\tau)} = \mp i |w(\gamma(\tau))| \, \overline{\dot{\gamma}(\tau)}.$$

Taking the square of this identity,

$$w^2(\gamma(\tau)) = -|w(\gamma(\tau))|^2 \, (\overline{\dot{\gamma}(\tau)})^2$$

and, as a result,

$$w^2(\gamma(\tau))\dot{\gamma}(\tau) = -|w(\gamma(\tau))|^2 \, (\overline{\dot{\gamma}(\tau)})^2 \dot{\gamma}(\tau) = -|w(\gamma(\tau))|^2 \, \overline{\dot{\gamma}(\tau)} \, |\dot{\gamma}(\tau)|^2$$
$$= -|w(\gamma(\tau))|^2 \, \overline{\dot{\gamma}(\tau)} = -|\nabla\psi(\gamma(\tau))|^2 \, \overline{\dot{\gamma}(\tau)}.$$

We thus combine this information with the complex conjugation of (9.16) and conclude that

$$\overline{\frac{2iF}{\rho}} = \int_0^L |\nabla\psi(\gamma(\tau))|^2 \, \dot{\gamma}(\tau) \, d\tau = \int_0^L |\nabla\psi(\gamma(\tau))|^2 \, \overline{\dot{\gamma}(\tau)} \, d\tau$$
$$= -\int_0^L w^2(\gamma(\tau))\dot{\gamma}(\tau) \, d\tau = -\oint_\gamma w^2(z) \, dz. \tag{9.19}$$

Also, using the harmonicity of ψ, we observe that

$$\partial_x(\Re w) = \partial_{xx}\psi = -\partial_{yy}\psi = \partial_y(\Im w)$$

and, moreover,

$$\partial_y(\Re w) = \partial_{xy}\psi(x, y) = -\partial_x(\Im w).$$

These observations give that w satisfies the Cauchy–Riemann equations; therefore, w is holomorphic outside Ω.

We can therefore take $R > 0$ large enough such that $\Omega \Subset B_R$ and employ Laurent's theorem (see e.g. [ST18, Theorem 11.1], applied here with $R_1 := R$, $R_2 := +\infty$ and $z_0 := 0$); in this way, we obtain that, for all z in the exterior of B_R, the following series representation holds true for suitable a_k, $b_k \in \mathbb{C}$:

$$w(z) = w_0(z) + w_\star(z)$$

with

$$w_0(z) := \sum_{k=0}^{+\infty} a_k z^k \quad \text{and} \quad w_\star(z) := \sum_{k=1}^{+\infty} \frac{b_k}{z^k},$$

where the power series representing w_0 converges in $\mathbb{C}$ and the one representing $w_\star$ converges in the exterior of B_R.

In particular, w_0 is holomorphic in the whole of $\mathbb{C}$, and the convergence of the power series defining $w_\star$ entails (see e.g. [ST18, Theorems 3.19 and 3.23]) that

$$\limsup_{k \to +\infty} |b_k|^{1/k} \leqslant 2R.$$

As a result, we can take $k_0 \in \mathbb{N} \cap [2, +\infty)$ sufficiently large such that $|b_k| \leqslant (3R)^k$ for every $k > k_0$ and, therefore,

$$\limsup_{|z| \to +\infty} |w_\star(z)| \leqslant \limsup_{|z| \to +\infty} \sum_{k=1}^{k_0} \frac{|b_k|}{|z|^k} + \limsup_{|z| \to +\infty} \sum_{k=k_0+1}^{+\infty} \frac{|b_k|}{|z|^k}$$

$$\leqslant 0 + \limsup_{|z| \to +\infty} \sum_{k=k_0+1}^{+\infty} \left(\frac{3R}{|z|}\right)^k$$

$$= \limsup_{|z| \to +\infty} \left(\frac{3R}{|z|}\right)^{k_0+1} \frac{1}{1 - \frac{3R}{|z|}}$$

$$= 0.$$

For this reason, recalling (9.4),

$$|v_0| = \limsup_{|(x,y)| \to +\infty} |\nabla\psi(x,y)| = \limsup_{|z| \to +\infty} |w(z)|$$

$$= \limsup_{|z| \to +\infty} |w_0(z) + w_\star(z)| = \limsup_{|z| \to +\infty} |w_0(z)|.$$

As a consequence, w_0 is bounded in the whole of $\mathbb{C}$; therefore, by Liouville's theorem (see e.g. [ST18, Theorem 10.8]), we have that w_0 is constant.

That is, by (9.4), $w_0(z) = i\overline{v_0}$ for all $z \in \mathbb{C}$, whence, for all z, in the exterior of B_R,

$$w(z) = i\overline{v_0} + w_\star(z) = i\overline{v_0} + \sum_{k=1}^{+\infty} \frac{b_k}{z^k}. \tag{9.20}$$

Moreover,

$$\limsup_{M \to +\infty} \left| \oint_{\partial B_M} \sum_{k=k_0+1}^{+\infty} \frac{b_k}{z^k} \, dz \right| \leqslant \limsup_{M \to +\infty} \sum_{k=k_0+1}^{+\infty} \frac{2\pi |b_k|}{M^{k-2}}$$

$$\leqslant \limsup_{M \to +\infty} \sum_{k=k_0+1}^{+\infty} \frac{2\pi M^2 (3R)^k}{M^k}$$

$$= \limsup_{M \to +\infty} 2\pi M^2 \left(\frac{3R}{M} \right)^{k_0+1} \frac{1}{1 - \frac{3R}{M}} = 0.$$

Thus, by (9.20) and Cauchy's theorem (see e.g. [ST18, Theorem 9.13]),

$$\oint_\gamma w(z) \, dz = \lim_{M \to +\infty} \oint_{\partial B_M} w(z) \, dz$$

$$= \lim_{M \to +\infty} \oint_{\partial B_M} \left(i\overline{v_0} + \sum_{k=1}^{+\infty} \frac{b_k}{z^k} \right) dz \tag{9.21}$$

$$= \lim_{M \to +\infty} \oint_{\partial B_M} \left(i\overline{v_0} + \sum_{k=1}^{k_0} \frac{b_k}{z^k} \right) dz.$$

We stress that the function in the latter integrand is meromorphic in B_M, and we can thereby employ Cauchy's residue theorem (see e.g. [ST18, Theorem 12.3]), thus deducing from (9.21) that

$$\oint_\gamma w(z) \, dz = \lim_{M \to +\infty} \oint_{\partial B_M} w(z) \, dz = 2\pi i b_1. \tag{9.22}$$

In addition,

$$\oint_\gamma w(z)\,dz$$

$$= \int_0^L (\partial_x\psi(\gamma(\tau)) - i\partial_y\psi(\gamma(\tau)))(\dot\gamma_1(\tau) + i\dot\gamma_2(\tau))\,d\tau$$

$$= \int_0^L (\partial_x\psi(\gamma(\tau))\dot\gamma_1(\tau) + \partial_y\psi(\gamma(\tau))\dot\gamma_2(\tau) - i\partial_y\psi(\gamma(\tau))\dot\gamma_1(\tau)$$

$$+ i\partial_x\psi(\gamma(\tau))\dot\gamma_2(\tau))\,d\tau$$

$$= \int_0^L \left(\frac{d}{d\tau}\psi(\gamma(\tau)) - i\partial_y\psi(\gamma(\tau))\dot\gamma_1(\tau) + i\partial_x\psi(\gamma(\tau))\dot\gamma_2(\tau)\right)\,d\tau$$

$$= \psi(\gamma(L)) - \psi(\gamma(0)) - i\int_0^L (\partial_y\psi(\gamma(\tau))\dot\gamma_1(\tau) - \partial_x\psi(\gamma(\tau))\dot\gamma_2(\tau))\,d\tau$$

$$= -i\int_0^L (\partial_y\psi(\gamma(\tau))\dot\gamma_1(\tau) - \partial_x\psi(\gamma(\tau))\dot\gamma_2(\tau))\,d\tau.$$

Therefore, recalling the definition of Γ in (9.1) and the relations in (9.10),

$$\oint_\gamma w(z)\,dz = i\int_0^L (v_1(\gamma(\tau))\dot\gamma_1(\tau) + v_2(\gamma(\tau))\dot\gamma_2(\tau))\,d\tau = i\Gamma.$$

As a result of this and (9.22), we deduce that

$$b_1 = \frac{\Gamma}{2\pi}.$$

This and (9.20) lead to

$$w^2(z) = \left(i\overline{v_0} + \frac{\Gamma}{2\pi z} + \sum_{k=2}^{+\infty}\frac{b_k}{z^k}\right)^2 = -\overline{v_0}^2 + \frac{i\overline{v_0}\Gamma}{\pi z} + \sum_{k=2}^{+\infty}\frac{c_k}{z^k},$$

for suitable $c_k \in \mathbb{C}$.

This, Cauchy's residue theorem (see e.g. [ST18, Theorem 12.3]) and (9.19) give that

$$\frac{2i\overline{F}}{\rho} = 2\overline{v_0}\Gamma$$

and, therefore,

$$\frac{2iF}{\rho} = \overline{2\overline{v_0}\Gamma} = 2v_0\Gamma.$$

Consequently,

$$F = -i\rho v_0\Gamma,$$

which establishes the claim in (9.1).

A natural question now, however, is whether it is possible to produce a nonzero circulation in (9.1). This is in general quite an intriguing problem: here, we only provide a simple and explicit[2] example. For this, we let

$$\Omega := B_1 \tag{9.23}$$

and

$$\psi(x,y) := y\left(1 - \frac{1}{x^2 + y^2}\right) + \ln(x^2 + y^2). \tag{9.24}$$

See Figure 9.5 for the level sets of ψ. These level sets will correspond to streamlines since we consider the vector field $v = (v_1, v_2)$, with

$$v_1(x,y) := -\partial_y\psi(x,y) = \frac{2x^2}{(x^2+y^2)^2} - \frac{1+2y}{x^2+y^2} - 1$$

$$\text{and} \quad v_2(x,y) := \partial_x\psi(x,y) = \frac{2x}{x^2+y^2} + \frac{2xy}{(x^2+y^2)^2}. \tag{9.25}$$

See Figure 9.6 for a sketch of the vector field v.

Complex analysis enthusiasts will also enjoy the fact that the setting in (9.18) is also available by posing

$$w := \frac{2}{z} + i\left(\frac{1}{z^2} - 1\right). \tag{9.26}$$

Note that w is holomorphic away from the origin. And the knowledge of this complex map is pretty much all that is needed to produce

---

[2]Though we do not really exploit this fact here, the example is inspired by a rotating cylinder producing a Magnus effect, which is what soccer players use to make the ball curve during flight. See https://www.youtube.com/watch?v=XdL 7EDKr_rk for a famous application of the Magnus effect in soccer.

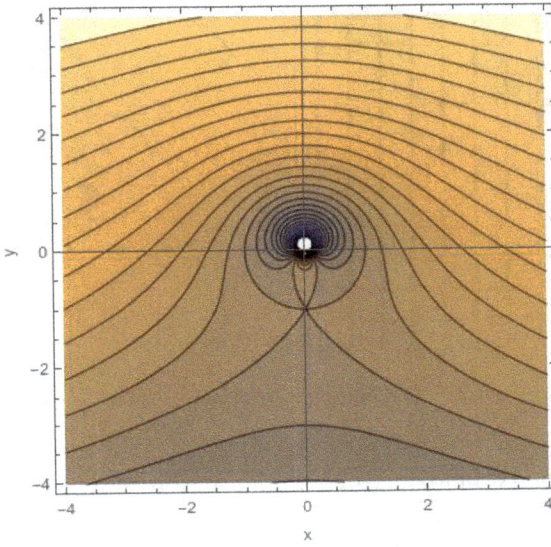

Figure 9.5. The level sets of the potential function in (9.24).

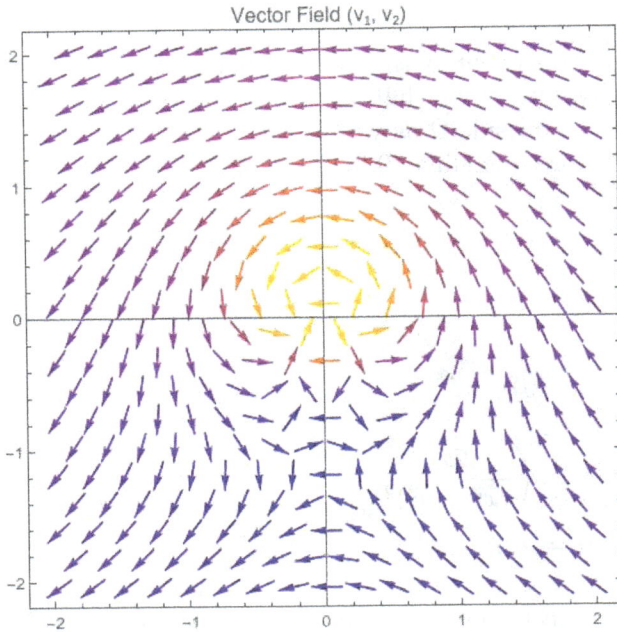

Figure 9.6. The vector field in (9.25).

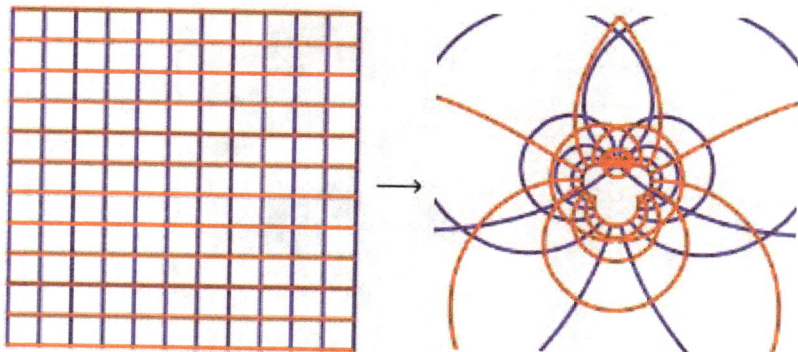

Figure 9.7.   The complex map w in Equation (9.26).

the fluid velocity field in (9.25). The action of the above map w is sketched in Figure 9.7.

We observe that v is tangential to $\partial\Omega$ since

$$v(\cos\theta, \sin\theta) \cdot (\cos\theta, \sin\theta) = 0,$$

and this is consistent with the impenetrability condition in (8.9). Additionally, we have that $\Delta\psi = 0$ outside Ω, which implies that $\operatorname{div} v = 0$ outside Ω as well.

Furthermore, recalling (9.3),

$$\operatorname{curl} v = (\partial_x v_2 - \partial_y v_1)e_3 = (\partial_x(\partial_x\psi) + \partial_y(\partial_y\psi))e_3 = 0 \qquad (9.27)$$

and

$$\lim_{|(x,y)|\to+\infty} v(x,y)$$

$$= \lim_{|(x,y)|\to+\infty} \left(\frac{2x^2}{(x^2+y^2)^2} - \frac{1+2y}{x^2+y^2} - 1, \; \frac{2x}{x^2+y^2} + \frac{2xy}{(x^2+y^2)^2} \right)$$

$$= (-1, 0) =: v_0.$$

Thus, to check that this example is indeed a solution of our fluid dynamics problem, it remains to show that the first equation in (9.2) holds true. For this, we choose $\rho := 1$ and, inspired by Bernoulli's principle in (9.5),

$$p(x, y) := \frac{|v_0|^2 - |v(x, y)|^2}{2} = \frac{1 - |v(x, y)|^2}{2},$$

and we combine the vectorial identity in (6.3) with (9.27) to calculate that

$$\rho(v \cdot \nabla)v + \nabla p = (v \cdot \nabla)v - \nabla \frac{|v|^2}{2} = -v \times (\text{curl } v) = 0.$$

This shows that the vector field in (9.25) is consistent with the fluid dynamics setting in (9.1). To confirm the practical importance and the physical consistency of (9.1) it thus remains to check that this vector field provides a nontrivial circulation Γ. To this end, we compute that

$$\Gamma = \int_0^{2\pi} v(\cos \theta, \sin \theta) \cdot (- \sin \theta, \cos \theta) \, d\theta$$

$$= 2 \int_0^{2\pi} (\cos^2 \theta - \sin \theta - 1, \ \cos \theta + \sin \theta \cos \theta) \cdot (- \sin \theta, \cos \theta) \, d\theta$$

$$= 2 \int_0^{2\pi} (1 + \sin \theta) \, d\theta$$

$$= 4\pi$$

$$\neq 0,$$

as desired.

But hold on a sec! The skeptical reader (who is always very welcome) will complain that we've been cheating on them: we had promised that this section was devoted to airfoils, but then we tried to sell in (9.23) the disk as an example of airfoils. Come on! Airfoils are objects like the ones shown in Figure 9.8, and nothing in Figures 9.5, 9.6 or 9.7 looks like an airfoil – after all, a disk is not an airfoil at all!

Well, in fact, to a certain degree, it is: this is the beauty of mathematics and of complex analysis in particular. Indeed, one of the brilliant ideas of the pioneers of aerodynamics is that the potential in (9.26) describes all the fluid dynamics outside the disk in the example that we have discussed explicitly in detail; therefore, other examples can be constructed via conformal transformations. In particular, using dilations, translations and maps of the form $z + \frac{1}{z}$, one can transform a disk into an airfoil shaped as in Figure 9.8.

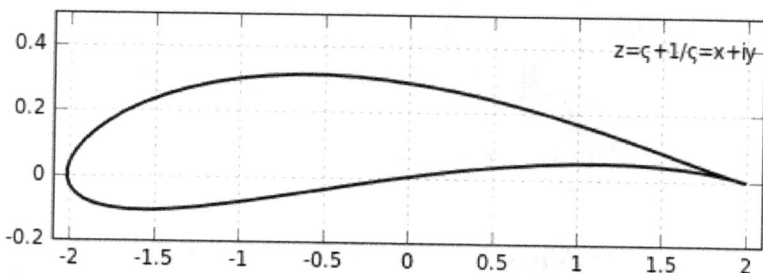

Figure 9.8.   Joukowski airfoil (image by Krishnavedala from Wikipedia, licensed under the Creative Commons Attribution-Share Alike 4.0 International license).

The set of conformal transformations linking circles and airfoils can be visualized, for instance, using the Wolfram Demonstrations Projects: https://demonstrations.wolfram.com/TheJoukowski MappingAirfoilsFromCircles/; https://demonstrations.wolfram.com/JoukowskiAirfoilGeometry/.
It can also be visualized using the GeoGebra application: https://www.geogebra.org/m/XwmqSR49.
A number of alternative computing resources are also available.

Interestingly, on https://complex-analysis.com/content/joukowsky_airfoil.html, one can also plot the resulting flow around the airfoil: the result that we have obtained using this application is shown in Figure 9.9, which nicely shows how a disk is mapped into an airfoil.

For further readings about the Kutta–Joukowski theory and the analysis of airfoils, see e.g. [MT60, HJ09, HJJ16] and the references therein. See also [CBV74, pp. 227–238] for an extensive treatment of fluid dynamics from a complex analysis perspective.

We cannot avoid mentioning that one of the most spectacular applications of fluid dynamics is probably showcased by the motion[3] of a boomerang.

---

[3]The origin of the name "boomerang" is a bit uncertain: some references link it to an extinct Aboriginal language of New South Wales while others link it to the language of the Turuwal people (a sub-group of the Darug) of the Georges River (Tucoerah River). Besides the well-known traditional employment by some Aboriginal Australian peoples for hunting (see Figure 9.10), it seems that boomerangs have also been used in ancient Europe, Egypt, and North America (ancient Egyptian boomerangs have been tested and seemed to work well as returning boomerangs, and a boomerang discovered in the Carpathian Mountains in Poland was dated to about 30,000 years ago.

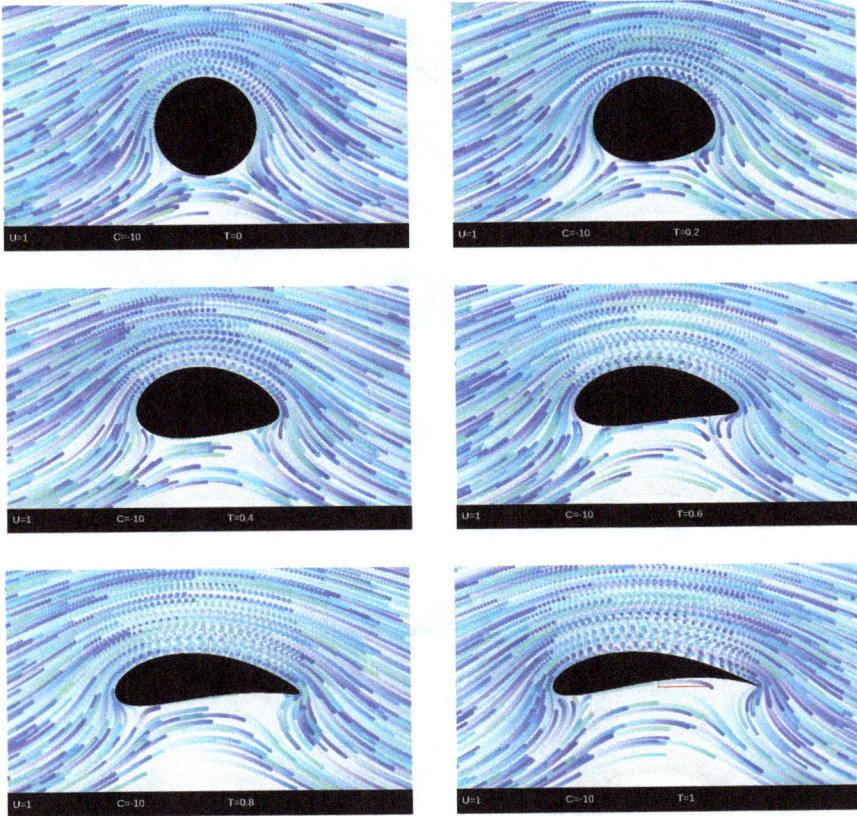

Figure 9.9. From a disk to an airfoil, and the corresponding fluid flow (images produced by the online application in https://complex-analysis.com/content/joukowsky_airfoil.html).

Interestingly, a boomerang was used to set the Guinness World Record for the longest throw of any object by a human: in 2005, at Murarrie Recreation Ground, in Queensland, David Schummy performed a throw of 427.2 m, see https://www.youtube.com/watch?v=ly3nCEbcQig.

Currently, long-distance boomerangs are mostly shaped as a question mark and often have a beveled edge to facilitate the pitch and to lower the drift since the boomerang, in this case, is usually thrown almost horizontally (whereas, in the above-mentioned record, the objective was not to make the boomerang return to the point of release; actually the boomerang ended up on a tree).

Figure 9.10. Otto Jungarryi Sims with a boomerang sitting inside a cave in the Northern Territory of Australia (photo by Ed Gold; image from Wikipedia for free use under ticket #2020101210010454).

In rough terms, each wing of a boomerang is shaped like an airfoil section, allowing the airflow over the wings to create significant lift. If the boomerang is thrown nearly upright, the rotating blades generate more lift at the top than at the bottom because at the top the speed of the rotation adds up to the forward speed, while at the bottom the speed of the rotation subtracts from the forward speed; see Figure 9.11, in which the forward speed is represented by the yellow arrow. This additional lift from the top produces a torque

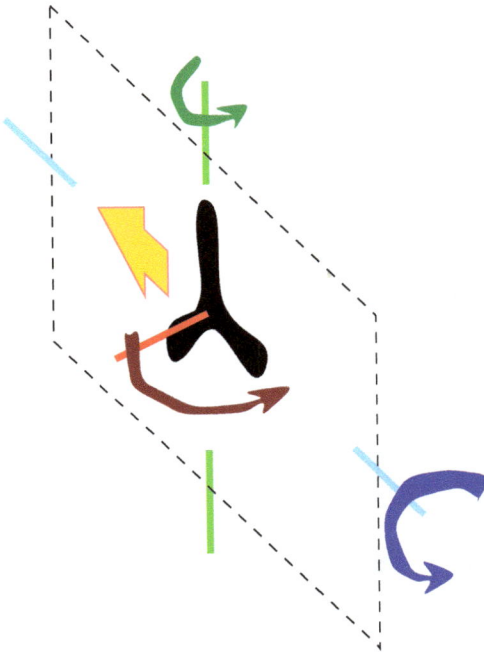

Figure 9.11.   Why do boomerangs return?

(represented by the blue arrow in Figure 9.11), which causes the rotation plane to turn around. We point out that this torque is typically not sufficient to tilt the boomerang around its axis of travel, given its high angular momentum (the spinning of the boomerang being represented by the red arrow in Figure 9.11); hence, the stability of the rotating plane, ensured by the gyroscopic precession, combines with the aerodynamic torque and leads to the curved trajectory sketched by the green arrow in Figure 9.11.

Chapter 10

Surfing the Waves

A topical problem in fluid dynamics consists of the description of waves in shallow waters since this is a typical case arising in the proximity of land. Though a large number of different models are available for solving this problem, in these pages we recall the classical equation

$$a\partial_t\phi + b\partial_x^3\phi + c\,\phi\,\partial_x\phi = 0 \qquad (10.1)$$

for a, b, $c \in \mathbb{R} \setminus \{0\}$, with $\phi = \phi(x,t)$, $x \in \mathbb{R}$ and $t \in (0,+\infty)$. Equation (10.1) is called[1] the Korteweg–de Vries equation (or KdV equation for short).

---

[1]Equation (10.1) is named after Diederik Johannes Korteweg and Gustav de Vries [KdV95], though it was introduced by Joseph Valentin Boussinesq [Bou77]. De Vries completed his PhD under Korteweg's supervision and then worked all his life as a high school teacher in Haarlem in the Netherlands.

His strong interest in the formation and propagation of waves in canals was possibly the outcome of a direct observation by John Scott Russell [Rus45] at the Union Canal (a canal in Scotland, running from Falkirk to Edinburgh). Russell's own words have been repeated in virtually every paper and books that even remotely discusses water wave problems, and we have no intention of breaking this consolidated tradition. So, here is his report on the astounding experience of meeting a traveling wave for the first time: "I was observing the motion of a boat which was rapidly drawn along a narrow channel by a pair of horses, when the boat suddenly stopped – not so the mass of water in the channel which it had put in motion; it accumulated round the prow of the vessel in a state of violent agitation, then suddenly leaving it behind, rolled forward with great velocity, assuming the form of a large solitary elevation, a rounded, smooth and well-defined heap of water, which continued its course along the channel apparently without change

Figure 10.1. Henry Espinoza Panta smashing a wave at Lobitos, Perú (photo by Freddy Sinarahua, image from Wikipedia, available under the Creative Commons Attribution-Share Alike 4.0 International license).

The physical content of equation (10.1) is that ϕ models the shape of a wave in a canal. The position on the canal is given by $x \in \mathbb{R}$, and t stands for the time variable. The amplitude of the wave and the depth of the channel are supposed to be small (as will be discussed as follows in the approximations leading to the derivation of the equation), hence the validity of (10.1) has to be limited to small-amplitude waves in shallow waters.

Though equation (10.1) should be considered as a very simplified model, lacking the ability to capture the complexity of oscillatory phenomena and waves in the real world, it is interesting to note that the Korteweg–de Vries equation can describe some interesting features, such as traveling waves, and provide interesting information about the shape of the waves, as we now discuss.

---

of form or diminution of speed. [...] Such, in the month of August 1834, was my first chance interview with that singular and beautiful phenomenon which I have called the Wave of Translation".

See Figure 10.2 for a historical picture of the Union Canal.

Figure 10.2. Students picnic on the Union Canal in 1922 (copyright The University of Edinburgh, available for public use; image from Wikipedia, licensed under the Creative Commons Attribution-Share Alike 3.0 Unported license).

First of all (recalling the presentation on p. 75), one can seek traveling wave solutions of (10.1) in the form

$$\phi(x, t) = \phi_0(x - vt) \tag{10.2}$$

for some velocity v. In this setting, equation (10.1) reduces to

$$
\begin{aligned}
0 &= -av\phi_0' + b\phi_0''' + c\,\phi_0\,\phi_0' = -av\phi_0' + b\phi_0''' + \frac{c}{2}\,(\phi_0^2)' \\
&= \left(-av\phi_0 + b\phi_0'' + \frac{c}{2}\,\phi_0^2\right)'.
\end{aligned}
\tag{10.3}
$$

We integrate this equation by assuming that at infinity, ϕ_0, and its derivatives converge to zero, finding that

$$0 = -av\phi_0 + b\phi_0'' + \frac{c}{2}\,\phi_0^2.$$

As a result,

$$0 = \left(-av\phi_0 + b\phi_0'' + \frac{c}{2}\,\phi_0^2\right)\phi_0' = \left(-\frac{av}{2}\phi_0^2 + \frac{b}{2}(\phi_0')^2 + \frac{c}{6}\,\phi_0^3\right)'.$$

Integrating this as above, we deduce that

$$0 = -\frac{av}{2}\phi_0^2 + \frac{b}{2}(\phi_0')^2 + \frac{c}{6}\phi_0^3.$$

This ordinary differential equation can be solved by separation of variables (assuming $\phi_0 \geqslant 0$, $c > 0$ and $abv > 0$), leading, up to a translation, to

$$\phi_0(r) = \frac{6av}{c\left(\cosh\left(r\sqrt{\frac{av}{b}}\right) + 1\right)}. \qquad (10.4)$$

See Figure 10.3 for a sketch[2] of this traveling wave (recall (10.2)) when $a := 1$, $b := 1$, $c := 1$ and $v := 1$.

It is also interesting to observe that the highest crest of the wave in (10.4) is attained for $r := 0$, and it is equal to

$$\frac{3av}{c}. \qquad (10.5)$$

This suggests that the fastest waves (i.e. waves with the largest velocity v) correspond to the highest ones.

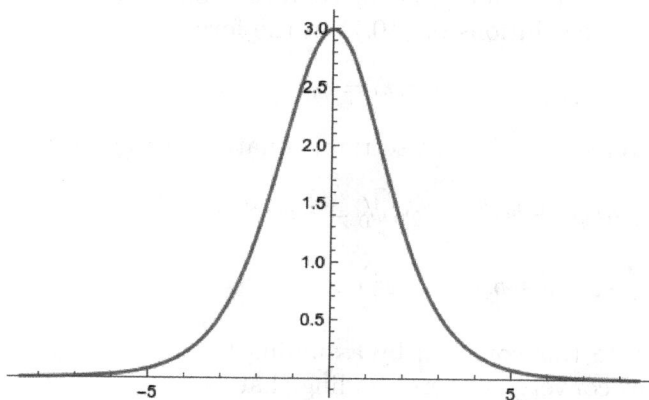

Figure 10.3.   The traveling wave solution for the Korteweg–de Vries equation, as found in (10.4).

---

[2]See also http://lie.math.brocku.ca/$\sim$sanco/solitons/kdv_solitons.php for several animations of (possibly interacting) traveling waves of the Korteweg–de Vries equation.

Another nice feature captured by the Korteweg–de Vries equation in (10.1) is a sufficiently accurate description of the shape of a set of waves: it is indeed a common experience that "real" waves exhibit sharper crests and flatter troughs than those of sine and cosine functions, see Figures 10.4–10.6.

To understand the shape of these waves, it is convenient to define, given $\varphi \in \mathbb{R}$ and $m \in [0, 1)$,

$$U(\varphi, m) := \int_0^\varphi \frac{d\theta}{\sqrt{1 - m \sin^2 \theta}}. \tag{10.6}$$

We observe that $\frac{\partial U}{\partial \varphi} > 0$. Hence, given $m \in (0, 1)$, the function $\varphi \mapsto U(\varphi, m)$ is invertible, and with a slight abuse of notation, we call $\varphi(U, m)$ the inverse function. In this setting, one defines the elliptic cosine (in Latin, *cosinus amplitudinis*) by

$$\mathrm{cn}(U, m) := \cos(\varphi(U, m)). \tag{10.7}$$

Figure 10.4. Wave trains crossing in front of Île de Ré, in the Atlantic Ocean (photo by Michel Griffon; image from Wikipedia, licensed under the Creative Commons Attribution-Share Alike 3.0 Unported license).

Figure 10.5.   Cnoidal wave solution to the Korteweg–de Vries equation (image by Kraaiennest from Wikipedia, licensed under the Creative Commons Attribution-Share Alike 3.0 Unported license).

Figure 10.6.   Cnoidal looking waves seen from North West Cape.

Now, to find periodic solutions of the Korteweg–de Vries equation in (10.1) and to compare their shapes with Figure 10.5, we return to (10.3) and integrate it, now without assuming that ϕ_0 decays at infinity: in this way, we have that

$$-av\phi_0 + b\phi_0'' + \frac{c}{2}\phi_0^2 = \kappa_1,$$

for some constant $\kappa_1 \in \mathbb{R}$.

As a consequence,

$$0 = \left(-av\phi_0 + b\phi_0'' + \frac{c}{2}\phi_0^2 - \kappa_1\right)\phi_0'$$

$$= \left(-\frac{av}{2}\phi_0^2 + \frac{b}{2}(\phi_0')^2 + \frac{c}{6}\phi_0^3 - \kappa_1\phi_0\right)',$$

from which a further integration produces

$$-\frac{av}{2}\phi_0^2 + \frac{b}{2}(\phi_0')^2 + \frac{c}{6}\phi_0^3 - \kappa_1\phi_0 = \kappa_2, \tag{10.8}$$

for some constant $\kappa_2 \in \mathbb{R}$.

We observe that we can write (10.8) as

$$(\phi_0')^2 = \mathcal{P}(\phi_0), \tag{10.9}$$

for a suitable polynomial $\mathcal{P}$ of degree 3. The setting that we consider now is that in which $\mathcal{P}$ possesses three distinct real roots $\beta_1 > \beta_2 > \beta_3$. With this, we rewrite (10.9) as

$$(\phi_0')^2 = -\kappa\,(\phi_0 - \beta_1)(\phi_0 - \beta_2)(\phi_0 - \beta_3), \tag{10.10}$$

where $\kappa := c/(3b)$, which we assume to be positive.

Now, it turns out to be useful to seek solutions in the form

$$\phi_0(r) = \beta_1 \cos^2(\Phi(r)) + \beta_2 \sin^2(\Phi(r)), \tag{10.11}$$

where the function Φ is supposed to be increasing and has to be determined.

We observe that

$$(\phi_0')^2 = (-2\beta_1 \cos\Phi\, \sin\Phi\, \Phi' + 2\beta_2 \sin\Phi\, \cos\Phi\, \Phi')^2$$
$$= 4(\beta_1 - \beta_2)^2 \sin^2\Phi\, \cos^2\Phi\, (\Phi')^2.$$

Additionally,

$$\phi_0 - \beta_1 = \beta_1(\cos^2\Phi - 1) + \beta_2 \sin^2\Phi = -(\beta_1 - \beta_2)\sin^2\Phi$$

and

$$\phi_0 - \beta_2 = \beta_1 \cos^2\Phi + \beta_2(\sin^2\Phi - 1) = (\beta_1 - \beta_2)\cos^2\Phi.$$

These observations and (10.10) yield that

$$4(\Phi')^2 = \kappa\,(\phi_0 - \beta_3).$$

Thus, since

$$\phi_0 - \beta_3 = \beta_1(1-\sin^2\Phi) + \beta_2 \sin^2\Phi - \beta_3 = (\beta_1 - \beta_3) - (\beta_1 - \beta_2)\sin^2\Phi,$$

we conclude that

$$\Phi' = \frac{\sqrt{\kappa}}{2}\sqrt{(\beta_1 - \beta_3) - (\beta_1 - \beta_2)\sin^2\Phi} = \kappa_0\sqrt{1 - m\sin^2\Phi}, \tag{10.12}$$

where

$$\kappa_0 := \frac{\sqrt{\kappa\,(\beta_1 - \beta_3)}}{2} \quad \text{and} \quad m := \frac{\beta_1 - \beta_2}{\beta_1 - \beta_3}.$$

We can thus utilize the separation of variable method in (10.12) and, in view of (10.6), obtain that

$$\kappa_0 r = \int_0^{\Phi(r)} \frac{d\vartheta}{\sqrt{1 - m \sin^2 \vartheta}} = U(\Phi(r), m),$$

where we normalized the picture so that $\Phi(0) = 0$.

From this and (10.7), it follows that

$$\mathrm{cn}(\kappa_0 r) = \mathrm{cn}(U(\Phi(r), m), m) = \cos(\Phi(r)),$$

where we have used the short setting $\mathrm{cn}(\cdot) := \mathrm{cn}(\cdot, m)$ to ease notation.

Returning to (10.11), we thereby conclude that

$$
\begin{aligned}
\phi_0(r) &= \beta_1 \cos^2(\Phi(r)) + \beta_2(1 - \cos^2(\Phi(r))) \\
&= (\beta_1 - \beta_2)\cos^2(\Phi(r)) + \beta_2 \quad\quad\quad (10.13) \\
&= (\beta_1 - \beta_2)\,\mathrm{cn}^2(\kappa_0 r) + \beta_2 = A\,\mathrm{cn}^2(\kappa_0 r) + B,
\end{aligned}
$$

where $A := \beta_1 - \beta_2$ and $B := \beta_2$. The profiles of these functions, for suitable choices of parameters, resemble the shape of waves in Figures 10.4 and 10.5 (which are often called "cnoidal waves" due to the presence of the elliptic cosine cn in their expression). To get a feeling for how the different parameters change the shape of a cnoidal waves see Figure 10.7.

It is also interesting to take note of the fact that, despite its (relative) simplicity, the Korteweg–de Vries equation in (10.1) is capable

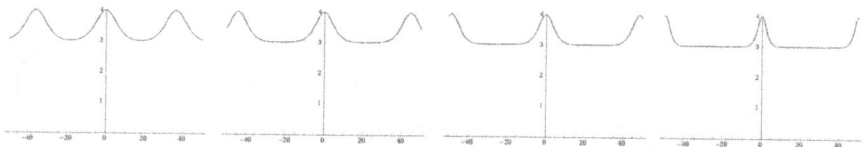

Figure 10.7. The cnoidal waves in (10.13) with $A := 1$, $B := 3$ and $(\kappa_0, m) \in \{(1/7, 0.9), (1/6, 0.99), (1/5, 0.999), (1/3, 0.999999)\}$.

of capturing complex phenomena, such as the formation of oscillatory waves, with crests traveling at different speeds according to their
heights; these complicated situations may even arise from very simple
initial data, see Figure 10.8.

Furthermore, we point out that the Korteweg–de Vries equation
often comes in handy in the description of natural phenomena that
seem to be well beyond its range of applicability. For example (see
e.g. [GY16] and the references therein), the Korteweg–de Vries equation has been exploited in the description[3] of tsunamis even though
the extreme depth of oceans may appear to be incompatible with
an equation which was originally designed for shallow waters. The
reason for the success of the Korteweg–de Vries equation even in this
setting is, at least, twofold. First of all, though oceans are very deep,
tsunami waves can reach a spatial extent of even larger size; therefore, when the depth of the ocean is sufficiently small with respect
to the breadth of the wave, the shallow water theory may still be
applied since what counts is the relative (and not absolute) sizes of
the length parameters. Secondly, while tsunamis in the deep ocean
consists typically of very long waves with quite small amplitudes, as
they approach shallow waters, the nonlinear effects become predominant, and they change significantly the shape and velocity of the
leading waves, for which the Korteweg–de Vries equation can also
provide an approximate, but often effective, model.

Now, we give a brief motivation for the Korteweg–de Vries equation in (10.1). While its expression may appear rather mysterious at
first glance, especially in view of the third derivative appearing as a
leading term, we will see now that equation (10.1) arises naturally
from the equations of fluid dynamics presented in Chapter 5.

More specifically, we consider the Euler fluid flow equation
in (5.7). Here, we take the coordinates $(x, z) \in \mathbb{R}^2$, where $x \in \mathbb{R}$
represents the direction of a canal and $z \in \mathbb{R}$ is the vertical displacement. The fluid velocity v is written in components as $v = (\overline{u}, \overline{w})$,
with $\overline{u}$ and $\overline{w}$ scalar functions. Hence, with ρ being the density of

---

[3]Although it is debatable whether the artists intended their works to be interpreted in that way, it is customary to refer to the artworks in Figure 10.9 in
connection with tsunamis.

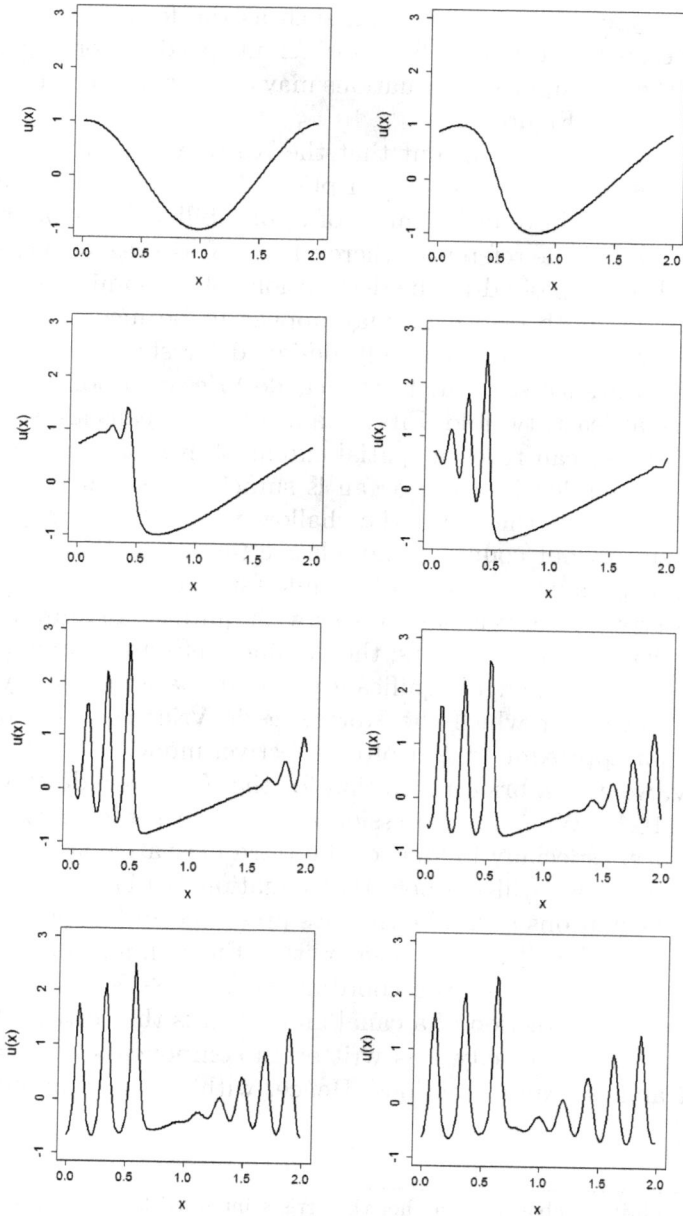

Figure 10.8. Numerical solution of the KdV equation $\partial_t u + u\partial_x u + (0.022)^2 \partial_x^3 u = 0$ with initial condition $u(x,0) = \cos(\pi x)$. The initial cosine wave evolves into a train of solitary-type waves [ZK65] (by Ta2o; image from Wikipedia, licensed under the Creative Commons Attribution-Share Alike 3.0 Unported license).

(a)   (b)

Figure 10.9. (a) The Great Wave off Kanagawa, famous woodblock print by the ukiyo-e artist Katsushika Hokusai (H. O. Havemeyer Collection, Bequest of Mrs. H. O. Havemeyer, 1929, Metropolitan Museum of Art; Public Domain image from Wikipedia). (b) Die Woge, sculpture by Tobias Stengel in Dresden, Germany (photo by Christoph Münch, Press Section of the city of Dresden's official homepage; image from Wikipedia, licensed under the Creative Commons Attribution-Share Alike 3.0 Unported license).

the fluid (assumed to be constant) and p being the pressure, equation (5.7) reads

$$\rho \partial_t (\overline{u}, \overline{w}) + \rho((\overline{u}, \overline{w}) \cdot (\partial_x, \partial_z))(\overline{u}, \overline{w}) = -\rho g(0, 1) - (\partial_x, \partial_z)p. \quad (10.14)$$

Hence, setting

$$\overline{P} := \frac{p}{\rho} + gz, \quad (10.15)$$

we rewrite (10.14) as the following system of two equations:

$$\begin{cases} \partial_t \overline{u} + \overline{u} \partial_x \overline{u} + \overline{w} \partial_z \overline{u} = -\partial_x \overline{P}, \\ \partial_t \overline{w} + \overline{u} \partial_x \overline{w} + \overline{w} \partial_z \overline{w} = -\partial_z \overline{P}. \end{cases} \quad (10.16)$$

We suppose that the bottom of the canal is $z = 0$ and the fluid at rest has height $h_0 \in (0, +\infty)$. The wave is thus described as a perturbation of the rest fluid surface $z = h_0$, described by the graph of a function $\overline{\eta} = \overline{\eta}(x, t)$, modulated by a small amplitude $\alpha \in (0, +\infty)$: namely, at every instant of time t, the fluid in the channel corresponds to the region of the points $(x, z) \in \mathbb{R}^2$, with $0 < z < h_0 + \alpha \overline{\eta}(x, t)$. That is, the equations in (10.16) are set in the domain

$$\Omega^{(t)} := \{(x, z) \in \mathbb{R}^2 \text{ s.t. } x \in \mathbb{R} \text{ and } z \in (0, h_0 + \alpha \overline{\eta}(x, t))\}. \quad (10.17)$$

We also assume that the fluid is incompressible, and hence, by (5.9),

$$0 = \text{div}(\overline{u}, \overline{w}) = \partial_x \overline{u} + \partial_z \overline{w}. \tag{10.18}$$

The equations in (10.16) and (10.18) are complemented by boundary conditions corresponding to the bottom of the canal and the free surface of the fluid. Indeed, at the bottom $z = 0$, we suppose that the floor of the canal is impenetrable; hence the vertical component of the fluid velocity vanishes, which reads

$$\overline{w}(x, 0, t) = 0. \tag{10.19}$$

As for the free surface, we assume that fluid parcels staying at the top remain at the top; hence, if $(x(t), z(t))$ represents the trajectory of a fluid particle on the upper surface of the fluid at time t, we have that

$$z(t) = h_0 + \alpha \overline{\eta}(x(t), t).$$

Thus, taking the time derivative of this expression and recalling that the fluid parcel velocity $(\dot{x}(t), \dot{z}(t))$ agrees with $(\overline{u}, \overline{w})$,

$$\overline{w}(x(t), h_0 + \alpha \overline{\eta}(x(t), t), t) = \dot{z}(t) = \alpha \partial_x \overline{\eta}(x(t), t)\dot{x}(t) + \alpha \partial_t \overline{\eta}(x(t), t)$$

$$= \alpha \partial_x \overline{\eta}(x(t), t)\overline{u}(x(t), h_0 + \alpha \overline{\eta}(x(t), t), t) + \alpha \partial_t \overline{\eta}(x(t), t). \tag{10.20}$$

The free surface $z = h_0 + \alpha \overline{\eta}$ also presents an additional boundary condition related to pressure. Indeed, the pressure on the top surface of the fluid is balanced by the atmospheric pressure, which we denote by p_0 and assume to be constant. Hence, by (10.15),

$$\overline{P}(x, h_0 + \alpha \overline{\eta}(x, t), t) = \frac{p_0}{\rho} + g\left(h_0 + \alpha \overline{\eta}(x, t)\right). \tag{10.21}$$

We now consider a regime of shallow waters and small amplitudes. That is, we pick $\Theta > 0$ and[4] we consider the quantities

$$\varepsilon := \frac{\alpha}{h_0} \quad \text{and} \quad \delta := \frac{h_0}{\Theta^2 g}. \tag{10.22}$$

---

[4]To simplify the computation, one can simply take $\Theta := 1$. The role of the "arbitrary" constant Θ is only to introduce a quantity with the dimension of time. In this way, we can think that $\overline{\eta}$, ε and δ are dimensionless, while α and h_0 have the dimension of space. In the end, the auxiliary dimensional constant Θ will simplify and will no longer appear in the final equation (10.42).

We suppose that ε and δ are small and also, up to normalizing constants, that

$$\delta = \varepsilon. \tag{10.23}$$

We define

$$u^\star(x,z,t) := \frac{\overline{u}(\sqrt{gh_0}x, h_0 z, t)}{\sqrt{gh_0}}, \quad w^\star(x,z,t) := \frac{\Theta\,\overline{w}(\sqrt{gh_0}x, h_0 z, t)}{h_0},$$

$$\eta^\star(x,t) := \overline{\eta}(\sqrt{gh_0}x, t) \quad \text{and} \quad P^\star(x,z,t) := \frac{\overline{P}(\sqrt{gh_0}x, h_0 z, t)}{gh_0}. \tag{10.24}$$

In view of (10.17), the above functions are defined in the region

$$\Omega_t^\star := \{(x,z) \in \mathbb{R}^2 \text{ s.t. } (\sqrt{gh_0}x, h_0 z) \in \Omega^{(t)}\}$$
$$= \{(x,z) \in \mathbb{R}^2 \text{ s.t. } x \in \mathbb{R} \text{ and } z \in (0, 1 + \varepsilon\eta^\star(x,t))\}. \tag{10.25}$$

We thus collect the equations in (10.16), (10.18), (10.19), (10.20) and (10.21), exploiting also (10.22) and (10.23), by writing

$$\begin{cases} \partial_t u^\star + u^\star \partial_x u^\star + \frac{w^\star \partial_z u^\star}{\Theta} = -\partial_x P^\star & \text{in } \Omega_t^\star, \\ \varepsilon(\Theta \partial_t w^\star + \Theta u^\star \partial_x w^\star + w^\star \partial_z w^\star) = -\partial_z P^\star & \text{in } \Omega_t^\star, \\ \Theta \partial_x u^\star + \partial_z w^\star = 0 & \text{in } \Omega_t^\star, \\ P^\star = \frac{p_0}{\rho g h_0} + 1 + \varepsilon\eta^\star & \text{on } z = 1 + \varepsilon\eta^\star, \\ w^\star = \varepsilon\Theta\,(\partial_t \eta^\star + u^\star \partial_x \eta^\star) & \text{on } z = 1 + \varepsilon\eta^\star, \\ w^\star = 0 & \text{on } z = 0. \end{cases} \tag{10.26}$$

We let $c_0 := \frac{p_0}{\rho g h_0} + 1$. It is now convenient to "surf the wave" and look at the new variable $\xi := x - t$ (in rough terms, using this variable, one is moving with a wave that travels at unit[5] speed). It is

---

[5]Interestingly, moving at unit speed for $\eta^\star$ corresponds to moving at speed $\sqrt{gh_0}$ for the original profile $\overline{\eta}$, thanks to the change in spatial variables in (10.24). The quantity $\sqrt{gh_0}$ happens indeed to be one of the characteristic velocities of small waves in shallow waters (relating the velocity to $\sqrt{gh_0}$ also tells us that the speed of these waves decreases when the height of the canal is smaller and, conversely, their speed increases when moving from very shallow to slightly deeper water).

also appropriate to look at large times by setting $\tau := \varepsilon t$. Thus, we rephrase our physical quantities with respect to these new variables by defining

$$u(\xi, z, \tau) := u^\star\left(\xi + \frac{\tau}{\varepsilon}, z, \frac{\tau}{\varepsilon}\right), \quad w(\xi, z, \tau) = w^\star\left(\xi + \frac{\tau}{\varepsilon}, z, \frac{\tau}{\varepsilon}\right),$$

$$\eta(\xi, \tau) = \eta^\star\left(\xi + \frac{\tau}{\varepsilon}, \frac{\tau}{\varepsilon}\right) \quad \text{and} \quad P(\xi, z, \tau) := P^\star\left(\xi + \frac{\tau}{\varepsilon}, z, \frac{\tau}{\varepsilon}\right).$$

$$(10.27)$$

By (10.25), these functions are defined in the region

$$\Omega_\tau := \{(\xi, z) \in \mathbb{R}^2 \text{ s.t. } \xi = x - t, \ \tau = \varepsilon t \text{ and } (x, z) \in \Omega_t^\star\}$$
$$= \{(\xi, z) \in \mathbb{R}^2 \text{ s.t. } \xi \in \mathbb{R} \text{ and } z \in (0, 1 + \varepsilon\eta(\xi, \tau))\}.$$

In this setting, one can rewrite (10.26) in the form

$$\begin{cases} -\partial_\xi u + \varepsilon\partial_\tau u + u\partial_\xi u + \frac{w\partial_z u}{\Theta} = -\partial_\xi P & \text{in } \Omega_\tau, \\ \varepsilon(-\Theta\partial_\xi w + \varepsilon\Theta\partial_\tau w + \Theta u\partial_\xi w + w\partial_z w) = -\partial_z P & \text{in } \Omega_\tau, \\ \Theta\partial_\xi u + \partial_z w = 0 & \text{in } \Omega_\tau, \\ P = c_0 + \varepsilon\eta & \text{on } z = 1 + \varepsilon\eta, \\ w = \varepsilon\Theta(\varepsilon\partial_\tau\eta + (u - 1)\partial_\xi\eta) & \text{on } z = 1 + \varepsilon\eta, \\ w = 0 & \text{on } z = 0. \end{cases}$$

$$(10.28)$$

It is now convenient to consider formal Taylor expansions of the physical quantities involved in the powers of ε by writing

$$u(\xi, z, \tau) = \sum_{k=0}^{+\infty} \varepsilon^k u_k(\xi, z, \tau), \quad w(\xi, z, \tau) = \sum_{k=0}^{+\infty} \varepsilon^k w_k(\xi, z, \tau)$$

$$P(\xi, z, \tau) = \sum_{k=0}^{+\infty} \varepsilon^k P_k(\xi, z, \tau) \quad \text{and} \quad \eta(\xi, \tau) = \sum_{k=0}^{+\infty} \varepsilon^k \eta_k(\xi, \tau).$$

Substituting these expressions into (10.28), we obtain that

$$
\begin{cases}
-\sum_{k=0}^{+\infty}\varepsilon^k\partial_\xi u_k + \sum_{k=0}^{+\infty}\varepsilon^{k+1}\partial_\tau u_k + \sum_{k,m=0}^{+\infty}\varepsilon^{k+m}u_k\partial_\xi u_m \\
\quad +\dfrac{1}{\Theta}\sum_{k,m=0}^{+\infty}\varepsilon^{k+m}w_k\partial_z u_m = -\sum_{k=0}^{+\infty}\varepsilon^k\partial_\xi P_k \qquad \text{in } \Omega_\tau, \\[2mm]
-\Theta\sum_{k=0}^{+\infty}\varepsilon^{k+1}\partial_\xi w_k + \Theta\sum_{k=0}^{+\infty}\varepsilon^{k+2}\partial_\tau w_k \\
\quad +\Theta\sum_{k,m=0}^{+\infty}\varepsilon^{k+m+1}u_k\partial_\xi w_m + \sum_{k,m=0}^{+\infty}\varepsilon^{k+m+1}w_k\partial_z w_m \quad \text{in } \Omega_\tau, \\
\quad = -\sum_{k=0}^{+\infty}\varepsilon^k\partial_z P_k \\[2mm]
\sum_{k=0}^{+\infty}\varepsilon^k(\Theta\partial_\xi u_k + \partial_z w_k)=0 \qquad \text{in } \Omega_\tau, \\[2mm]
P\left(\xi,1+\sum_{k=0}^{+\infty}\varepsilon^{k+1}\eta_k,\tau\right)=c_0+\sum_{k=0}^{+\infty}\varepsilon^{k+1}\eta_k, \\[2mm]
\dfrac{1}{\Theta}w\left(\xi,1+\sum_{k=0}^{+\infty}\varepsilon^{k+1}\eta_k,\tau\right)=\sum_{k=0}^{+\infty}\varepsilon^{k+2}\partial_\tau\eta_k \\
\quad +\sum_{k,m=0}^{+\infty}\varepsilon^{k+m+1}u_k\partial_\xi\eta_m - \sum_{k=0}^{+\infty}\varepsilon^{k+1}\partial_\xi\eta_k, \\[2mm]
w(\xi,0,\tau)=0.
\end{cases}
\tag{10.29}
$$

The strategy is now to approximate (10.29) by neglecting the terms of size ε^k with $k\geqslant 3$. For this, we consider the orders in ε^k in (10.29) with $k\in\{0,1,2\}$, and we formally equate the corresponding coefficients. Namely, by formally considering the coefficients related to ε^0 in (10.29), we see that

$$
\begin{cases}
-\partial_\xi u_0 + u_0\partial_\xi u_0 + \frac{w_0\partial_z u_0}{\Theta} = -\partial_\xi P_0 & \text{in } \Omega_\tau, \\
0 = \partial_z P_0 & \text{in } \Omega_\tau, \\
\Theta\partial_\xi u_0 + \partial_z w_0 = 0 & \text{in } \Omega_\tau, \\
P_0(\xi,1,\tau)=c_0, \\
w_0(\xi,1,\tau)=0, \\
w_0(\xi,0,\tau)=0.
\end{cases}
$$

Since we are interested in small perturbations of steady states, we also take $u_0 := 0$ and $w_0 := 0$; therefore,

$$\begin{cases} 0 = -\partial_\xi P_0 & \text{in } \Omega_\tau, \\ 0 = \partial_z P_0 & \text{in } \Omega_\tau, \\ P_0(\xi, 1, \tau) = c_0. \end{cases} \tag{10.30}$$

From the second and third lines in (10.30), it follows that

$$P_0(\xi, z, \tau) = c_0. \tag{10.31}$$

Now, we consider the coefficients related to ε^1 in (10.29), and we obtain that

$$\begin{cases} -\partial_\xi u_1 = -\partial_\xi P_1 & \text{in } \Omega_\tau, \\ 0 = -\partial_z P_1 & \text{in } \Omega_\tau, \\ \Theta \partial_\xi u_1 + \partial_z w_1 = 0 & \text{in } \Omega_\tau, \\ P_1(\xi, 1, \tau) + \partial_z P_0(\xi, 1, \tau)\eta_0 = \eta_0, \\ w_1(\xi, 1, \tau) = -\Theta \partial_\xi \eta_0, \\ w_1(\xi, 0, \tau) = 0. \end{cases} \tag{10.32}$$

Owing to (10.31), the fourth line in (10.32) can be written as $P_1(\xi, 1, \tau) = \eta_0(\xi, \tau)$, which, combined with the second line, gives that

$$P_1(\xi, z, \tau) = \eta_0(\xi, \tau). \tag{10.33}$$

Hence, from the first line,

$$\partial_\xi u_1 = \partial_\xi \eta_0. \tag{10.34}$$

This and the third line give that

$$\partial_z \left(\frac{w_1}{\Theta} + z\partial_\xi \eta_0 \right) = \frac{\partial_z w_1}{\Theta} + \partial_\xi \eta_0 = -\partial_\xi u_1 + \partial_\xi \eta_0 = 0;$$

therefore, from the last line in (10.32), we infer that

$$\frac{w_1}{\Theta} = -z\partial_\xi \eta_0, \tag{10.35}$$

which is also in agreement with the second-last line in (10.32).

Additionally, from (10.34), we have that $u_1(\xi,\tau,z) = \eta_0(\xi,\tau) + v(\tau,z)$, for an auxiliary function v. As a matter of fact, if we have a solution of (10.32) and we add to u_1 a function of (τ,z), we obtain another solution. Therefore, we choose the "simplest possible" solution by picking v identically equal to zero, thus obtaining

$$u_1(\xi,\tau,z) = \eta_0(\xi,\tau). \tag{10.36}$$

Now, we consider the order related to ε^2 in the expansion in (10.29) (the higher orders being formally neglected). In this way, we find that

$$\begin{cases} -\partial_\xi u_2 + \partial_\tau u_1 + u_1 \partial_\xi u_1 + \frac{w_1 \partial_z u_1}{\Theta} = -\partial_\xi P_2 & \text{in } \Omega_\tau, \\ -\Theta \partial_\xi w_1 = -\partial_z P_2 & \text{in } \Omega_\tau, \\ \Theta \partial_\xi u_2 + \partial_z w_2 = 0 & \text{in } \Omega_\tau, \\ \partial_z P_0(\xi,1,\tau)\eta_1 + \frac{1}{2}\partial_{zz} P_0(\xi,1,\tau)\eta_0^2 \\ \quad + \partial_z P_1(\xi,1,\tau)\eta_0 + P_2(\xi,1,\tau) = \eta_1, \\ \partial_z w_1(\xi,1,\tau)\eta_0 + w_2(\xi,1,\tau) = \Theta\partial_\tau\eta_0 + \Theta u_1\partial_\xi\eta_0 - \Theta\partial_\xi\eta_1, \\ w_2(\xi,0,\tau) = 0. \end{cases} \tag{10.37}$$

Recalling (10.31) and (10.33), the fourth line in (10.37) becomes

$$P_2(\xi,1,\tau) = \eta_1(\xi,\tau).$$

Also, from the second line in (10.37) and (10.35),

$$\partial_z P_2 = -\Theta^2\,\partial_\xi(z\partial_\xi\eta_0) = -z\Theta^2\,\partial_{\xi\xi}\eta_0.$$

These observations entail that, for all $z \in (0,1)$,

$$\eta_1(\xi,\tau) - P_2(\xi,z,\tau) = P_2(\xi,1,\tau) - P_2(\xi,z,\tau) = \int_z^1 \partial_z P_2(\xi,z',\tau)\,dz'$$

$$= -\Theta^2 \int_z^1 z'\partial_{\xi\xi}\eta_0(\xi,\tau)\,dz' = -\frac{\Theta^2\,(1-z^2)}{2}\partial_{\xi\xi}\eta_0(\xi,\tau)$$

and, therefore,

$$P_2(\xi,z,\tau) = \eta_1(\xi,\tau) + \frac{\Theta^2\,(1-z^2)}{2}\partial_{\xi\xi}\eta_0(\xi,\tau).$$

From this and the first equation in (10.37), we infer that

$$\partial_\xi u_2 = \partial_\tau u_1 + u_1 \partial_\xi u_1 + \frac{w_1 \partial_z u_1}{\Theta} + \partial_\xi P_2$$

$$= \partial_\tau u_1 + u_1 \partial_\xi u_1 + \frac{w_1 \partial_z u_1}{\Theta} + \partial_\xi \eta_1 + \frac{\Theta^2 (1 - z^2)}{2} \partial_{\xi\xi\xi} \eta_0.$$

Note the appearance of a third derivative for η_0. Thus, recalling the third equation in (10.37),

$$-\frac{\partial_z w_2}{\Theta} = \partial_\xi u_2 = \partial_\tau u_1 + u_1 \partial_\xi u_1 + \frac{w_1 \partial_z u_1}{\Theta} + \partial_\xi \eta_1$$

$$+ \frac{\Theta^2 (1 - z^2)}{2} \partial_{\xi\xi\xi} \eta_0.$$

Combining this with the last two equations in (10.37), we conclude that

$$\partial_\tau \eta_0(\xi, \tau) + u_1(\xi, 1, \tau) \partial_\xi \eta_0(\xi, \tau) - \partial_\xi \eta_1(\xi, \tau) - \frac{\partial_z w_1(\xi, 1, \tau) \eta_0(\xi, \tau)}{\Theta}$$

$$= \frac{w_2(\xi, 1, \tau)}{\Theta}$$

$$= \frac{w_2(\xi, 1, \tau) - w_2(\xi, 0, \tau)}{\Theta}$$

$$= \frac{1}{\Theta} \int_0^1 \partial_z w_2(\xi, z, \tau) \, dz$$

$$= -\int_0^1 \Big(\partial_\tau u_1(\xi, z, \tau) + u_1(\xi, z, \tau) \partial_\xi u_1(\xi, z, \tau)$$

$$+ \frac{w_1(\xi, z, \tau) \partial_z u_1(\xi, z, \tau)}{\Theta} + \partial_\xi \eta_1(\xi, \tau)$$

$$+ \frac{\Theta (1 - z^2)}{2} \partial_{\xi\xi\xi} \eta_0(\xi, \tau) \Big) \, dz.$$

This and (10.35) yield that

$$\partial_\tau \eta_0(\xi, \tau) + u_1(\xi, 1, \tau) \partial_\xi \eta_0(\xi, \tau) - \partial_\xi \eta_1(\xi, \tau) + \partial_\xi \eta_0(\xi, \tau) \eta_0(\xi, \tau)$$

$$= -\int_0^1 (\partial_\tau u_1(\xi, z, \tau) + u_1(\xi, z, \tau) \partial_\xi u_1(\xi, z, \tau)$$

$$- z \partial_\xi \eta_0(\xi, \tau) \partial_z u_1(\xi, z, \tau)) \, dz - \partial_\xi \eta_1(\xi, \tau) - \frac{\Theta^2}{3} \partial_{\xi\xi\xi} \eta_0(\xi, \tau).$$

Then, we can simplify the term $-\partial_\xi \eta_1(\xi, \tau)$ and exploit (10.36), finding that

$$\partial_\tau \eta_0(\xi, \tau) + \eta_0(\xi, \tau)\partial_\xi \eta_0(\xi, \tau) + \partial_\xi \eta_0(\xi, \tau)\eta_0(\xi, \tau)$$

$$= -\int_0^1 (\partial_\tau \eta_0(\xi, \tau) + \eta_0(\xi, \tau)\partial_\xi \eta_0(\xi, \tau))\, dz - \frac{\Theta^2}{3}\partial_{\xi\xi\xi}\eta_0(\xi, \tau)$$

$$= -\partial_\tau \eta_0(\xi, \tau) - \eta_0(\xi, \tau)\partial_\xi \eta_0(\xi, \tau) - \frac{\Theta^2}{3}\partial_{\xi\xi\xi}\eta_0(\xi, \tau),$$

which is the Korteweg–de Vries equation

$$\partial_\tau \eta_0 + \frac{3}{2}\partial_\xi \eta_0 \eta_0 + \frac{\Theta^2}{6}\partial_{\xi\xi\xi}\eta_0 = 0, \tag{10.38}$$

to be compared with (10.1), as desired.

It is now instructive to recover equation (10.38) in terms of the original variables. For this, owing to (10.24) and (10.27),

$$\eta(\xi, \tau) = \eta^\star\left(\xi + \frac{\tau}{\varepsilon}, \frac{\tau}{\varepsilon}\right) = \bar\eta\left(\sqrt{gh_0}\left(\xi + \frac{\tau}{\varepsilon}\right), \frac{\tau}{\varepsilon}\right). \tag{10.39}$$

Moreover, since $\bar\eta(x, t)$ was the profile of the free surface of the fluid in the canal and $\eta_0(\xi, \tau)$ the leading order of this profile in the new variables (ξ, τ) in shallow water and low-amplitude approximation, in light of (10.39), we can consider the leading order of the original profile $\bar\eta_0$ as defined implicitly by the relation

$$\eta_0(\xi, \tau) = \bar\eta_0\left(\sqrt{gh_0}\left(\xi + \frac{\tau}{\varepsilon}\right), \frac{\tau}{\varepsilon}\right).$$

Therefore, (10.38) becomes

$$\frac{\sqrt{gh_0}}{\varepsilon}\partial_x \bar\eta_0 + \frac{1}{\varepsilon}\partial_t \bar\eta_0 + \frac{3\sqrt{gh_0}}{2}\partial_x \bar\eta_0 \bar\eta_0 + \frac{(gh_0)^{3/2}\,\Theta^2}{6}\partial_x^3 \bar\eta_0 = 0$$

and thus, recalling (10.22),

$$\partial_t \bar\eta_0 + \frac{\sqrt{g}h_0^{5/2}}{6}\partial_x^3 \bar\eta_0 + \frac{3h_0^{3/2}}{2\Theta^2 \sqrt{g}}\partial_x \bar\eta_0 \bar\eta_0 + \sqrt{gh_0}\partial_x \bar\eta_0 = 0. \tag{10.40}$$

In this expression, one can get rid of the term $\partial_x \bar\eta_0$ by a vertical translation of $\bar\eta_0$, that is, if

$$\zeta_0(x, t) := \alpha\left(\bar\eta_0(x, t) + \frac{2\Theta^2 g}{3h_0}\right) = \frac{h_0^2}{\Theta^2 g}\left(\bar\eta_0(x, t) + \frac{2\Theta^2 g}{3h_0}\right), \tag{10.41}$$

it follows that

$$\partial_t \zeta_0 + \frac{\sqrt{g} h_0^{5/2}}{6} \partial_x^3 \zeta_0 + \frac{3\sqrt{g}}{2\sqrt{h_0}} \partial_x \zeta_0 \zeta_0 = 0, \qquad (10.42)$$

which corresponds[6] to (10.1), with $a := 1$, $b := \frac{\sqrt{g} h_0^{5/2}}{6}$ and $c := \frac{3\sqrt{g}}{2\sqrt{h_0}}$.

It is also worth reconsidering the solitary waves of the Korteweg–de Vries equation found in (10.4) and note that these correspond, in principle, to waves with an arbitrarily large velocity v and an arbitrarily large amplitude (proportional to v, see (10.5)): the large-amplitude solutions however are, in strict terms, not consistent with the technical derivation of the model, which relies on expansions that are valid only for small-amplitude waves (recall the setting

---

[6]Reconsidering (10.5), we know that the velocity of the traveling wave ϕ_0 is proportional to $\frac{c}{a}$ times the amplitude of the wave. It is suggestive to compare this information with the numerical properties of the coefficients and the relation with the physical shape of the wave. Namely, in this case, we have that $\frac{c}{a} = c = \frac{3\sqrt{g}}{2\sqrt{h_0}}$, and by (10.41), the amplitude of the traveling wave would formally correspond to $\frac{h_0^2}{\Theta^2 g} \left(\max \overline{\eta}_0 + \frac{2\Theta^2 g}{3 h_0} \right) = \frac{h_0^2}{\Theta^2 g} \max \overline{\eta}_0 + \frac{2h_0}{3}$. The velocity of the traveling wave would thus be proportional to $\frac{3\sqrt{g}}{2\sqrt{h_0}} \left(\frac{h_0^2}{\Theta^2 g} \max \overline{\eta}_0 + \frac{2h_0}{3} \right) = \frac{3 h_0^{3/2}}{2\Theta^2 \sqrt{g}} \max \overline{\eta}_0 + \sqrt{g h_0}$. Once again, this suggests that the speed of small-amplitude waves in a shallow canal is dictated by a quantity proportional to $\sqrt{g h_0}$, plus an increasing term in the amplitude of the wave.

The leading term proportional to $\sqrt{g h_0}$ for the speed of the wave, which becomes smaller in shallower water near the coast, also suggests that waves have the tendency to slow down as they approach the shore because of the force exerted on them by the seabed.

This phenomenon is somewhat related to the fact that, in our experience, the waves coming toward the shore have a strong tendency to follow the shape of the coastline, see e.g. Figures 10.10 and 10.11.

This effect can be explained in terms of the slowing process induced by shallow waters. Indeed, the waves offshore can well be oblique with respect to the coastline, but when approaching the shoreline, the part of the wave closest to the coast is affected sooner by the slowing effect of shallow waters, while the farthest wavefront continues its run for a while with the same speed. This relative variation in speed causes the ridges of the waves to gradually arrange themselves in an almost parallel fashion to the coastline, see Figures 10.12 and 10.13.

Of course, we do not aim here at an exhaustive study of the complicated topic of water waves' speed. See e.g. [GQ18, SKSC21] and the references therein for further details on this subject.

(a)   (b)

Figure 10.10.   (a) The coast of Fengbin, Taiwan; (b) satellite view of the Shark Bay, Western Australia (images from Wikipedia, Photo by Wu Pei Hsuan licensed under the Creative Commons Attribution-Share Alike 4.0 International license for the first, public Domain photo by Jeff Schmaltz for the second).

Figure 10.11.   Wave fronts approaching the shore at North West Cape.

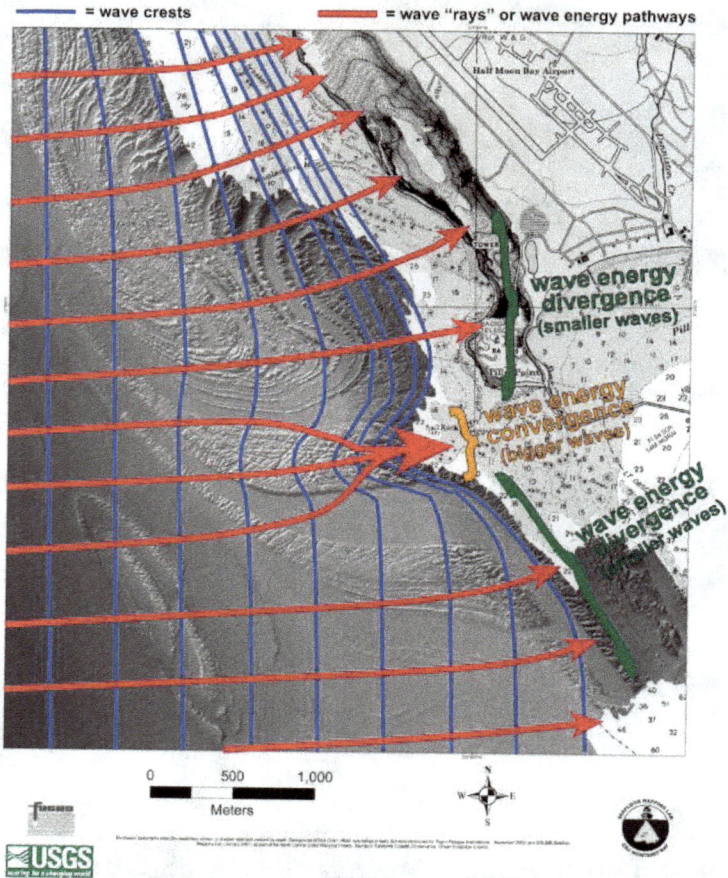

Figure 10.12.   Wavefronts at Mavericks, California (Public Domain image from Wikipedia).

in (10.22)). We also remark that, since the coefficient $\sqrt{\frac{av}{b}}$ in (10.4) is monotone increasing in v and since this coefficient modulates the hyperbolic cosine in the denominator, faster and higher waves in (10.4) are also narrower.

The fact that higher waves (at least in this setting) need to be narrower can be guessed also from the fact that, at least for waves "vanishing at infinity" and when $a \neq 0$, the total mass of the

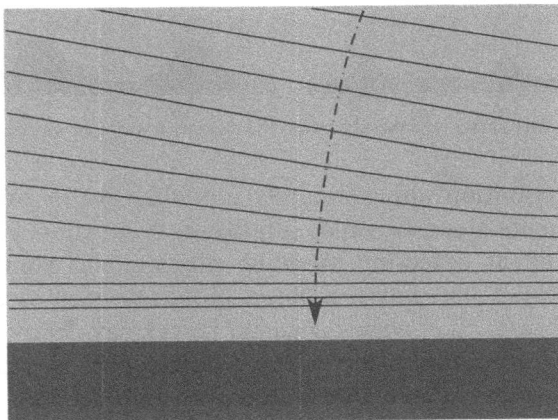

Figure 10.13.   Waterfronts approaching the coastline.

wave $m(t) := \int_{\mathbb{R}} \phi(x,t)\,dx$ is a conserved quantity for the Korteweg–de Vries equation (10.1) since

$$\dot{m}(t) = \int_{\mathbb{R}} \partial_t \phi(x,t)\,dx$$

$$= -\frac{b}{a}\int_{\mathbb{R}} \partial_x^3 \phi(x,t)\,dx - \frac{c}{a}\int_{\mathbb{R}} \phi(x,t)\,\partial_x \phi(x,t)\,dx$$

$$= -\frac{b}{a}\int_{\mathbb{R}} \partial_x(\partial_x^2 \phi(x,t))\,dx - \frac{c}{2a}\int_{\mathbb{R}} \partial_x(\phi^2(x,t))\,dx$$

$$= -\frac{b}{a}(\partial_x^2 \phi(+\infty,t) - \partial_x^2 \phi(-\infty,t)) - \frac{c}{2a}(\phi^2(+\infty,t) - \phi^2(-\infty,t))$$

$$= 0.$$

For further information[7] about the Korteweg–de Vries equation and related topics see e.g. [SG69, Kru74, Mil81, KN86, DJ89, Pal97, Joh97] and the references therein.

Since we are discussing here about waves in a shallow canal, let us spend a few, certainly not exhaustive, words about the phenomenon of wave breaking, which is the case in which a solution

---

[7]For simplicity, we did not discuss the role of the surface tension in the derivation of the Korteweg–de Vries equation, but we mention that this additional term can also be easily included in the coefficient b in (10.1).

Figure 10.14.   Breaking waves: sharp crests versus waves which break into bores.

of a certain equation approximately modeling water waves remains bounded[8] but its slope becomes infinite in finite time. In particular, one would aim at detecting breaking phenomena[9] that correspond to "sharp crests" (i.e. waves with cuspidal tops) or to waves which "break into bores" (i.e. waves u which travel, say, from left to right, starting from a configuration with u' sufficiently negative at some point in the space, whose profile steepens until u' becomes $-\infty$ in finite time), see Figure 10.14. However, the existence and the properties of breaking waves heavily depend on the specific partial differential equation chosen to model the phenomenon. For instance, the Korteweg–de Vries equation does not allow breaking waves, as proved in [CKS+01, CKS+03]. The heuristic reason for which the Korteweg–de Vries equation prevents wave braking is the third derivative term in (10.1), which provides[10] a smoothing effect. As a matter of fact, by dropping this term, one obtains another equation, namely

---

[8]On a different note, we mention that other possible singularities are the ones in which crests exhibit a corner. These may occur for a set of equations similar to, but structurally different from, the ones presented here. The theory of some of these equations go back to Stokes, and several remarkable properties hold true about the maximal periodic deep-water waves arising in this context, e.g. the wave height is about one-seventh of the wavelength, and the maximum-height waves have a 120° corner. See [Tol78, SF82] for further details about these types of problems.

Traveling waves with corners at their top edge (that is, with discontinuous first derivatives) are sometimes called "peakons".

Another type of equation which develops peakons and singularities (i.e. moving waves with corners as well as waves whose slopes become vertical in finite time, provided they are initially sufficiently negative) is the so-called Camassa–Holm equation, see [CH93].

[9]The classification of breaking waves is not homogeneous across the scientific literature, and other types of wave breaking have been classified as well, see e.g. [Lem92, p. 42] and [Lin08, p. 60].

[10]Quite surprisingly, this smoothing effect will play a fundamental role in the evolution phenomena of chain of oscillators, see the forthcoming discussion on p. 266 for full details.

the Bateman–Burgers equation, which will be introduced as follows in (26.8), and which develops singularities in finite time; however, as clarified in [Whi99, p. 457], the theory of the Bateman–Burgers equation (that is, the Korteweg–de Vries equation without third derivative term) "goes too far: it predicts that all waves carrying an increase of elevation break. Observations have long since established that some waves do not break. So an invalid theory seems to be right sometimes and wrong sometimes!"

To develop a mathematical theory of breaking waves, several modifications of the Korteweg–de Vries equation have been introduced in the literature (for instance, by replacing the third derivative term with a nonlocal or fractional term, see [Whi67, HT14, Hur17, HT18]). Yet, the matter does not appear to be completely settled. For instance, several attempts have been made to determine the breaking wave height α with respect to the water depth h and[11] the "numerology" does not always match conjectures, theory and experiments. As an example, in [Sho05, p. 14], one finds that the water depth corresponds to 1.5 times the breaking wave height, giving $\frac{\alpha}{h} = \frac{2}{3} = 0.\overline{6}$. The problem has an ancient tradition: already in [McC91, equation (52)], John McCowan, a pioneer in the study of the fluid mechanics behind surfing, calculated $\frac{\alpha}{h} \simeq 0.9$, concluding that "the wave will break for an elevation rather less than the mean depth", believing this condition for breaking to be "not be far wrong. Scott Russell's experiments confirm this: he found that the wave broke when the elevation was about equal to the depth; but from some experiments of my own I am inclined to think that 3/4 is a closer approximation for the elevation at the breaking-point". This experiment, determining $\frac{\alpha}{h} \simeq 0.75$, was reconsidered from a theoretical point of view in [McC94, equation (34)], proposing at this stage the value $\frac{\alpha}{h} \simeq 0.78$. This value has been commented about in [Lem92, p. 39] as "appropriate for estimating the breaker height in mild-sloped surf-zones".

The reliability of these numbers is possibly a bit controversial, see [KT14], according to which "[God10] mentions that a value of 0.8261 is more accurate [YKO68]. From the field observations, it is found to be between 0.78 and 0.86". But [IK54, p. 29] lists

---

[11] The ratio between the height of the breaking wave and the depth of the bottom is sometimes referred to by the name "breaker index".

five different theoretical findings for the ratio between the breaking wave elevation and the depth of the bottom, ranging from 0.73 to 1.03. Numerous laboratory experiments have been performed to empirically quantify the wave breaking phenomenon, including by installing offshore structures resembling the actual coastal terrain, but most of the formulas found depend significantly on the specific laboratory experiment data, see e.g. [Mun49, Table 1] for the effect of the beach characteristics on break height. See also [LC21, YKD22] for a machine learning viewpoint to the problem. Other approaches to describe wave breaking relate to equations with variable[12] depth, instead of flat bottoms, see e.g. [Isr10]. In general, a common belief is that the phenomena related to wave breaking are highly complex and are linked to many additional properties of the environment, such as the shape of the land (sometimes referred to by the name "topography") and the specific features of the bottom (such as impermeability, presence of irregularities and slope oscillations).

In essence, about two centuries after the solitary wave of translation observed by John Scott Russell, a unified and all-encompassing theory of water waves with an outright match between rigorous results and experimental findings is, perhaps, still not available.

---

[12]It is possible that the steepness of the bottom also plays a role in the determination of the height of a breaking wave: according to [CML07], "most formulas show quite good predictions for cases including gentle slopes, however, the predictions are typically not satisfactory for breaking waves on steep slopes". This paper also presents a comparison of different approaches, methodologies and formulas on the topic of breaking waves.

Chapter 11

Plasma Physics

A state of matter is one of the distinct forms in which matter can exist in nature. Classically, nature was classified into three states of matter: solid, liquid and gas. Nowadays, it is common[1] to consider also a fourth state of matter, namely plasma.

Though our everyday experience with plasma appears to be rather limited (our experience with plasma is restricted to neon tubes, plasma lamps and lightnings, see Figure 11.1), plasma is allegedly the most abundant form of ordinary matter in the Universe (excluding dark matter and dark energy) and likely amounts to about 99% of the

---

[1] Also, a large number of intermediate states are nowadays known to exist, such as liquid crystals, Bose–Einstein condensates, Fermionic condensates, neutron-degenerate matter, quark-gluon plasma, superfluids, and superconductors, but we do not investigate these states here.

From a historical point of view, plasma was first identified in the laboratory by Sir William Crookes in 1879 (though more systematic studies of plasma only began in the twentieth century).

Besides discovering plasma, Crookes is also credited with discovering thallium, inventing the radiometer and, quite conveniently, a 100% ultraviolet-blocking sunglass lens.

He was also interested in spiritualism and became the president of the Society for Psychical Research, whose purpose is to understand psychic and paranormal events.

See Figure 11.2 for a caricature of Sir William Crookes (by caricaturist Sir Leslie Ward).

Figure 11.1. (a) a plasma lamp (photo by Luc Viatour, posted on Wikipedia under the Creative Commons Attribution-Share Alike 3.0 Unported license). (b) neon lights in Osaka, Japan (photo by Pedro Szekely, posted on Wikipedia under the Creative Commons Attribution-Share Alike 2.0 Generic license). (c) a large bolt strikes west of downtown Denver on June 8, 2004 (photo by refractor, posted on Wikipedia under the Creative Commons Attribution 2.0 Generic license).

total mass of the visible universe. It consists of a gas of ions and free electrons, usually produced by very high temperatures (10^4 degrees or more), which make electrons leave their orbit surrounding the corresponding nuclei. In this way, if the initial status of the gas is overall in an electrically neutral state, the plasma consists of a mixture of charged particles, namely free electrons and the corresponding ions.

The mathematical description of a plasma is a highly advanced subject, but in a nutshell, there is a hierarchy of models accounting for the evolution of a plasma.

The most intuitive one would be to describe the plasma by detecting the positions and velocities of all its particles (ions and electrons) at a given time.

This approach is highly precise but often impractical, given the high number of particles involved; therefore, intermediate models (often called "kinetic models") have been introduced to describe plasma "on average" through statistical analyses of particle distribution.

Moreover, at a large scale, efficient models for a plasma leverage the knowledge of the fluid dynamics equations (somewhat close in spirit to the ones we presented in Chapters 5 and 6); namely, at a macroscopic level and close to a thermodynamic equilibrium, one can identify each species of particle of a plasma (ions and electrons) with

Figure 11.2. Caricature of Sir William Crookes (Public Domain image from Wikipedia).

a fluid and describe its corresponding density, velocity and energy via a set of[2] partial differential equations.

To make this model concrete, we denote by an index $j \in \{1, 2\}$ the different species of particles of a given plasma (e.g. $j = 1$ corresponding to ions and $j = 2$ corresponding to electrons). We suppose that the plasma, being neutral on average but composed of charged particles at a small scale, is subject to its own magnetic and electric fields (denoted by B and E, respectively). We recall that charged particles are influenced by both the electric and magnetic fields; more specifically, the so-called Lorentz force acting on a single charged particle

---

[2]To confirm the importance of partial differential equations in our understanding of plasma, let us mention for instance that the laboratory specialized in plasma, located in Toulouse, France, is named LAboratoire PLAsma et Conversion d'Energie (LAPLACE).

(say, with charge q_j and velocity v_j) is of the form $q_j(E + v_j \times B)$. Therefore, the total contribution of the Lorentz force acting on a portion of the plasma with particle density μ_j takes the form

$$\mu_j q_j (E + v_j \times B). \tag{11.1}$$

Additionally, variations in density exert a force on the plasma similar to that of pressure. Thus, we denote by

$$-\nabla p_j \tag{11.2}$$

this type of pressure (the minus sign indicating the fact that higher densities oppose the motion).

Differently from the case of classical fluids, in the description of a plasma, we should also account for the effect of possible particle collisions. Since we are interested here in a macroscopic description of plasma, we suppose that the total momentum of the particles of species j comes from the average over all particles of that species; therefore collisions between like particles do not change the total momentum (say, if two electrons collide, one might slow down after the collision, while the other is accelerated, but the total momentum is preserved, thanks to Newton's law; the case of two ions colliding is similar). We should therefore focus on collisions between unlike particles, which allow momentum to be exchanged between species. In this situation, when an electron and a ion collide, the momentum loss of one particle entails a corresponding momentum gain of the other particle; that is, if the first particle experiences a variation in momentum given by P_1, the second particle experiences a variation in momentum given by $P_2 = -P_1$.

Computing the momentum exchange of colliding particles can be, in general, a rather complicated task. A simplifying assumption in this framework is that the particles are reduced to points. An additional simplification arises if we are willing to suppose that the collisions are perfectly elastic (i.e. no dissipation of energy due to particle collisions). Furthermore, we can also reduce the problem to a simpler one if we are willing to use the fact that the mass of an electron is typically much smaller than that of an ion and, therefore, consider the latter as the leading term of a more complex approximation. To make these ideas work, at least in this simplified setting, we recall that the elastic collision of two point masses, say with masses m_1

and m_2, initial (before collision) velocities v_{1i} and v_{2i} and final (after collision) velocities v_{1f} and v_{2f}, is fully described by the equations

$$\begin{cases} v_{1f} = \dfrac{(m_1 - m_2)v_{1i} + 2m_2 v_{2i}}{m_1 + m_2}, \\ v_{2f} = \dfrac{(m_2 - m_1)v_{2i} + 2m_1 v_{1i}}{m_1 + m_2}, \end{cases}$$

see e.g. [LL60, equation (17.2)].

Hence, since we consider the index $j = 1$ as corresponding to the electrons and we assume that $m_1 \ll m_2$, we have that the momentum exchange for the colliding electron is

$$\begin{aligned} m_1(v_{1f} - v_{1i}) &= m_1 \left(\frac{(m_1 - m_2)v_{1i} + 2m_2 v_{2i}}{m_1 + m_2} - v_{1i} \right) \\ &= \frac{2m_1 m_2 (v_{2i} - v_{1i})}{m_1 + m_2} \\ &\simeq 2m_1(v_{2i} - v_{1i}). \end{aligned}$$

In the framework of plasma collisions, this would give $P_1 = 2\mu_1 m_1(v_2 - v_1)$ and, accordingly, $P_2 = 2\mu_1 m_1(v_1 - v_2)$. Thus, if we denote by ϕ the collision frequency between the two species (for simplicity, we are disregarding here the collisions between ions, or electrons, with the remaining neutral gas), the species corresponding to $j = 1$ changes its momentum by a quantity $\phi \mu_1 m_1(v_2 - v_1)$ and the species corresponding to $j = 2$ changes its momentum by a quantity $\phi \mu_2 m_2(v_1 - v_2)$ (where constants are omitted for the sake of simplicity).

From this, (11.1) and (11.2), by Newton's law of momentum balance, we have that

$$\begin{aligned} \mu_j m_j \partial_t v_j + \mu_j m_j (v_j \cdot \nabla) v_j &= \mu_j m_j \frac{d}{dt}(v_j) \\ &= \mu_j q_j (E + v_j \times B) \\ &\quad - \nabla p_j + \phi \mu_j m_j (v_{k_j} - v_j), \end{aligned}$$

where

$$k_j := \begin{cases} 1 & \text{if } j = 2, \\ 2 & \text{if } j = 1. \end{cases}$$

Figure 11.3. The first durable color photographic image (Public Domain image from Wikipedia).

This equation is usually complemented by a continuity equation (corresponding to the conservation of the total number of particles, which can be seen here as the counterpart of the mass transport equation in (5.4)), an equation of state (that is, a constitutive law relating pressure and particle density, which can be seen as the counterpart of the barotropic flow description given on p. 67, usually taken to be of the form $p_j = \kappa \mu_j^\gamma$, for suitable positive constants κ and γ) and the classical Maxwell's equations[3] for the electric and magnetic

---

[3]James Clerk Maxwell founded the theory of electromagnetism, proving that electric and magnetic fields travel through space as waves moving at the speed of light, and he argued that light itself is an undulation causing electric and magnetic phenomena. This great conceptual unification of light and electrical phenomena made Maxwell one of the 19th-century scientists who had the greatest impact on the subsequent developments of relativity and quantum field theory.

Maxwell also worked on the kinetic theory of gases and pioneered the early developments of control theory. Also, he established that the rings of Saturn were made of numerous small particles, and he is credited with presenting the first durable color photograph. For this, he had the idea of superimposing on a screen shots taken with red, green and blue filters; his result, depicting a tartan ribbon, is shown in Figure 11.3. The photographic plates available at the time were almost insensitive to red and barely sensitive to green; hence, the results in themselves were perhaps far from perfect, but Maxwell clearly knew this was after all only a minor detail in the big picture of science. He wrote, "by finding photographic materials more sensitive to the less refrangible rays, the representation of the colors of objects might be greatly improved".

fields; omitting structural constants, this set of equations thus takes the form[4]

$$\begin{cases} \mu_j m_j \partial_t v_j + \mu_j m_j (v_j \cdot \nabla) v_j = \mu_j q_j (E + v_j \times B) \\ \quad - \nabla p_j + \phi \mu_j m_j (v_k - v_j), \\ \partial_t \mu_j + \mathrm{div}(\mu_j v_j) = 0, \\ p_j = \kappa \mu_j^\gamma, \\ \mathrm{div}\, B = 0, \\ \mathrm{div}\, E = \mu_1 q_1 + \mu_2 q_2, \\ \mathrm{curl}\, E = -\partial_t E, \\ \mathrm{curl}\, B = \mu_1 q_1 v_1 + \mu_2 q_2 v_2 + \partial_t E, \end{cases} \qquad (11.3)$$

with $j \in \{1, 2\}$.

For additional information about plasma physics, see e.g. [SM95, BS03, Jar10, Sen14] and the references therein.

---

[4]Sometimes the setting in (11.3) is called in jargon the "Euler–Maxwell's equations" since it mixes the classical Euler's formulation of fluid dynamics presented in Chapter 5 (after suitable corrections to deal with particle collision) with Maxwell's equations for electromagnetism.

Chapter 12

Galaxy Dynamics

Interestingly, a suitable modification of the model used to describe the motion of fluids presented in Chapter 5 can represent the dynamics of galaxies and globular clusters (see e.g. Figure 12.1 for a fascinating picture). The *ansatz* of the model described here is that the stars forming the galaxies interact only through the gravitational field collectively created by them, and they can be described by parcels of a self-gravitating gas. Neglecting collisional and relativistic effects and disregarding the physical and chemical reactions that continuously modify the internal structure of stars, one can derive a suitable set of equations to describe large-scale galaxy motion. These are named Jeans equations after Sir James Hopwood Jeans (though they were probably first derived by James Clerk Maxwell and can be seen as a collisionless version of the Boltzmann equations, which describe the statistical behavior of a thermodynamic system not in equilibrium) and are given as follows:

$$\begin{cases} \partial_t f + v \cdot \nabla_x f - \nabla u \cdot \nabla_v f = 0, \\ \Delta u = \rho, \end{cases} \qquad (12.1)$$

where $f = f(x, t, v)$ denotes the phase-space[1] density of stars, that is, the density of stars corresponding to time t, position $x \in \mathbb{R}^3$ and velocity $v \in \mathbb{R}^3$; $u = u(x, t)$ is the gravitational potential induced

---

[1] By "phase-space" here we mean the collection of position and velocity variables.

147

Figure 12.1. Hubble Space Telescope picture of the galaxies NGC 2207 (left) and IC 2163 (right) (Public Domain image from Wikipedia).

collectively by the stars; and

$$\rho = \rho(x,t) := \int_{\mathbb{R}^3} f(x,t,v)\, dv$$

is the density of the stars corresponding to time t and position x.

To understand the rationale behind equation (12.1), we argue as follows. In analogy with the dynamics of fluid parcels described in (5.1), we assume that star dust trajectories follow the law $\dot{x}(t) = v(x(t),t)$ and that their acceleration follows the gravitational law $\ddot{x}(t) = -\nabla u(x(t),t)$.

One can therefore obtain an analogous form of the continuity equation in (5.4) by assuming that, in the absence of encounters, collisions and collapses, the amount of stars is preserved by the flow. Namely, the total quantity of stars occupying a region Z of the phase-space $\mathbb{R}^3 \times \mathbb{R}^3$ at a given time t_0 is given by the quantity $\int_Z f(x,t_0,v)\, dx\, dv$ and, in a small time τ, possibly occupying a different region, named Z_τ, which collects all the evolution trajectories in the phase space given by

$$\Big(x(t_0 + \tau), v\big(x(t_0 + \tau), t_0 + \tau\big)\Big) = \big(x(t_0 + \tau), \dot{x}(t_0 + \tau)\big)$$

$$= (x(t_0), \dot{x}(t_0)) + \tau(\dot{x}(t_0), \ddot{x}(t_0)) + o(\tau)$$

$$= \Big(x(t_0), v\big(x(t_0), t_0\big)\Big) + \tau\Big(v\big(x(t_0), t_0\big), -\nabla u(x(t_0), t_0)\big)\Big) + o(\tau).$$

Thus, if $(x_0, v_0) \in Z$ and $(x(\tau), z(\tau)) = \Phi^\tau(x_0, v_0)$ denote the integral flow starting at (x_0, v_0), it follows that

$$\left| \det D_{(x_0, v_0)} \Phi^\tau(x_0, v_0) \right|$$
$$= \left| \det D_{(x_0, v_0)} \left((x_0, v_0) + \tau(v_0, -\nabla u(x_0, t_0)) \right) \right| + o(\tau)$$
$$= 1 + o(\tau),$$

leading to

$$\int_{Z_\tau} f(x, t_0 + \tau, v) \, dx \, dv = \int_Z f(x_0 + \tau v_0, t_0 + \tau, v_0$$
$$- \tau \nabla u(x_0, t_0)) \, dx_0 \, dv_0 + o(\tau).$$

Hence, the assumption that stars are conserved thus translates into

$$0 = \frac{d}{d\tau} \int_{Z_\tau} f(x(t_0 + \tau), t_0 + \tau, v(t_0 + \tau)) \, dx \, dv \bigg|_{\tau = 0}$$
$$= \frac{d}{d\tau} \left(\int_Z f(x_0 + \tau v_0, t_0 + \tau, v_0 - \tau \nabla u(x_0, t_0)) \, dx_0 \, dv_0 + o(\tau) \right) \bigg|_{\tau = 0}$$
$$= \int_Z \left[\nabla_x f(x_0, t_0, v_0) \cdot v(x_0, t_0) + \partial_t f(x_0, t_0, v_0) \right.$$
$$\left. - \nabla_v f(x_0, t_0, v_0) \cdot \nabla u(x_0, t_0) \right] dx_0 \, dv_0.$$

The arbitrariness of the phase-space domain Z thus leads to

$$\nabla_x f(x, t_0, v) \cdot v(x, t_0) + \partial_t f(x, t_0, v) - \nabla_v f(x, t_0, v) \cdot \nabla u(x, t_0) = 0,$$

which is the first equation in (12.1).

Additionally, by Gauß's flux law for gravity, we know that the flux of the gravitational field through a surface balances the total mass enclosed within the surface: namely, neglecting dimensional constants, for every domain $\Omega \subset \mathbb{R}^3$,

$$\int_{\partial \Omega} \nabla u(x) \cdot \nu(x) \, d\mathcal{H}_x^{n-1} = \int_\Omega \rho(x, t) \, dx.$$

From this and the divergence theorem, we arrive at

$$\int_\Omega \Delta u(x) \, dx = \int_\Omega \rho(x, t) \, dx,$$

and consequently, since Ω is arbitrary, we find that $\Delta u = \rho$, which is the second equation in (12.1).

For additional information about galaxy dynamics, see e.g. [BT08] and the references therein.

Chapter 13

How to Count What We Cannot See

In 1812, Lorenzo Romano Amedeo Carlo Avogadro, Count of Quaregna and Cerreto, hypothesized that the volume of a gas at a given pressure and temperature is proportional to the number of atoms or molecules. That is, equal volumes of gases at the same temperature and pressure have the same number of molecules, regardless of the nature of the gas. In particular, by Avogadro's law, the number of molecules or atoms in a given volume of an ideal gas is independent of their size. While this prescription only holds true for ideal gases, and real gases show instead small deviations from this ideal configuration, Avogadro's law is often a very useful approximation and always provides a great conceptual tool since it detects a universal quantity that only depends on reasonable macroscopic parameters, such as temperature and pressure, and is independent of the specific situation under consideration, namely the type of gas that one is studying. See Figure 13.1 for a portrait of Avogadro (by C. Sentier).

It is perhaps worth stressing that Avogadro's law is not quite intuitive. For instance, if one has three units of volumes of hydrogen and one unit of a volume of nitrogen, after they combine to produce ammonia, how many units of volume of ammonia do we expect (assuming that pressure and temperature are maintained constant)? One unit? Three units? Four units?

The correct answer according to Avogadro's law is *two* units of volume. This is because three volumes of hydrogen contain $3k$ molecules of hydrogen H_2 (for some $k \in \mathbb{N}$), and one volume of nitrogen contains $1k$ molecules of nitrogen N_2 (and the proportionality factor k is the same for both hydrogen and nitrogen by

Figure 13.1. Portrait of Amedeo Avogadro (Public Domain image from Wikipedia).

Avogadro's law). Thus, from the reaction

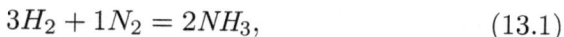

$$3H_2 + 1N_2 = 2NH_3, \tag{13.1}$$

the combination of $3k$ molecules of hydrogen and $1k$ molecules of nitrogen produces $2k$ molecules of ammonia NH_3. And again by Avogadro's law, this corresponds to two volumes of ammonia. See Figure 13.2 for a sketch of this phenomenon.

Furthermore, in chemistry, an obvious practical issue is given by the fact that molecules are small, while in the lab one has to deal with macroscopic quantities. Thus, while it is desirable from a theoretical point of view to base the concept of an "amount of a given substance" on the number of "elementary entities", i.e. atoms or molecules present in the substance (because these elementary entities are the ones which will play a role in chemical reactions), for practical purposes it is often convenient to relate the notion of the "amount of a given substance" to the ratio of measured macroscopic quantities (because measuring a huge number of microscopic entities is typically unfeasible).

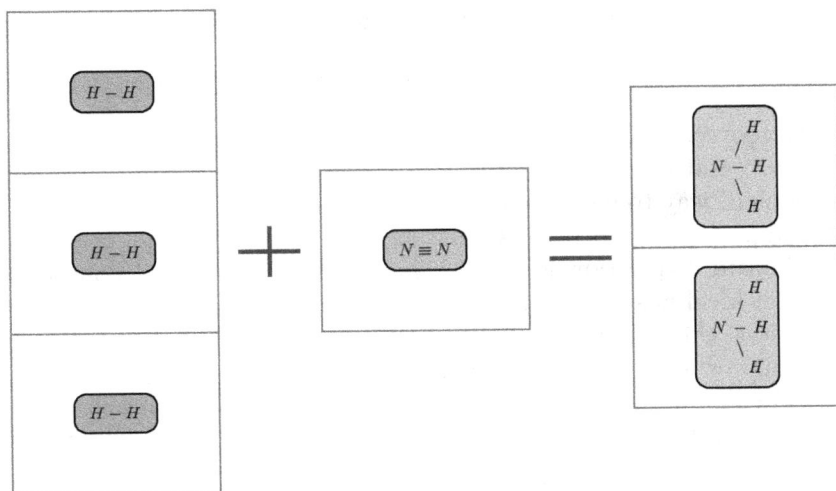

Figure 13.2. The algebra in (13.1) with the corresponding gas volumes according to Avogadro's law.

We remark that the equivalence between these two approaches is a notable consequence of Avogadro's law. Indeed, if, at a given pressure and temperature, we take, say, one gram of atomic hydrogen H (or, more simply, two grams of the common molecule of hydrogen H_2) and we measure its volume, and then we consider the same volume of another gas, such as carbon dioxide CO_2, we know from Avogadro's law that the latter contains the same number of molecules of CO_2 as the number of atoms of hydrogen H (or the number of molecules of H_2) in the former. If we measure the weight of the carbon dioxide CO_2 in the second container, we find that it amounts to 44 grams: we can accordingly conclude that the mass of a molecule of carbon dioxide CO_2 is 44 times that of an atom of hydrogen H (or 22 times that of a molecule of hydrogen H_2). That is, we have measured, with minimal effort, the microscopical weight of any molecule of any gase in terms of a given "unit of measure", such as the atomic weight of H (or the molecular weight of H_2).

Hence, it has become a common practice to adopt the notion of "mole" to denote the mass of a given substance which contains the same number of molecules (or atoms in the case of pure atomic elements) as one gram of atomic hydrogen H. In the previous example, since the mass of a molecule of carbon dioxide CO_2 happens to be 44 times that of an atom of hydrogen H, we infer that

$$\text{a mole of carbon dioxide weights 44 grams.} \qquad (13.2)$$

Moreover, by Avogadro's law, a mole of any gas at a given temperature and pressure occupies the same volume, making it a quite convenient way to measure the (relative) amount of substance simply by comparing volumes of gases.

A natural question is thus, given a certain chemical compound (say, an ideal gas), how many molecules of it correspond to a mole? How does this number depend on the specific chemical compound? The answer is easy: by construction, this number is the same for all (ideal) substances and equals the number of hydrogen atoms H necessary to form one gram of hydrogen, and this is called[1] Avogadro's number $\mathcal{N}_A$.

Note that the knowledge of $\mathcal{N}_A$ also permits "passing from ratios to real quantities". For instance, from (13.2), we obtain that each molecule of carbon dioxide weighs $\frac{44}{\mathcal{N}_A}$ grams: in general, the knowledge of Avogadro's number entails the knowledge of atomic and molecular masses as a byproduct of macroscopic measurements.

Therefore, it is highly desirable to have a precise[2] quantification of $\mathcal{N}_A$. This is not a mathematical problem per se because, to

---

[1] We stress that Avogadro's number is not a "pure number" but a physical constant of dimension "one over moles". See also [Whe10] for additional information on Avogadro's number.

[2] It turns out (possibly with some slight approximations) that $\mathcal{N}_A = 6.02214076 \times 10^{23}$. Several researchers have observed a similarity between this number and 2^{79} (which is half of the so-called yobibyte 2^{80}); indeed, it is a rather surprising coincidence that

$$\frac{\mathcal{N}_A - 2^{79}}{\mathcal{N}_A + 2^{79}}$$

is almost zero. There is nothing special in this – coincidences happen. For instance, it is a mere, and quite remarkable, coincidence that the most important physical constant, namely the speed of light, is almost equal to 300.000.000 m/s. It is also a coincidence that the angular diameter of the Sun seen from Earth is quite close to that of the Moon, making it possible to see a total solar eclipse. It is also a coincidence that

$$\left(\frac{\pi^e - e^\pi}{\pi^e + e^\pi} \right)^{\pi + e}$$

turns out to be almost equal to zero.

quantify $\mathcal{N}_A$, one can only measure it through experiments. However, one needs a brilliant piece of mathematics to contrive a practical procedure for experimentally calculating $\mathcal{N}_A$, and we recall here the method devised by Albert Einstein in [Ein05] (at that time, he was working as a patent clerk in Bern, Switzerland, and the very notion of atoms and molecules was still a subject of controversy and intense scientific debate since individual atoms and molecules were, for the instruments available in 1905, simply "too small" and "too fast").

To measure Avogadro's number, Einstein utilized a molecular theory of heat more or less in the lines of Brownian motion, described in Chapter 2, and leading to the heat equation in (2.7). For this, Einstein thought of some ideal spherical particles of radius a subject to Brownian motion in an ideal (for simplicity, one-dimensional and infinitely long) pipe containing some fluid with a viscosity coefficient equal to μ. In this setting, according to Einstein's calculation,

$$\mathcal{N}_A = \frac{RTt}{3\pi\mu a\, X^2(t)}, \tag{13.3}$$

where R is the universal[3] gas constant, T denotes the temperature (or, more precisely, the thermodynamic temperature, which is the temperature measured in Kelvin), t is any given time and $X^2(t)$ is the averaged squared distance traveled by a Brownian particle in an interval of time equal to t.

Equation (13.3) is quite remarkable since it equates Avogadro's number with quantities which can be measured experimentally, thus leading to a precise determination of $\mathcal{N}_A$ (that's the power of mathematics!).

The experiment suggested by Einstein was indeed conducted by Jean Baptiste Perrin (see Figure 13.3) and his team of research students in 1909. The test setup consisted of a camera lucida with a microscope to observe and record the motion of suspended gamboge[4]

---

[3]Note that, in principle, one can measure R via the ideal gas prescription $PV = nRT$ in terms of macroscopic quantities (though more precise measures of R typically rely on a finer analysis). Thus, for the purposes of this notes, R is considered simply a "known" constant.

[4]Gamboge is a yellowish pigment. The word "gamboge" comes from *gambogium*, the Latin word for pigment, which in turns derives from "Gambogia", the Latin word for Cambodia.

Figure 13.3.   Jean Perrin (Public Domain image from Wikipedia).

particles in a liquid of a given viscosity and constant temperature. Marking the particle's position on a piece of graph paper at
set time intervals, and repeating the test under different conditions
(such as different viscosities, temperatures and radii of the gamboge
grains), Perrin obtained consistent results about the right-hand side
of (13.3), thus producing a rather accurate measurement of Avogadro's number $\mathcal{N}_A$ through a wealth of measurements that could
not be contested.

Besides its strong laxative properties, which did not play a role in Perrin's
experiment, gamboge also possess the convenient feature of producing, with
appropriate alcoholic additives and after a selective centrifuge, almost perfect
spherules with equal radius, thus providing the ideal suspension particles for the
experiment proposed by Einstein (this was not a cheap preparation; about one
kilo of gamboge and several months of work produced a few decigrams of useful
spherules).

This experiment also provided an experimental confirmation of Einstein's equation (13.3) and raised atoms and molecules from the status of hypothetical objects to concrete and observable entities, thus offering the ultimate confirmation of the atomic nature of matter and resolving the debate regarding the physical reality of molecules. For these achievements, Perrin was awarded the Nobel Prize for Physics in 1926.

It is now time to return to Einstein's equation (13.3) to understand its roots in the theory of partial differential equations. To this end, we recall that the first ingredient in the derivation of (13.3) is Stokes' law, which pertains to the viscosity force acting on a small sphere moving through a viscous fluid. Namely, by studying the incompressible steady flow of the Navier–Stokes equation (see Chapter 5), George Gabriel Stokes in 1851 had proposed the formula

$$F = 6\pi\mu a v \tag{13.4}$$

to describe the frictional force F acting on a spherical particle of radius a with a relative velocity v in a fluid with a viscous coefficient equal to μ.

The second ingredient in the derivation of (13.3) is the van 't Hoff equation for[5] the osmotic pressure p of a solution containing n moles of solute particles in a solution of volume V, given by

$$p = \frac{nRT}{V}, \tag{13.5}$$

in which R is the universal gas constant.

The third ingredient in the derivation of (13.3) is the assumption that the Brownian motion of suspended particles can be effectively described by the heat equation, as discussed in Chapter 2. Hence, we suppose that the density u of the particles satisfies the equation

$$\partial_t u = c\partial_{xx} u, \tag{13.6}$$

---

[5]Equation (13.5) is named after Jacobus Henricus van 't Hoff Jr., the first winner of the Nobel Prize in Chemistry, see Figure 13.4. Allegedly, van 't Hoff chose to study chemistry against the wishes of his father. Also, it seems he came up with equation (13.5) after a chance encounter and a quick chat with a botanist friend during a walk in a park in Amsterdam, thinking about an analogy to the law for ideal gases.

Figure 13.4.   Henry van 't Hoff (Public Domain image from Wikipedia).

for some constant diffusion coefficient $c > 0$ (here, we assume that the motion of the particles is confined to a linear pipe, and hence the variable x is one-dimensional).

We can now retake van 't Hoff equation (13.5) by considering that the number of floating particles N over the volume V equals the density u, and by definition, a mole contains $\mathcal{N}_A$ particles. These observations lead to

$$p = \frac{NRT}{\mathcal{N}_A V} = \frac{RTu}{\mathcal{N}_A}. \tag{13.7}$$

Also, the force F acting on each floating particle arises from a pressure gradient; namely, the gradient pressure force with respect to a unit of mass is $\frac{\partial_x p}{u}$. This remark and (13.7) give that

$$Fu = \partial_x p = \frac{RT \, \partial_x u}{\mathcal{N}_A}.$$

At equilibrium, the force must correspond to the frictional force in (13.4); therefore, we obtain

$$6\pi\mu a v u = \frac{RT\, \partial_x u}{N_A}. \tag{13.8}$$

Now, we compute the flux of floating particles through an ideal cross-section of the pipe by using two possible strategies. On one hand, we can consider the model of particles traveling with velocity v; hence, we can quantify the flux using the product of v and the particle density u. On the other hand, the flux at some cross-section x can also be obtained from the variation in time of the mass preceding x, that is, $\partial_t \int_{-\infty}^x u(\xi, t)\, d\xi$. These observations and (13.6) give that

$$v(x,t)u(x,t) = \partial_t \int_{-\infty}^x u(\xi, t)\, d\xi = \int_{-\infty}^x \partial_t u(\xi, t)\, d\xi$$

$$= c \int_{-\infty}^x \partial_{xx} u(\xi, t)\, d\xi = c\partial_x u(x,t),$$

where, as usual, we have performed formal calculations and assumed some convenient decay of the solution at $-\infty$.

From this and (13.8), we arrive at

$$N_A = \frac{RT\, \partial_x u}{6\pi\mu a v u} = \frac{RT}{6\pi\mu a c}. \tag{13.9}$$

Now, we observe that

$$g(x,t) := \frac{1}{\sqrt{4\pi c t}} \exp\left(-\frac{x^2}{4ct}\right)$$

is a solution of (13.6), concentrating at the origin when $t = 0$; hence, we can suppose that the general solution u of (13.6) occurs as a superposition[6] of translations of g of the form $g(x - x_0, t)$ (note that each of these translations would correspond to a particle located at x_0 when $t = 0$).

---

[6] The formalization of this superposition method is the core of the notion of fundamental solution, and in an elliptic setting, it will play a decisive role in Section 2.7 of the companion book [DV23], particularly in Proposition 2.7.4.

The average squared displacement at time t of the particle starting at the origin is therefore provided by the quantity

$$\int_{\mathbb{R}} x^2 g(x,t)\, dx = \int_{\mathbb{R}} \frac{x^2}{\sqrt{4\pi ct}} \exp\left(-\frac{x^2}{4ct}\right) dx = 2ct.$$

Since this displacement should not depend on the original position, we infer that it is the same for all possibilities of the initial location of the particles, namely

$$X^2(t) = 2ct. \tag{13.10}$$

This is conceptually an important result since it suggests that the velocity (namely the ratios of $X(t)$ to t) of Brownian motion behaves like $\frac{\sqrt{2ct}}{t} = \frac{\sqrt{2c}}{\sqrt{t}}$ and hence quite irregularly[7] for short durations. That is, measuring the velocity of the floating particles in an extremely short interval of time produces a result that approaches infinity (in a sense, experiments attempted in this direction would simply end up measuring the wrong quantity). One of the merits of Einstein's approach was that it bypassed this hindrance by detecting the relevant physical quantities that are both consistent from a mathematical point of view and experimentally measurable in laboratories. Note indeed that by plugging (13.10) into (13.9), we obtain (13.3), as desired.

---

[7]For a formalization of this heuristic and imprecise discussion, see e.g. [Eva13, Section 3.4].

Chapter 14

Good Vibrations

Another classical occurrence of partial differential equations arises in the description of vibrating strings (e.g. the string of a guitar, see Figure 14.1). Different models are possible, taking into account different physical assumptions (see e.g. [PR05, Remark 10.5]), but we focus here on the simplest possible (indeed linear) equation. In our description, the string at rest is modeled by the segment $[0, L] \times \{0\}$ for some $L > 0$, which represents the length of the string in the absence of additional external forces. One can assume that the string is constrained at the extrema and can be deformed into a shape given by the graph of a function u; namely, the position of the vibrating string at each moment of time t is described by the graph $(x, u(x, t)) \in [0, L] \times \mathbb{R}$, and the constraints for the string correspond to the boundary prescription $u(0) = u(L) = 0$.

We suppose that the string is subject to gravity (acting downward in the vertical direction) and to its own "tension" (which is the internal force of the string acting between its elements, providing an elastic reaction to the external forces). The magnitude of this tension at each point of the string $(x, u(x, t))$ is denoted by $T(x, t)$.

We also consider the unit tangent vector to the string, as given by

$$\tau(x, t) := \frac{(1, \partial_x u(x, t))}{\sqrt{1 + (\partial_x u(x, t))^2}}.$$

Figure 14.1. Jimi Hendrix (photo by Steve Banks, image from Wikipedia, licensed under the Creative Commons Attribution-Share Alike 4.0 International license).

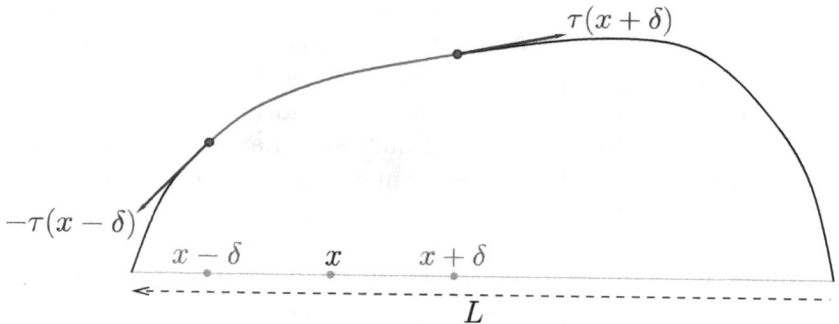

Figure 14.2. A string subject to its own tension.

The vibrating string model then consists of the assumption that, given a small $\delta > 0$, the tension forces acting at the string point $(x, u(x,t))$ are the byproducts of a force F_- acting on $(x - \delta, u(x - \delta, t))$ and a force F_+ acting on $(x + \delta, u(x + \delta, t))$ whose magnitude is proportional to the tension $T(x,t)$ and is opposite to the tangential directions (namely, the direction of F_- is $-\tau(x - \delta, t)$ and the direction of F_+ is $\tau(x + \delta, t)$), see Figure 14.2.

Therefore, the total tension force acting on the string at the point $(x, u(x,t))$ takes the form

$$
\begin{aligned}
F_+ + F_- &= \kappa T(x)\big(\tau(x+\delta,t) - \tau(x-\delta,t)\big) \\
&= 2\kappa\delta T(x)\partial_x \tau(x,t) + o(\delta) \\
&= 2\kappa\delta T(x)\partial_x \left(\frac{(1,\partial_x u(x,t))}{\sqrt{1+\big(\partial_x u(x,t)\big)^2}} \right) + o(\delta).
\end{aligned}
\tag{14.1}
$$

The constant $\kappa > 0$ depends on the material with which the string is made and accounts for the elastic properties of the string.

The total force acting on the string, taking into account gravity, is therefore

$$
\begin{aligned}
F = (F_1, F_2) &:= F_+ + F_- - (0, m_\delta g) \\
&= 2\kappa\delta T(x)\partial_x \left(\frac{(1,\partial_x u(x,t))}{\sqrt{1+\big(\partial_x u(x,t)\big)^2}} \right) - (0, m_\delta g) + o(\delta),
\end{aligned}
$$

where m_δ is the mass of the string located between $x-\delta$ and $x+\delta$ and g denotes the gravitational acceleration.

If we suppose that the density of the string is constant, say equal to some $\rho > 0$, the mass m_δ is the product of ρ and the length of the string located between $x-\delta$ and $x+\delta$, namely

$$
m_\delta = \rho \int_{x-\delta}^{x+\delta} \sqrt{1+\big(\partial_x u(\zeta,t)\big)^2}\, d\zeta = 2\rho\delta\sqrt{1+\big(\partial_x u(x,t)\big)^2} + o(\delta).
\tag{14.2}
$$

Furthermore, by Newton's second law, the vertical component of the force is equal to the mass times the vertical acceleration of the string, that is,

$$
F_2 = m_\delta\, \partial_{tt} u(x,t).
$$

In light of these considerations, we find that[1] for small δ,

$$\partial_{tt}u = \frac{F_2}{m_\delta}$$

$$= \frac{1}{m_\delta}\left[2\kappa\delta T\partial_x\left(\frac{\partial_x u}{\sqrt{1+\left(\partial_x u\right)^2}}\right) - m_\delta g + o(\delta)\right]$$

$$= \frac{2\kappa\delta T}{m_\delta}\partial_x\left(\frac{\partial_x u}{\sqrt{1+\left(\partial_x u\right)^2}}\right) - g + o(1) \tag{14.3}$$

$$= \frac{\kappa T}{\rho\sqrt{1+\left(\partial_x u\right)^2}+o(1)}\partial_x\left(\frac{\partial_x u}{\sqrt{1+\left(\partial_x u\right)^2}}\right) - g + o(1)$$

$$= \frac{\kappa T}{\rho\sqrt{1+\left(\partial_x u\right)^2}}\partial_x\left(\frac{\partial_x u}{\sqrt{1+\left(\partial_x u\right)^2}}\right) - g + o(1).$$

By formally sending $\delta \searrow 0$, we thus obtain the partial differential equation

$$\partial_{tt}u = \frac{\kappa T}{\rho\sqrt{1+\left(\partial_x u\right)^2}}\partial_x\left(\frac{\partial_x u}{\sqrt{1+\left(\partial_x u\right)^2}}\right) - g. \tag{14.4}$$

Several different models can be considered to describe explicitly the tension T and thus find an expression for (14.4) that solely depends on the shape of the string, on its density and elastic constant and on the gravitational acceleration. One commonly accepted model is to consider the string as "inextensible" and T as a constant (the tension being distributed uniformly along the whole string). Another possibility is to consider the string as a superposition of infinitesimal

---

[1]From a geometric point of view, it is interesting to observe that the term $\partial_x\left(\frac{\partial_x u}{\sqrt{1+\left(\partial_x u\right)^2}}\right)$ in (14.3) detects the curvature of the string (remarks of this type will be formalized in a general setting in Theorem 2.3.7 in the companion book [DV23]).

elastic springs and thus take T to be proportional to the string's elongation. Both these models however share the common trait that, for small elongations, T is constant. Therefore, in the small-elongation approximation, we take $T := 1$, and we disregard the quadratic terms in u in (14.4) (the *ansatz* being that for small u, the linear terms will prevail over the quadratic ones). With this simplification in mind, one can reduce (14.4) to

$$\partial_{tt}u = \frac{\kappa}{\rho}\partial_{xx}u - g. \tag{14.5}$$

In particular, in the absence of gravity, the linear approximation of the vibrating string can be seen as a particular case of the wave equation in (7.2). Interestingly, the speed of propagation c in (7.2) corresponds here to $\sqrt{\frac{\kappa}{\rho}}$. Namely, the higher the elastic parameter of the string, the higher the speed of propagation; the higher the density (or, equivalently, the heavier the string), the slower the speed of propagation.

The dependence of the solution of (14.5) on the structural parameters of the string is not a mere mathematical curiosity; on the contrary, it has a deep impact on music since these parameters precisely determine the pitch of the sound that the string generates. Let us see, for instance, how understanding partial differential equations may turn out to be useful when tuning a guitar. For accomplishing this goal, we neglect the gravity effect and summarize (14.5) and the boundary conditions of the string in the following mathematical setting:

$$\begin{cases} \partial_{tt}u(x,t) = \dfrac{\kappa}{\rho}\partial_{xx}u(x,t) & \text{for all } (x,t) \in (0,L) \times (0,+\infty), \\[2mm] u(x,0) = \eta\sin\dfrac{\pi x}{L}, \\[2mm] \partial_t u(x,0) = 0, \\[2mm] u(0,t) = u(L,t) = 0 & \text{for all } t \in (0,+\infty). \end{cases} \tag{14.6}$$

The prescriptions in (14.6) model a guitar string with a rest length of L and a null initial velocity which is initially displaced (say by expert fingerpicking) as a sinusoidal graph (for the sake of simplicity, though more complicated initial situations could be considered).

The small parameter $\eta > 0$ has been introduced here only to ensure consistency with the small-elongation approximation.

It can be readily checked that the function

$$u(x,t) = \eta \cos \frac{\sqrt{\kappa}\,\pi t}{\sqrt{\rho}\,L} \sin \frac{\pi x}{L} \qquad (14.7)$$

solves (14.6). This solution is quite telling regarding the pitch produced by the string. Indeed, in light of (14.7), the string vibrates with a frequency of

$$\omega := \frac{\sqrt{\kappa}\,\pi}{\sqrt{\rho}\,L}. \qquad (14.8)$$

In particular:

- Strings with a higher elastic coefficient produce a higher pitch (and we already know this by experience because tightening the tuning pegs of a guitar enhances the string's tension and correspondingly raises its pitch).
- Longer strings produce lower pitches, while shorter strings produce higher pitches (and we already know this by experience too, since, while playing, one changes the length of the vibrating string by holding it firmly against the fingerboard with a finger and shortening the string, that is, stopping it on a higher fret, gives higher pitch).
- Strings with a higher density produce lower pitches, that is, more massive strings vibrate more slowly (and this is also in agreement with our experience since, for instance, on classical guitars, low-density nylon strings are used for the high pitches, while higher-density wire-wound strings are employed for low pitches).

To appreciate even more the impact of mathematics on guitar playing, we recall that an octave[2] is the distance between one pitch and

---

[2]The name "octave" comes from the Latin adjective "octava", meaning "eighth". This etymology, however, is possibly a bit confusing since it only represents the interval between one musical pitch and another with double its frequency; hence, an octave is more related to the number 2 than to the number 8. Also, in common Western musical compositions, the octave is composed of 7 notes, or 12 semitones, and hence the octave seems to relate more to the numbers 7 and 12 rather than

another with half or double its frequency; thus, from (14.8), holding the string against the fingerboard and thus halving its length has the effect of doubling the frequency of the root note (consistent with the fact that each fret on the guitar's fretboard corresponds[3] to one semitone, the distance of 12 semitones is the octave, and the position of the 12th fret corresponds to a string half as long as the original one).

Let us also mention a typical question asked by many beginners of (say, classical) guitar:

can I replace the steel strings with nylon strings?   (14.9)

This question is often motivated by a practical reason:

nylon strings are usually "softer" compared to steel strings;
                                                              (14.10)

---

to 8. The fact is that when talking about intervals of notes, the tradition is to count from the first note to the final note included (which makes 8 notes, somewhat justifying the name of octave).

[3]The choice of dividing the octave into 12 semitones is also mathematically well grounded since it corresponds to frequency ratios of the form $2^{\frac{j}{12}}$ for $j \in \{0, \dots, 12\}$. On the one hand, these are irrational numbers (except for $j = 0$ and $j = 12$). On the other, notes sound harmonious to our ear if the frequency of the notes is close to a simple interval, for instance, the frequency ratios such as $\frac{3}{2}$ (the "perfect fifth" in musical jargon), $\frac{4}{3}$ (the "fourth") and $\frac{5}{4}$ (the "major third"). These two observations seem contradictory until we realize how "close to rationals" the irrational numbers of the form $2^{\frac{j}{12}}$ are (for $j \in \{1, \dots, 11\}$, these numbers are "almost" $\frac{16}{15}, \frac{9}{8}, \frac{6}{5}, \frac{5}{4}, \frac{4}{3}, \frac{7}{5}, \frac{3}{2}, \frac{8}{5}, \frac{5}{3}, \frac{16}{9}$ and $\frac{15}{8}$). For example,

$$\frac{3}{2} - 2^{\frac{7}{12}} = 1.5 - 1.49830707688\ldots = 0.00169292312\ldots,$$

showing how close the seventh semitone is to the perfect fifth.

While other tuning scales are possible and are indeed employed in several contexts, the equal temperament one (i.e. the one splitting the scale into equal intervals) presents several advantages, such as the possibility of transposing a tune into other keys, thus allowing instruments to be played in all keys (or singers to sing a song in a key which is more congenial to their voice) with minimal flaws in intonation. The 12-tone equal temperament based on the division of the octave into 12 equally spaced parts on a logarithmic scale appears the most widespread system in music today. Likely, the common adoption of this specific temperament also influenced the composition of music in order to accommodate the system and minimize dissonance coming from irrational approximations of rational ratios.

therefore, they are less likely to cause pain to the fingers (whereas professional guitarists develop callouses on their fingertips with practice, so they do not perceive this problem). Moreover, steel strings are generally perceived by beginners to be more uncomfortable and may require more effort to play in terms of muscle tension in the hand and wrist.

Well, from a musical theory point of view, there are technical issues about how the timbre of the instrument would be affected by the change proposed in (14.9): many guitarists would argue that messing up with the proper setting of the guitar may well compromise its unique[4] tone quality (nylon strings tend to have a "rounder" and "mellower" tone, while inappropriate materials may make the strings buzz excessively and slow down the playing action, and some strings lend themselves better to certain techniques than others, e.g. strumming with a pick may not work too well with a nylon string, which are instead often more appreciated by experienced fingerstylists, while ordinarily finger picking the steel strings is considered to be easier, etc.).

But if you are going to tolerate these performance issues, in principle the answer to (14.9) is yes, you can: *placing nylon strings on a guitar built for steel strings will (more or less) work.*

But, beware! The other way around does not work: *steel strings on a guitar built for nylon strings can break the neck of the instrument due to the increased tension.*

Indeed, by the discussions after (14.8), we know that the higher the density, the lower the pitch. Thus, putting a steel (denser) string in place of a nylon (less dense), one would bring down the pitch. To compensate for the pitch drop, again by the discussions after (14.8), one will need to increase the tension of the string (since we know that this would raise the pitch). So, all in all, to adjust the tuning,

---

[4]And here, we confine ourselves to a rough dichotomy between steel and nylon strings, without going into finer distinctions and specific materials, such as nickel, brass and bronze, on the metals side, catgut, on the natural fibers side, and fluorocarbon polymers, on the synthetic materials side. Interestingly, an important development in the production of modern nylon strings occurred in the late 1940s, thanks to close interactions between luthier Albert Augustine, the chemical company DuPont and virtuoso classical guitarist Andrés Segovia (see Figure 14.3 for a portrait of the latter, by Hilda Wiener).

Figure 14.3.   Andrés Segovia Torres, 1st Marquis of Salobreña, during a recital in Brussels on December 15, 1932 (Public Domain image from Wikipedia, source Sotheby's).

replacing nylon with steel strings would require an increased tension, which, on the long run, can damage the instrument.

This fact, showcasing that nylon strings have lower tension than their steel counterparts, also explains the phenomenon described in (14.10) since, due to their reduced tension, nylon strings are softer and more gentle on players' fingers.

The difference in tension between nylon and steel strings also justifies why they affect differently the timbre of the instruments. In rough terms, on the one hand, with less tension, the nylon strings may move more freely, thus producing more overtones by dislodging some energy onto adjacent or resonant modes, while the higher tension of steel strings may tend to produce sounds with a greater clarity and a brighter effect. On the other hand, denser strings have in principle more potential energy, which gets converted into kinetic

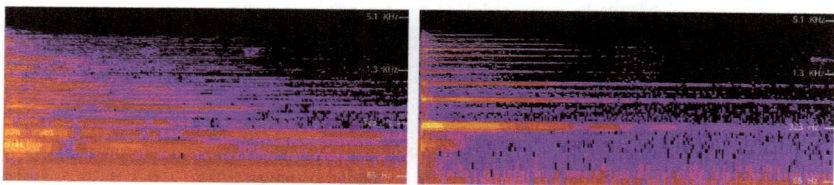

Figure 14.4.   Frequency plot, over time, of a metal E2 string (left) and a synthetic E4 string (right).

energy upon releasing the string. This can also enhance some high-frequency harmonics in a steel string, providing its characteristic "metallic" sound (and, if one relates high-frequency modes with more irregular behaviors of the function that they describe, this also justifies the fact that steel strings often produce a vibration which may sound less "round", or "harder" than their nylon counterparts). For a diagram comparing the sound spectrum of steel and nylon strings, see e.g. Figure 3.10 in [Pho17].

It is also instructive to compare the frequency diagrams obtained from two simple pick strokes on a cheap guitar: one on a metal E2 string and one on a synthetic E4 string, see Figure 14.4. In this setting, both the strings correspond to the same note E, but the synthetic string plays two octaves higher than the metallic one (specifically, E4 should correspond approximately to 329.6276 Hz, while E2 should have a frequency of approximately 82.40689 Hz, and note the ratio of approximately 4 between the two frequencies, in agreement with their distance of two octaves). We note from Figure 14.4 that, indeed, the metal strings not only tend to excite additional harmonics of higher frequencies (and possibly too many of them, indicating a poor quality of the instrument and an excessively forceful pitch on the string), but also their vibration is more persistent in time; this fact is also in agreement with our intuition since heavier metal strings possess more (potential, hence kinetic) energy than lighter synthetic ones, whence they keep oscillating longer since longer is the time needed to dissipate energy due to friction.

Moreover, it is experienced by some guitarists[5] that stiffer strings play "faster"; namely, steel strings with high tension have a "quicker

---

[5]Other musical instruments can also be understood through the lens of mathematics, see e.g. http://newt.phys.unsw.edu.au/jw/flutes.v.clarinets.html for flutes and clarinets.

Figure 14.5.   Gustav Kirchhoff (Public Domain image from Wikipedia, source Smithsonian Libraries).

attack", while their nylon counterparts have a slower one (the attack being the time it takes for the pick of the string to reach the loudest peak of the sound, i.e. the time when one begins to hear the note after the string is released).

For completeness, let us also mention a slightly different model for vibrating strings, proposed[6] by Gustav Robert Kirchhoff in [Kir76], see Figure 14.5. The gist of this model is that the elastic coefficient κ

---

[6]Besides the nonlocal equations for vibrating strings related to (14.11), Gustav Kirchhoff made fundamental contributions to the theory of electrical circuits, spectroscopy and black-body radiation. This creates quite a confusion because the names of Kirchhoff's law and Kirchhoff equation end up being used in all these topics, with different meanings.

in (14.5) is assumed to be a positive constant; however, in reality, the elastic properties of the string may depend on its elongation (according to the material, the string could either become "stiff" or "slack off" for large elongations). To account for such a possibility, Kirchhoff proposed that κ instead depends on the length of the deformed string, namely on

$$\mathscr{L}(t) := \int_0^L \sqrt{1 + \big(\partial_x u(x,t)\big)^2}\, dt. \qquad (14.11)$$

In this model, κ in (14.5) must be interpreted as $\kappa(\mathscr{L}(t))$, that is, a function of the length of the string (the corresponding equation is thus called the Kirchhoff equation). Note that not only does κ vary with time in this situation, but also it depends on a "nonlocal" quantity. Indeed, the length of the deformed string in (14.11) is a "global" object: to calculate it, it is not sufficient to know the shape of the string at a given point; instead, complete information about its large-scale geometry is required. And, of course, mathematical problems requiring the knowledge of nonlocal quantities become structurally harder (but often quite interesting since they aim at capturing the big picture).

By the way, Kirchhoff's home town was Königsberg, a historic Prussian city that is now Kaliningrad, in Russia's Kaliningrad Oblast (a small exclave of the vast Russian state, providing an interesting counterexample to the conjecture that political states are connected regions, up to additional islands). The reason for which we mention Königsberg is because one classical and very famous mathematical problem that was settled by Euler is related to the bridges of Königsberg (see e.g. [HW04] to know more about the Königsberg bridges).

Chapter 15

Elastic Membranes

The description of elastic membranes can be seen as a multi-dimensional version of the one of vibrating strings presented in Chapter 14. As a simple model, we could consider a membrane composed of a web of infinitesimal strings in the coordinate directions. In this way, one may think that the vertical force exerted at a point $(x, u(x)) \in \mathbb{R}^n \times \mathbb{R}$ on the membrane is the sum of the forces produced by the tension of the infinitesimal strings in each direction, plus gravity. Recalling (14.1), we may think that, for each $i \in \{1, \ldots, n\}$, the vertical force exerted by the tension of the infinitesimal string located in direction e_i (say, located from $x - \delta e_i$ and $x + \delta e_i$) is equal to

$$2\kappa\delta T(x)\partial_i \left(\frac{\partial_i u(x,t)}{\sqrt{1 + \left(\partial_i u(x,t)\right)^2}} \right),$$

up to higher orders in δ.

Hence, the total vertical force acting on the membrane at $(x, u(x)) \in \mathbb{R}^n \times \mathbb{R}$ is given by

$$2\kappa\delta T(x) \sum_{i=1}^{n} \partial_i \left(\frac{\partial_i u(x,t)}{\sqrt{1 + \left(\partial_i u(x,t)\right)^2}} \right) - m_\delta g,$$

where, in view of (14.2),

$$m_\delta = 2\rho\delta \sum_{i=1}^{n} \sqrt{1 + \left(\partial_i u(x,t)\right)^2},$$

up to higher orders in δ, with ρ being the linear densities of the infinitesimal strings in each direction (which we suppose to be constant and independent of the direction). In this setting, Newton's second law gives that

$$\partial_{tt} u = \frac{2\kappa\delta T}{m_\delta} \sum_{i=1}^{n} \partial_i \left(\frac{\partial_i u}{\sqrt{1 + (\partial_i u)^2}} \right) - g$$

$$= \frac{\kappa T}{\rho \sum_{i=1}^{n} \sqrt{1 + (\partial_i u)^2}} \sum_{i=1}^{n} \partial_i \left(\frac{\partial_i u}{\sqrt{1 + (\partial_i u)^2}} \right) - g, \tag{15.1}$$

which can be considered the higher-dimensional[1] version of (14.4). In the small elongation approximation, we thus consider $T := 1$ and disregard the terms that are quadratic in the displacement u, thus reducing (15.1) to

$$\partial_{tt} u = \frac{\kappa}{\rho} \Delta u - g, \tag{15.2}$$

which is the higher-dimensional version of (14.5).

In particular, an elastic membrane in equilibrium represents a stationary solution of (15.2) and thus satisfies

$$\frac{\kappa}{\rho} \Delta u = g, \tag{15.3}$$

which is a form of the so-called Poisson's equation (compare with (1.8)).

In the absence of gravity, (15.3) reduces to the so-called Laplace's equation (compare with (1.9))

$$\Delta u(x) = 0.$$

This means that (at least in the linear approximation) elastic membranes constrained at their boundaries are described by the graphs

---

[1]From a geometric point of view, it is interesting to observe that the term $\sum_{i=1}^{n} \partial_i \left(\dfrac{\partial_i u}{\sqrt{1 + (\partial_i u)^2}} \right)$ in (15.1) detects the mean curvature of the membrane (remarks of this type will be formalized in a general setting in Theorem 2.3.7 in the companion book [DV23]).

of harmonic functions. Once again, this fact brings us a useful hint: it suggests a smoothing effect for harmonic functions, perhaps inherited from the fact that elastic membranes do not develop spikes (the elasticity would indeed reduce the peak to find a balance): this intuition will be mathematically formalized, developed and consolidated in Chapter 2 of the companion book [DV23].

Chapter 16

Elasticity Theory and the Torsion of Bars

A topical argument in material sciences deals with the determination of the shape of objects subject to external forces, as well as the understanding of the relations between the forces applied and the resulting deformations. Providing a comprehensive account of this elasticity theory is well beyond the scope of these notes (see e.g. [HI11] and the references therein for further readings). Here, we only introduce, in a rather simplified form, some basic concepts which will lead us to the study of an interesting[1] elliptic equation (which will in fact be studied in detail in Sections 7.3 and 8.2 of the companion book [DV23]). For concreteness, we focus on the three-dimensional case, and we consider an infinitesimal portion of the material under consideration, modeled as tiny cubes with faces oriented along the coordinate axes, see Figure 16.1. If the cubes have the form $(0, \delta)^3$, for some small $\delta > 0$, we denote by S_1, S_2 and S_3 the three facets of the cubes in the directions e_1, e_2 and e_3, respectively, i.e.

$$S_1 := \{\delta\} \times (0, \delta) \times (0, \delta),$$

$$S_2 := (0, \delta) \times \{\delta\} \times (0, \delta)$$

and $\quad S_3 := (0, \delta) \times (0, \delta) \times \{\delta\}.$

---

[1]Actually, the final equation that we will obtain in (16.14) is precisely the same as the one arising from fluid dynamics in (5.14). This coincidence shows once again the extraordinary unifying power of mathematics.

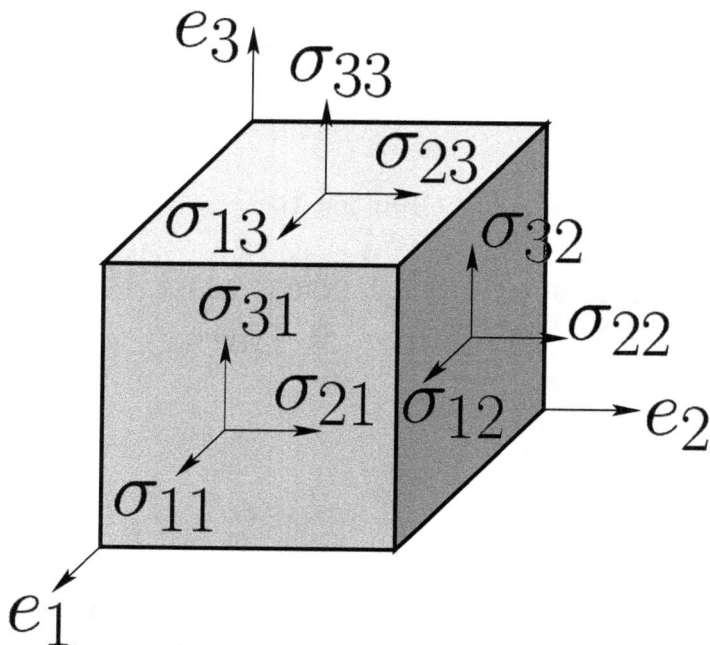

Figure 16.1. The stress tensor.

For each $i,\ j\ \in\ \{1,2,3\}$, we denote by σ_{ij} the force per unit area applied in the direction e_i to the surface S_j. In jargon, the matrix $\{\sigma_{ij}\}_{i,j\in\{1,2,3\}}$ is sometimes called the "stress tensor".

By Newton's third law, at equilibrium, the forces acting on the above ideal infinitesimal cube must attain a balance; therefore, the forces acting on the facets of the cubes opposite to S_1 (or S_2, or S_3) are the ones discussed above but with a change in sign. In particular, given $i \in \{1,2,3\}$, at equilibrium, the total forces per unit area applied in the direction e_i must balance out and therefore (noting that if $x \in S_j$, then $x - \delta e_j$ belongs to the facet opposite to S_j)

$$0 = \sum_{j=1}^{3}\Big(\sigma_{ij}(x) - \sigma_{ij}(x - \delta e_j)\Big) = \delta\sum_{j=1}^{3}\partial_j\sigma_{ij}(x) + o(\delta),$$

leading to the following balance prescription for the stress tensor:

$$\sum_{j=1}^{3}\partial_j\sigma_{ij} = 0 \qquad \text{for all } i \in \{1,2,3\}. \tag{16.1}$$

Additionally, the angular momentum contribution ϑ_1 coming from S_1 is proportional to the vector product between the vectorial force per unit area on S_1, corresponding to $\sigma_{11}e_1 + \sigma_{21}e_2 + \sigma_{31}e_3$, and the vector joining normally the center of the cube with S_1, namely e_1. Hence, up to normalizing constants,

$$\vartheta_1 = \det \begin{pmatrix} e_1 & e_2 & e_3 \\ \sigma_{11} & \sigma_{21} & \sigma_{31} \\ 1 & 0 & 0 \end{pmatrix} = \sigma_{31}e_2 - \sigma_{21}e_3.$$

Similarly, the angular momentum contribution ϑ_2 coming from S_2 is proportional to the vector product between the vectorial force per unit area on S_2, corresponding to $\sigma_{12}e_1 + \sigma_{22}e_2 + \sigma_{32}e_3$, and the vector joining normally the center of the cube with S_2, namely e_2, whence

$$\vartheta_2 = \det \begin{pmatrix} e_1 & e_2 & e_3 \\ \sigma_{12} & \sigma_{22} & \sigma_{32} \\ 0 & 1 & 0 \end{pmatrix} = -\sigma_{32}e_1 + \sigma_{12}e_3.$$

In addition, the angular momentum contribution ϑ_3 coming from S_3 is proportional to the vector product between the vectorial force per unit of area on S_3, corresponding to $\sigma_{13}e_1 + \sigma_{23}e_2 + \sigma_{33}e_3$, and the vector joining normally the center of the cube with S_2, namely e_3, whence

$$\vartheta_3 = \det \begin{pmatrix} e_1 & e_2 & e_3 \\ \sigma_{13} & \sigma_{23} & \sigma_{33} \\ 0 & 0 & 1 \end{pmatrix} = \sigma_{23}e_1 - \sigma_{13}e_2.$$

On this account, the conservation of the total angular momentum entails that, at equilibrium,

$$0 = \vartheta_1 + \vartheta_2 + \vartheta_3 = (\sigma_{23} - \sigma_{32})e_1 + (\sigma_{31} - \sigma_{13})e_2 + (\sigma_{12} - \sigma_{21})e_3.$$

This establishes that the stress tensor is symmetric, namely

$$\sigma_{ij} = \sigma_{ji} \qquad \text{for all } i, j \in \{1, 2, 3\}. \tag{16.2}$$

Now, in order to model the distortion of the material caused by these forces, we consider a displacement vector v that describes at any point the change in the configuration of the given body. The displacement vector itself may not be the most interesting object to take into account in an elasticity theory since translations and rotations do produce significant displacements without affecting the shape of the body under consideration. It is therefore common to try to detect the deformation effects that displacements produce by accounting for relative displacements. Namely, if a point x is moved to $S(x) := x + v(x)$, given $\omega \in \partial B_1$, the point $y := x + \delta\omega$ is moved to $\mathcal{P}(y) = y + v(y) = x + \delta\omega + v(x + \delta\omega)$. To compute the corresponding relative displacement, we check how $y - x$ is affected by this transformation. Namely, we point out that

$$\mathcal{P}(y) - \mathcal{P}(x) = \mathcal{P}(x + \delta\omega) - \mathcal{P}(x)$$
$$= \delta\, D\mathcal{P}(x)\omega + O(\delta^2) = D\mathcal{P}(x)(y - x) + O(\delta^2).$$

In other words, the "deformation gradient" $D\mathcal{P}$ is a linear approximation measuring the relative displacements.

If we want to measure the change in square length, we have that

$$|\mathcal{P}(y) - \mathcal{P}(x)|^2 - |y - x|^2 = |D\mathcal{P}(x)(y - x) + O(\delta^2)|^2 - |y - x|^2$$
$$= \big(D\mathcal{P}(x)(y - x) + O(\delta^2)\big) \cdot \big(D\mathcal{P}(x)(y - x) + O(\delta^2)\big) - |y - x|^2$$
$$= \big(D\mathcal{P}(x)(y - x)\big) \cdot \big(D\mathcal{P}(x)(y - x)\big) - |y - x|^2 + O(\delta^3)$$
$$= \big(D\mathcal{P}(x)\big)^\top D\mathcal{P}(x)(y - x) \cdot (y - x) - |y - x|^2 + O(\delta^3)$$
$$= \Big(\big(D\mathcal{P}(x)\big)^\top D\mathcal{P}(x) - \mathrm{Id}\Big)(y - x) \cdot (y - x) + O(\delta^3),$$

$$(16.3)$$

where Id is the identity matrix and the superscript "$\top$" denotes matrix transposition.

It is also useful to observe that

$$D\mathcal{P}^\top D\mathcal{P} - \mathrm{Id} = \big(\mathrm{Id} + Dv\big)^\top \big(\mathrm{Id} + Dv\big) - \mathrm{Id} = Dv + Dv^\top + Dv^\top Dv.$$

In particular, if one assumes that the displacements are small, the quadratic term $Dv^\top Dv$ is negligible with respect to the "symmetric gradient" $Dv + Dv^\top$, which is of the linear type. Therefore, it is

customary to consider the approximation $DP^\top DP - \mathrm{Id} \simeq Dv + Dv^\top$
and rewrite (16.3) as

$$|P(y) - P(x)|^2 - |y - x|^2 \simeq \left(Dv(x) + (Dv(x))^\top\right)(y - x) \times (y - x)$$

or, equivalently,

$$|P(y) - P(x)|^2 - |y - x|^2 \simeq \sum_{i,j=1}^{3} \varepsilon_{ij}(x)(y_i - x_i)(y_j - x_j), \quad (16.4)$$

where

$$\varepsilon_{ij}(x) := \left(Dv(x) + (Dv(x))^\top\right)_{ij} = \partial_i v_j(x) + \partial_j v_i(x). \quad (16.5)$$

In jargon, ε_{ij} is called the "strain tensor" (and note the symmetry feature that $\varepsilon_{ij} = \varepsilon_{ji}$), and (16.4) expresses in a quantitative way (though after some simplifying approximations) the relative change in the position of points under the deformation.

We now relate strain and stress. To this end, once again, the simplest possible *ansatz* leading to a linear theory of elasticity is that deformations (or, more precisely, strains, to avoid translations and rotations) are directly proportional to the forces exerted (this is indeed a possible rephrasing of Hooke's law for linear elastic materials). That is, neglecting proportionality constants, we suppose that

$$\varepsilon_{ij} = \sigma_{ij}. \quad (16.6)$$

It is interesting to remark that this identity is also algebraically compatible with the fact that both the tensors above are symmetric.

We now apply the general framework of the linear theory of elasticity described so far to the special situations of straight bars subject to torsion. This will indeed be one of the founding motivations in Sections 7.3 and 8.2 of the companion book [DV23].

To model a straight bar, we consider a body of the form

$$\Omega \times [0, +\infty), \quad (16.7)$$

with $\Omega \subset \mathbb{R}^2$ smooth and contractible, see Figure 16.2.

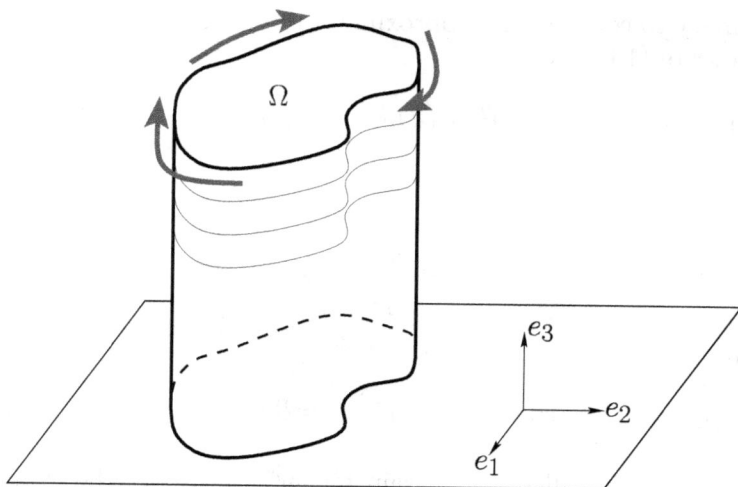

Figure 16.2.   A bar subject to torsion.

We assume that such a bar is constrained at the level $\{x_3 = 0\}$, but it gets "twisted" from the top. The effect of the twist would be to make the cross-section rotate by some angle, depending on the level that we consider. Assuming again linear behavior, we suppose that the rotation performed at the level $\{x_3 = \zeta\}$ is proportional to ζ (i.e. the angle of rotation increases linearly with the height, which is not so unreasonable, at least for small heights, since the bar is constrained at ground zero). Accordingly, we model the twist at the level $\{x_3 = \zeta\}$ by a rotation of angle $\Theta_\zeta = c_0\zeta$, for some $c_0 > 0$. The corresponding displacement therefore sends the point $x = (x_1, x_2, \zeta)$ to the point $z := x + v(x)$, with the horizontal component of v being prescribed by the above rotation by the angle Θ_ζ; namely, for small heights ζ,

$$
\begin{pmatrix} z_1 \\ z_2 \end{pmatrix} = \begin{pmatrix} \cos\Theta_\zeta & \sin\Theta_\zeta \\ -\sin\Theta_\zeta & \cos\Theta_\zeta \end{pmatrix} \begin{pmatrix} x_1 \\ x_2 \end{pmatrix}
$$

$$
= \begin{pmatrix} \cos(c_0\zeta) & \sin(c_0\zeta) \\ -\sin(c_0\zeta) & \cos(c_0\zeta) \end{pmatrix} \begin{pmatrix} x_1 \\ x_2 \end{pmatrix}
$$

$$
= \begin{pmatrix} 1 + O(\zeta^2) & c_0\zeta + O(\zeta^2) \\ -c_0\zeta + O(\zeta^2) & 1 + O(\zeta^2) \end{pmatrix} \begin{pmatrix} x_1 \\ x_2 \end{pmatrix}.
$$

This gives that

$$\begin{pmatrix} v_1(x) \\ v_2(x) \end{pmatrix} = \begin{pmatrix} z_1 \\ z_2 \end{pmatrix} - \begin{pmatrix} x_1 \\ x_2 \end{pmatrix} = \begin{pmatrix} O(\zeta^2) & c_0\zeta + O(\zeta^2) \\ -c_0\zeta + O(\zeta^2) & O(\zeta^2) \end{pmatrix} \begin{pmatrix} x_1 \\ x_2 \end{pmatrix}$$

$$= \begin{pmatrix} c_0\zeta x_2 \\ -c_0\zeta x_1 \end{pmatrix} + O(\zeta^2).$$

Hence, by sloppily[2] neglecting the higher-order terms and using the fact that $x = (x_1, x_2, x_3) \in \{x_3 = \zeta\}$, we can simply write that

$$v_1(x) = c_0 x_2 x_3 \qquad \text{and} \qquad v_2(x) = -c_0 x_1 x_3. \qquad (16.8)$$

As for the vertical component of the displacement, a common *ansatz* is to consider it to be independent of the height. In roughly terms, the torsion of the bar may exhibit a distortion of the cross-section out of its own plane (known in jargon as "warping deformation"), but usually this effect is less visible than other types of deformations, and we assume here that each layer of the bar (i.e. each horizontal cross-section) "equally pushes the next one", regardless of its height (once again, this is not a completely unrealistic assumption, especially if we restrict our attention to the layers close to ground zero). To model that this warping effect does not depend on the height, we assume that the vertical component of the displacement vector is independent of x_3, that is,

$$v_3(x) = w(x_1, x_2), \qquad (16.9)$$

for some function w.

Now, from (16.5), (16.6), (16.8) and (16.9), we arrive at

$$\sigma_{13} = \varepsilon_{13} = \partial_1 v_3 + \partial_3 v_1 = \partial_1 w + c_0 x_2.$$

Similarly,

$$\sigma_{23} = \varepsilon_{23} = \partial_2 v_3 + \partial_3 v_2 = \partial_2 w - c_0 x_1$$

and

$$\sigma_{33} = \varepsilon_{33} = 2\partial_3 v_3 = 0.$$

---

[2]This simply because here we are playing around with motivations. From the following chapter onward, this kind of sloppiness will no longer be tolerated!

Interestingly, the above expressions for σ_{i3}, with $i \in \{1,2,3\}$, are all independent of the height x_3 (which was probably not obvious from the beginning). Furthermore, recalling the balance prescription in (16.1) and the symmetry property in (16.2),

$$0 = \sum_{j=1}^{3} \partial_j \sigma_{3j} = \sum_{j=1}^{3} \partial_j \sigma_{j3} = \partial_1 \sigma_{13} + \partial_2 \sigma_{23}.$$

That is, the differential form $\sigma_{23}\,dx_1 - \sigma_{13}\,dx_2$ is closed and, therefore exact, thanks to the Poincaré lemma (see e.g. [Con93, Theorem 8.3.8]). This yields that there exists a function $\phi = \phi(x_1, x_2)$, sometimes refereed to by the name "warping potential", such that $d\phi = \sigma_{23}\,dx_1 - \sigma_{13}\,dx_2$.

This gives that $\partial_1 \phi = \sigma_{23}$ and $\partial_2 \phi = -\sigma_{13}$. Consequently,

$$\begin{aligned}\Delta\phi &= \partial_1 \sigma_{23} - \partial_2 \sigma_{13} = \partial_1 (\partial_2 w - c_0 x_1) - \partial_2 (\partial_1 w + c_0 x_2) \\ &= \partial_{12} w - c_0 - \partial_{12} w - c_0 = -2c_0.\end{aligned} \tag{16.10}$$

Another notion of interest for engineering purposes is the "traction" $\mathcal{T}$ along the boundary of the bar, which is defined as the normal component of the vertical stress. More explicitly, if we define $\sigma_3 := (\sigma_{31}, \sigma_{32}, \sigma_{33})$ and consider the outer unit normal $\nu^\star$ of the bar, we set

$$\mathcal{T} := \sigma_3 \cdot \nu^\star. \tag{16.11}$$

As a matter of fact, since the straight bar has the form given in (16.7), we can identify $\nu^\star \in \mathbb{R}^3$ with the vector $(\nu, 0)$, where $\nu \in \mathbb{R}^2$ is the outer unit normal at $\partial\Omega$, and consequently,

$$\mathcal{T} = \sigma_{31}\nu_1 + \sigma_{32}\nu_2 = \sigma_{13}\nu_1 + \sigma_{23}\nu_2 = -\partial_2\phi\nu_1 + \partial_1\phi\nu_2 = \nabla\phi \cdot \tau,$$

where $\tau := (-\nu_1, \nu_2)$ can be taken as the unit tangent vector along $\partial\Omega$ (say, clockwise oriented). In this spirit, we find that the traction is the tangential derivative of the warping potential.

In the special situation in which the bar is not subject to traction, it follows that $\nabla\phi \cdot \tau = 0$ on $\partial\Omega$. This gives that ϕ is constant along $\partial\Omega$ because, if $\partial\Omega$ is locally parameterized by a curve $\gamma : (-1,1) \to \mathbb{R}^2$ with $\dot\gamma$ proportional to τ, we have that

$$\frac{d}{dt}\phi(\gamma(t)) = \nabla\phi(\gamma(t)) \cdot \dot\gamma(t) = 0.$$

Thus, if the traction of the bar vanishes, we can write that

$$\phi = c_2 \text{ on } \partial\Omega, \tag{16.12}$$

for some $c_2 \in \mathbb{R}$.

Recalling (16.11), we call the magnitude of the traction on the surface of the bar $\partial \Omega$ the scalar $|\sigma_3|$. We remark that

$$|\nabla \phi| = \sqrt{(\partial_1 \phi)^2 + (\partial_2 \phi)^2} = \sqrt{\sigma_{23}^2 + \sigma_{13}^2} = \sqrt{\sigma_{23}^2 + \sigma_{13}^2 + \sigma_{33}^2} = |\sigma_3|;$$

therefore, the magnitude of the traction also coincides with the norm of the gradient of the warping potential.

Thus, a consequence of (16.12) is that $|\partial_\nu \phi| = |\nabla \phi|$, and hence in this situation, the magnitude of the traction also coincides (possibly up to a sign) with the normal derivative of the warping potential along the surface of the bar.

Figure 16.3. Bending steel beams with bare hands by the power of ε.

As a consequence, in the special situation in which the bar is not subject to traction and the magnitude of the traction is constant, we have that

$$\partial_\nu \phi = c_3 \text{ on } \partial\Omega, \qquad (16.13)$$

for some $c_3 \in \mathbb{R}$.

Accordingly, if we define

$$u := \frac{c_2 - \phi}{2c_0},$$

we deduce from (16.10), (16.12) and (16.13) that for a straight bar subject to torsion and no traction, if the magnitude of the traction along the surface of the bar is constant, then there exists a solution of

$$\begin{cases} \Delta u = 1 & \text{in } \Omega, \\ u = 0 & \text{on } \partial\Omega, \\ \partial_\nu u = c & \text{on } \partial\Omega. \end{cases} \qquad (16.14)$$

Note the perfect coincidence[3] of this set of prescriptions with that in (5.14), which also reveals a possibly unexpected connection between the dynamics of viscous fluids and the elastic reactions of bars subject to torsion.

---

[3] As remarked in footnote 6 on p. 69, equation (16.14) possesses a solution when Ω is a disk, which is when the bar has a circular cross-section. We will discuss in Sections 7.3 and 8.2 of the companion book [DV23] whether bars of other shapes maintain the same property of presenting no traction and constant traction magnitude along their surface.

Chapter 17

Bending Beams and Plates

Now, we deal with deflection of beams, which is a classical topic in all superhero comics, see Figure 16.3. Besides, the question has obvious applications[1] in engineering, see e.g. Figure 17.1. The beam equation

$$u'''' = q \qquad (17.1)$$

describes[2] the small deflection u of a homogeneous beam at equilibrium in terms of the applied load q.

---

[1]Sometimes mathematics plays an important role even in tragic events, such as the collapse of structures. One of the most famous events of this type was the collapse of the Tacoma Bridge in 1940, only four months after its opening, see Figure 17.2.

Fortunately, the Tacoma Bridge disaster resulted in only one fatality: Tubby, the dog (a black male Cocker Spaniel) belonging to a journalist, who was left in a car on the bridge. People attempted to rescue Tubby; however, he was too terrified to leave the car and bit one of the rescuers. Tubby died when the bridge fell, and neither his body nor the car was ever recovered, probably because the swift tides quickly moved the car away from the ruins of the bridge. See https://en.wikipedia.org/wiki/File:Tacoma_Narrows_Bridge_destructionogv for an impressive video of the Tacoma Bridge collapse, as well as [GJS23] and the references therein for a mathematical investigation of the causes of the disaster.

[2]The linear model of elasticity accounting for the load-carrying capacity and deflection of beams is often called the "Euler–Bernoulli beam theory". No confusion should arise between (17.1) and the model presented in Chapter 16, where the torsion theory of bars rather than deformation under a load was taken into account.

Figure 17.1. Narrows Bridge, Perth (image by Speddie23 from Wikipedia, licensed under the Creative Commons Attribution-Share Alike 4.0 International license).

In (17.1), u and q are functions of one variable $x \in \mathbb{R}$ (the beam is assumed to be infinitely long). In a nutshell, the derivation of (17.1) from first principles relies on the balance of forces and moments. In rough terms, the bending of the beam produces a compression force on one side of the beam and a tensile stress on the other side. These deformation stresses will be modeled via an elastic Hooke's law and produce turning forces which, at equilibrium, together with the corresponding moments, need to be balanced with the external load on the beam.

The details are as follows. We assume that the beam is displayed along the horizontal axis and slightly deformed into a shape given by the graph of the form $z = u(x)$ (the beam may well be three-dimensional, in which case we assume that its transversal sections possess some given shape that remains essentially invariant under small bending, and the three-dimensional coordinates are denoted, as usual, by (x, y, z)).

More precisely, the graph $z = u(x)$ describes the "neutral fiber" of the beam, that is, a fiber which maintains infinitesimally the original length that it possessed at rest, see Figure 17.3. The transversal sections of the beam that were perpendicular to the neutral

Figure 17.2. The Tacoma Bridge falling into the strait (Public Domain image from Wikipedia).

fiber before the beam deforms are assumed to remain perpendicular to the neutral fiber after the bending. We stress that this neutral fiber may not be located at the mid-height of the beam. In any case, we assume that, on one side of this neutral, fiber compression takes place, while on the other side, tension occurs (specifically, in Figure 17.3, the upper fibers are in compression and the lower fibers are under tension).

To efficiently describe the elastic forces produced by this interplay of compression and tension, it is beneficial to consider the osculating circle at a given point of the neutral fiber, see Figure 17.3. Considering an infinitesimal quantity $\varepsilon > 0$ and an infinitesimal element of the beam between x and $x + \varepsilon$, we thereby replace the graph of u with a small portion of the osculating circle of the graph of u tangent to $(x, u(x))$. We denote by $\varrho(x)$ the radius of this osculating circle and by ϑ the corresponding polar angle infinitesimally joining $(x, u(x))$ to $(x + \varepsilon, u(x + \varepsilon))$.

We consider an upper fiber located at distance z above the neutral fiber, and we try to quantify the force produced by compression.

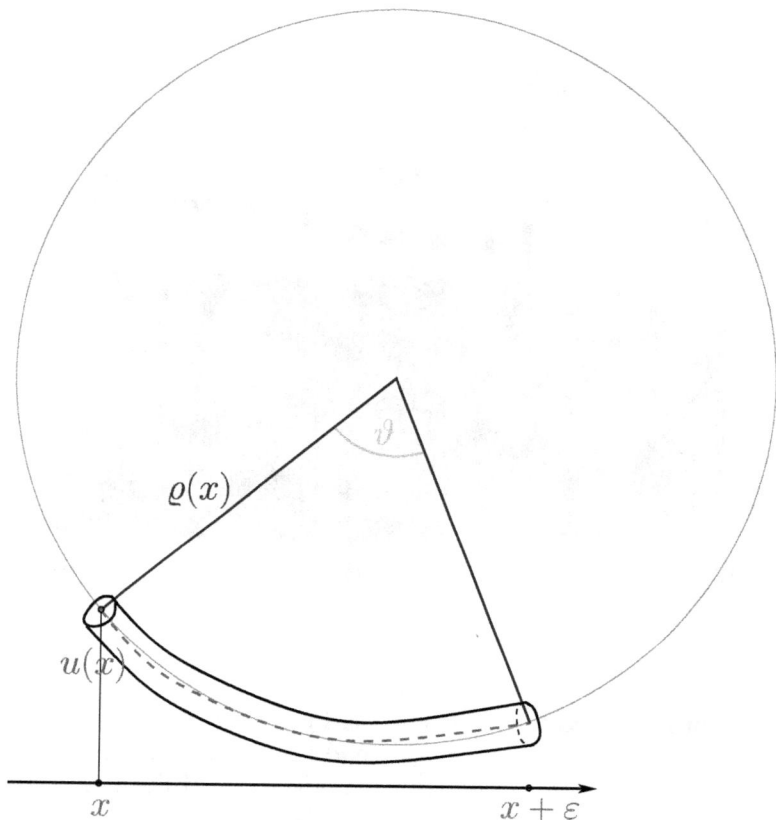

Figure 17.3.   A deflected beam in the (x, z)-plane with an osculating circle.

To this end, we consider the ratio $v(z)$ between the variation in the infinitesimal length of the fiber and its original length before bending, and we suppose that the corresponding compression force is linear with respect to $v(z)$ (linearity with respect to length variations is indeed the main ingredient of elasticity according to Hooke's law, and note that the bigger the distance z from the neutral fiber, the greater the compression). We also observe that the infinitesimal length of this fiber after bending (in the osculating circle approximation) is given by $(\varrho(x) - z)\vartheta$, while its original length before bending was equal to that of the neutral fiber, which is $\varrho(x)\vartheta$. Therefore,

$$v(z) = \frac{\varrho(x)\vartheta - (\varrho(x) - z)\vartheta}{\varrho(x)\vartheta} = \frac{z\vartheta}{\varrho(x)\vartheta} = \frac{z}{\varrho(x)},$$

and accordingly the compression force related to the upper fiber at a distance z from the neutral fiber is proportional to $\frac{z}{\varrho(x)}$.

Similarly, the lower fiber located at a distance z below the neutral one would produce by tension a force which is proportional to $\frac{z}{\varrho(x)}$ (note that the forces produced by tension are opposed to the ones produced by compression). Thus, if we suppose that the elastic modulus of the material is the same for compression and tension (which is another classical *ansatz* in the elasticity theory according to Hooke's law), and we use the convention that a positive z corresponds to an upper fiber and a negative z corresponds to a lower fibers, the resulting elastic force F is approximately oriented along the tangential axis and equals $\frac{Ez}{\varrho(x)}$, where $E > 0$ denotes the elastic modulus of the material (which we suppose to be constant).

We stress the fact that the force F changes sign with z (accordingly, that compression and tension produce forces in opposite directions above and below the neutral fiber) produces a total torque $M(x)$. Indeed, to compute this torque at the point $(x, u(x))$, for each z, we take into consideration the vector product of the position vector and the corresponding force; namely, if (y, z) belongs to the cross-section A of the beam corresponding to $(x, u(x))$, the magnitude of the torque at (y, z) is approximately given by $zF = \frac{Ez^2}{\varrho(x)}$.

The magnitude of the total torque $M(x)$ is thereby the corresponding integral for $(y, z) \in A$, that is,

$$M(x) = \iint_A \frac{Ez^2}{\varrho(x)} \, dy \, dz. \tag{17.2}$$

It is also customary[3] to define the axial second moment of area as

$$I := \iint_A z^2 \, dy \, dz.$$

---

[3]Of course, the axial second moment of area depends on the shape of the cross-section of the beam. For instance, for a circular cross-section we have that $A = \{(y, z) \in \mathbb{R}^2 \text{ s.t. } y^2 + z^2 < r^2\}$, for some $r > 0$; therefore, in this case,

$$I = \iint_{\{y^2 + z^2 < r^2\}} z^2 \, dy \, dz = 2 \int_{-r}^r z^2 \sqrt{r^2 - z^2} \, dz = \frac{\pi r^4}{4}.$$

Instead, if the cross-section is a rectangle, $(-a, a) \times (-b, b)$ for some $a, b > 0$,

$$I = \iint_{(-a, a) \times (-b, b)} z^2 \, dy \, dz = \frac{4ab^3}{3}.$$

With this notation, we can write (17.2) as

$$M(x) = \frac{EI}{\varrho(x)}. \tag{17.3}$$

We also recall that the osculating radius $\varrho(x)$ can be written as $\frac{(1+(u'(x))^2)^{3/2}}{u''(x)}$, see e.g. [Ami00]. Hence, for small deformations, the osculating radius $\varrho(x)$ can be approximated by $\frac{1}{u''(x)}$ and then, with this approximation, equation (17.3) reduces to

$$M(x) = EI \, u''(x). \tag{17.4}$$

To proceed with the derivation of (17.1), we now take into consideration the balance of moments, see Figure 17.4. For this, we assume that the beam is subject to a vertical distributed load of magnitude $q = q(x)$, and we compute its torque at the reference point corresponding to $x + \varepsilon$. For this, given $\ell \in (0, \varepsilon)$, we calculate the magnitude of the vector product of the distributed load at the point $x + \ell$, which is $(0, -q(x + \ell))$, and the vector joining the reference point to the point corresponding to $x + \ell$, which is $(x + \ell, u(x + \ell)) - (x + \varepsilon, u(x + \varepsilon)) = (\ell - \varepsilon, u(x + \ell) - u(x + \varepsilon))$, thus finding the quantity $q(x+\ell)(\ell-\varepsilon)$. Correspondingly, the magnitude of the total torque produced by the external load on the beam between x and $x + \varepsilon$ corresponds to

$$\int_0^\varepsilon q(x+\ell)(\ell-\varepsilon)\, d\ell = \int_0^\varepsilon \big(q(x)+O(\varepsilon)\big)(\ell-\varepsilon)\, d\ell = -\frac{\varepsilon^2 q(x)}{2}+O(\varepsilon^3).$$
$$\tag{17.5}$$

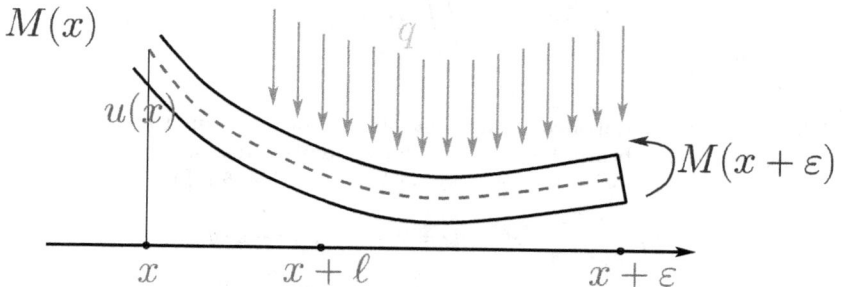

Figure 17.4.   Loads and moments of a beam.

Moreover,

$$M(x + \varepsilon) - M(x) = \varepsilon \partial_x M(x) + \frac{\varepsilon^2}{2} \partial_x^2 M(x) + O(\varepsilon^3). \qquad (17.6)$$

The balance of moments at the second order in ε between (17.5) and (17.6) leads[4] to

$$\partial_x^2 M(x) = q(x).$$

By combining this with (17.4), we obtain

$$q(x) = \partial_x^2(EI \, u''(x)) = EI \, u''''(x),$$

which corresponds to the beam equation in (17.1) (up to normalizing constants). See e.g. [Öch21] for further details on the beam equation and related topics.

It is interesting to generalize the model discussed so far to comprise the case of[5] bending plates. For instance, one could model a thin plate as a graph $x_{n+1} = u(x_1, \ldots, x_n)$, with $x = (x_1, \ldots, x_n) \in \mathbb{R}^n$ (of course, bearing in mind that concretely $n = 2$ and the corresponding plate is two-dimensional). In this setting, the beam equation in (17.1) leads to the plate equation

$$\Delta^2 u = q, \qquad (17.7)$$

where Δ^2 is the Laplace operator applied twice.

To deduce the plate equation in (17.7) from the beam equation in (17.1), one can argue as follows. Let us suppose that the vertical

---

[4]The first order in ε in (17.6) reveals the existence of a shear force acting on the faces of the beam; namely, the balance of moments in first order in ε would produce that this shear force equates to the term $\partial_x M$ (possibly up to a sign convention).

[5]The mathematical model commonly used to describe stresses and deformations in thin plates is sometimes called the "Kirchhoff–Love theory", after Gustav Robert Kirchhoff and Augustus Edward Hough Love. We have already met Kirchhoff on p. 154. Besides his work in elasticity theory, Love dedicated himself to the study of the spin-orbit locking, which is the phenomenon occurring when an astronomical body always has the same face toward the object it is orbiting (for example, up to some minor variability, the same side of the Moon always faces the Earth). See Figure 17.5 for a wood-engraved print representing Love taken from the British Newspaper The Graphic.

MR. E. H. LOVE
(St. John's College)
Second Wrangler

Figure 17.5. Augustus Edward Hough Love (Public Domain image from Wikipedia).

bending force of a horizontal plate is described, for small deformations, by a linear differential operator $\mathcal{L}$ of the form

$$\mathcal{L} := \sum_{\substack{\alpha \in \mathbb{N}^n \\ |\alpha| \leqslant m}} c_\alpha \partial^\alpha, \tag{17.8}$$

for some $m \in \mathbb{N}$ and $c_\alpha \in \mathbb{R}$. Under this assumption, the force balance would lead to the equation $\mathcal{L}u = q$, with q being the external load.

Then, to check (17.7), one needs to show that

$$\mathcal{L} = \Delta^2. \tag{17.9}$$

For this, we assume that, consistently with the model of beam deformation in (17.1), the operator $\mathcal{L}$ acts as a fourth derivative on any one-dimensional function. Namely, let us suppose that if $u_\omega(x) := u_0(\omega \cdot x)$ for some $u_0 : \mathbb{R} \to \mathbb{R}$ and $\omega \in \partial B_1$. Then,

$$\mathcal{L}u_\omega(x) = u_0''''(\omega \cdot x). \tag{17.10}$$

We observe that, for all $j \in \{1, \dots, n\}$,

$$\partial_j u_\omega(x) = \omega_j \, u_0'(\omega \cdot x)$$

and, therefore, if $\alpha = (\alpha_1, \dots, \alpha_n) \in \mathbb{N}^n$,

$$\partial^\alpha u_\omega(x) = \partial_1^{\alpha_1} \dots \partial_n^{\alpha_n} u_\omega(x) = \omega_1^{\alpha_1} \dots \omega_n^{\alpha_n} u_0^{(|\alpha|)}(\omega \cdot x) = \omega^\alpha u_0^{(|\alpha|)}(\omega \cdot x).$$

Hence, by (17.8) and (17.10),

$$u_0''''(\omega \cdot x) = \mathcal{L}u_\omega(x) = \sum_{\substack{\alpha \in \mathbb{N}^n \\ |\alpha| \leq m}} c_\alpha \, \omega^\alpha \, u_0^{(|\alpha|)}(\omega \cdot x).$$

Since u_0 is arbitrary, we infer that $c_\alpha = 0$ unless $|\alpha| = 4$ and

$$1 = \sum_{\substack{\alpha \in \mathbb{N}^n \\ |\alpha| = 4}} c_\alpha \, \omega^\alpha.$$

Now, given $X \in \mathbb{R}^n \setminus \{0\}$, we let $\rho := |X|$ and $\omega := \frac{X}{|X|}$, and we deduce that

$$\sum_{i,j=1}^{n} X_i^2 X_j^2 = \left(\sum_{i=1}^{n} X_i^2 \right)^2 = \left(|X|^2 \right)^2 = |X|^4 = |X|^4 \sum_{\substack{\alpha \in \mathbb{N}^n \\ |\alpha| = 4}} c_\alpha \, \omega^\alpha$$

$$= \sum_{\substack{\alpha \in \mathbb{N}^n \\ |\alpha| = 4}} c_\alpha \, \rho^{|\alpha|} \omega^\alpha = \sum_{\substack{\alpha \in \mathbb{N}^n \\ |\alpha| = 4}} c_\alpha \, (\rho\omega)^\alpha = \sum_{\substack{\alpha \in \mathbb{N}^n \\ |\alpha| = 4}} c_\alpha \, X^\alpha.$$

Consequently, by the identity principle for polynomials,

$$c_\alpha = \begin{cases} 1 & \text{if } \alpha = 2e_i + 2e_j \text{ for some } i, j \in \{1, \dots, n\}, \\ 0 & \text{otherwise.} \end{cases}$$

From this and (17.8), we arrive at

$$\mathcal{L} = \sum_{i,j=1}^{n} \partial_i^2 \partial_j^2 = \sum_{i=1}^{n} \partial_i^2 \Delta = \Delta^2,$$

which establishes (17.9), as desired.

Chapter 18

Gravitation and Electrostatics

One of the reasons why the Laplace operator became very popular, especially in the eighteenth and nineteenth centuries, is its pivotal role in the description of gravity and electromagnetic potentials. The physical intuition of these phenomena was actually quite helpful for the beautiful minds of those ages in understanding a number of very deep concepts, which paved the way to the modern theory of partial differential equations and which are now widely used in many branches of mathematics, physics and engineering.

We give here a couple of motivations[1] relating the gravity field and harmonic functions (similar arguments would link the electrostatic field and harmonic functions as well, simply by changing the notion of mass with that of electric charge, and possibly allowing a sign change, given the fact that unlike charges attract each other but like charges repel).

One observation is that the gravity force (in $\mathbb{R}^3$) is inversely proportional to the square of the distance; namely, a point mass located at x is attracted by a permanent center of gravity located at the origin via a force $f(x)$ equal (up to a dimensional constant) to $-\frac{x}{|x|^3}$, with the minus sign stressing that this force is attractive and hence tends to reduce the distance between the mass located at x and the origin.

---

[1] Also, a different but related perspective on electrostatics and magnetism will be given in Chapter 19.

It is readily seen that f originates from a gravity potential u: more precisely, if we set $u(x) := \frac{1}{|x|}$, we have that $\nabla u(x) = -\frac{x}{|x|^3} = f(x)$. Remarkably, the gravity potential is harmonic away[2] from the origin (here, it is important to work in three dimensions) since, if $x = (x_1, x_2, x_3) \in \mathbb{R}^3 \setminus \{0\}$,

$$\Delta u(x) = -\sum_{i=1}^{3} \partial_i \frac{x_i}{|x|^3} = -\sum_{i=1}^{3} \left(\frac{1}{|x|^3} - \frac{3x_i^2}{|x|^5} \right) = -\frac{3}{|x|^3} + \frac{3|x|^2}{|x|^5} = 0.$$

(18.1)

These considerations can be recast in a possibly more general, and more "geometric", framework while also allowing a higher-dimensional presentation. For this, we first employ the fact that the work done by the gravitational force depends only on the initial and final positions and not on the path between them: this gives that the gravitational vector field f is conservative and thus can be written as a gradient of some function u, which is usually called the gravitational potential.

Furthermore, Gauß's flux law states that the flux of the gravity field through an arbitrary closed surface enclosing no mass (or electric) charges is necessarily zero (or, in rough terms, the field lines going into the region enclosed by the surface balance exactly the ones coming out). Thus, from the divergence theorem, for every bounded region $\Omega \subset \mathbb{R}^n$ (say, with a smooth boundary),

$$0 = \int_{\partial\Omega} f(x) \cdot \nu(x) \, d\mathcal{H}_x^{n-1} = \int_{\Omega} \operatorname{div} f(x) \, dx$$

$$= \int_{\Omega} \operatorname{div}(\nabla u(x)) \, dx = \int_{\Omega} \Delta u(x) \, dx.$$

Since Ω is arbitrary, we thus conclude that the gravity potential (as well as the electrostatic potential) is harmonic away from the masses that are present in the environment.

The study of functions that are harmonic away from a point singularity will be the main topic of Section 2.7 of the companion book [DV23]. There, we will also appreciate how different dimensions affect the precise expression of the above potentials.

---

[2] We will learn from equation (2.5.3) of the companion book [DV23] a more handy way to perform the calculation in (18.1).

Chapter 19

Classical Electromagnetism

The theory of classical electromagnetism has its roots in the so-called Maxwell's equations, which describe the behavior and mutual interaction of electric and magnetic fields. These equations are somewhat a unified version of several specific information arising from Gauß's flux law, Faraday's law and Ampère's law. In a vacuum (and hence in a region of $\mathbb{R}^3$ with no electric charges and no electric currents), Maxwell's equations take the form

$$\text{div}\, E = 0, \qquad \text{curl}\, E = -\frac{\partial B}{\partial t}, \qquad \text{div}\, B = 0 \qquad \text{and}$$

$$\text{curl}\, B = \frac{1}{c^2}\frac{\partial E}{\partial t}. \tag{19.1}$$

where E is the electric field, B is the magnetic field and c is a physical constant (corresponding to the speed of light in vacuum).

We also observe that, for all smooth vector fields $v : \mathbb{R}^3 \to \mathbb{R}^3$,

$$\nabla(\text{div}\, v) - \text{curl}(\text{curl}\, v) = \Delta v. \tag{19.2}$$

To check this classical identity in an elementary way, given $v = (v_1, v_2, v_3)$, we write $v = v^{(1)} + v^{(2)} + v^{(3)}$, where $v^{(1)} := (v_1, 0, 0)$, $v^{(2)} := (0, v_2, 0)$ and $v^{(3)} := (0, 0, v_3)$, and we remark that, thanks to the linear structure of (19.2), it suffices to check it for the vector fields $v^{(1)}$, $v^{(2)}$ and $v^{(3)}$. We can focus our attention on $v^{(1)}$ up to reordering coordinates. In this situation, since, for every smooth vector field $V : \mathbb{R}^3 \to \mathbb{R}^3$,

$$\operatorname{curl} V = \left(\frac{\partial V_3}{\partial x_2} - \frac{\partial V_2}{\partial x_3}, \frac{\partial V_1}{\partial x_3} - \frac{\partial V_3}{\partial x_1}, \frac{\partial V_2}{\partial x_1} - \frac{\partial V_1}{\partial x_2} \right),$$

we have that

$$W := \operatorname{curl} v^{(1)} = \left(0, \partial_3 v_1, -\partial_2 v_1 \right)$$

and subsequently

$$
\begin{aligned}
&\nabla(\operatorname{div} v^{(1)}) - \operatorname{curl}(\operatorname{curl} v^{(1)}) \\
&\quad = \nabla(\partial_1 v_1) - \operatorname{curl} W \\
&\quad = (\partial_{11} v_1, \partial_{12} v_1, \partial_{13} v_1) - \left(\frac{\partial W_3}{\partial x_2} - \frac{\partial W_2}{\partial x_3}, -\frac{\partial W_3}{\partial x_1}, \frac{\partial W_2}{\partial x_1} \right) \\
&\quad = (\partial_{11} v_1, \partial_{12} v_1, \partial_{13} v_1) - (-\partial_{22} v_1 - \partial_{33} v_1, \partial_{12} v_1, \partial_{13} v_1) \\
&\quad = (\partial_{11} v_1 + \partial_{22} v_1 + \partial_{33} v_1, \, 0, \, 0) \\
&\quad = \Delta v^{(1)},
\end{aligned}
$$

thus establishing (19.2).

Now, in light of (19.1) and (19.2),

$$
\begin{aligned}
\partial_{tt} E = c^2 \partial_t(\operatorname{curl} B) &= c^2 \operatorname{curl}(\partial_t B) = -c^2 \operatorname{curl}(\operatorname{curl} E) \\
&= c^2 \big(\Delta E - \nabla(\operatorname{div} E) \big) = c^2 \Delta E
\end{aligned}
\tag{19.3}
$$

and, similarly,

$$
\begin{aligned}
\partial_{tt} B = -\partial_t(\operatorname{curl} E) &= -\operatorname{curl}(\partial_t E) = -c^2 \operatorname{curl}(\operatorname{curl} B) \\
&= c^2 \big(\Delta B - \nabla(\operatorname{div} B) \big) = c^2 \Delta B.
\end{aligned}
$$

These observations show that both the electric and magnetic fields propagate like waves do since they satisfy the wave equation in (7.2).

Chapter 20

Quantum-Mechanical Systems

A popular equation governing the wave function of a quantum-mechanical system was proposed by Erwin Schrödinger (this equation formed the basis for the work that resulted in Schrödinger's 1933 Nobel Prize).

In a nutshell, the Schrödinger equation reads

$$-i\hbar\partial_t\psi = \frac{\hbar^2\Delta\psi}{2m} - V\psi, \tag{20.1}$$

where $i = \sqrt{-1}$, $\psi = \psi(x,t)$ is the wavefunction[1] of a particle of mass m subject to a potential V, and $\hbar$ is a suitable positive physical constant (called the "reduced Planck constant").

Particularly interesting solutions of (20.1) are the so-called standing waves, namely solutions which oscillate in time but whose amplitude profiles do not change in space. From a mathematical point of view, these solutions are of the form

$$\psi(x,t) = u(x)\,e^{i\phi t},$$

where u is a real-valued function and $\phi > 0$ is the time frequency of oscillation.

---

[1]In rough terms, a wavefunction is a function that assigns a complex number to each point $x \in \mathbb{R}^3$ at each time $t \in \mathbb{R}$. The square of the magnitude of this function represents the probability density of measuring the particle as being at the point x at a given time t.

Since for standing waves we have that

$$\partial_t \psi(x,t) = i\phi\, u(x)\, e^{i\phi t} \qquad \text{and} \qquad \Delta\psi(x,t) = \Delta u(x)\, e^{i\phi t},$$

these special solutions of (20.1) are actually solutions[2] of

$$\hbar\phi u = \frac{\hbar^2 \Delta u}{2m} - V u. \tag{20.2}$$

Going back to the initial discussion of this section, one may wonder why and how Schrödinger introduced the equation in (20.1) above. This would be a rather long and complicated story, and a short answer was given in [FLS65, Chapter 16]: "Where did we get that from? Nowhere. It's not possible to derive it from anything you know. It came out of the mind of Schrödinger, invented in his struggle to find an understanding of the experimental observations of the real world". However, several quite convincing motivations for the Schrödinger equation are possible (see e.g. [Eft06, WV06, DdPDV15, Yan21] and the references therein), and we present one of them here in the following, which is based on the notion of canonical quantization.

To this end, we recall that the energy $\mathcal{E}$ of a photon is proportional to its temporal frequency ϕ, according to the so-called Planck–Einstein energy-frequency relation

$$\mathcal{E} = \hbar\phi.$$

---

[2]According to the classification presented in footnote 13 on p. 9, equation (20.2) is of the elliptic type. When $V = V(x)$, equation (20.2) is linear in u. More complex situations arise when V also depends on the magnitude of the wavefunction, say $V = V(x, u(x))$ or even $V = V(u(x))$, since in this situation, equation (20.2) is no longer linear in u (though it is linear with respect to the Hessian of u). A study of these "semilinear" equations will be provided in the companion book [DV23]. Further comments on semilinear equations will be given in footnote 2 on p. 277.

Also, Einstein's relativistic energy formula reads

$$\mathcal{E} = \sqrt{(|p|\,c)^2 + (mc^2)^2},\tag{20.3}$$

where p is the relativistic momentum, m is the mass at rest and c is the speed of light. For massless particles, such as photons, this reduces to

$$\mathcal{E} = |p|\,c,\tag{20.4}$$

from which we obtain

$$|p| = \frac{\mathcal{E}}{c} = \frac{\hbar\phi}{c}.\tag{20.5}$$

Now, we relate a photon to its electric field E traveling through space and manifesting itself as a solution of the wave equation (recall (19.3)). That is, we consider a direction of propagation $\varpi \in \partial B_1$, a spatial frequency κ and a temporal frequency ϕ, and we suppose that, for large distances away from the photon's source, the field is modeled using a simple plane wave, say

$$E = E_0 e^{i(\kappa\varpi \cdot x - \phi t)},$$

where $E_0 \in (0, +\infty)$. More precisely, by (19.3),

$$-\phi^2 E_0 e^{i(\kappa\varpi \cdot x - \phi t)} = \partial_{tt} E = c^2 \Delta E = -c^2 \kappa^2 E_0 e^{i(\kappa\varpi \cdot x - \phi t)}$$

and therefore

$$\phi = c\kappa,\tag{20.6}$$

thus relating the temporal and spatial frequencies of the photon. We thus obtain from (20.5) that

$$|p| = \hbar\kappa\tag{20.7}$$

and therefore

$$|p|E = \hbar\kappa E = -i\hbar\nabla E \cdot \varpi.\tag{20.8}$$

Taking the direction of the momentum to coincide with the spatial direction ϖ of the wave, i.e. taking $p = |p|\varpi$, we rewrite (20.8) in the form

$$pE = -i\hbar \nabla E. \qquad (20.9)$$

We can also reconsider (20.4) in view of (20.6) and (20.7) and write that

$$\mathcal{E} = |p|\, c = \hbar c \kappa = \hbar \phi,$$

consequently,

$$\mathcal{E} E = \hbar \phi E = i\hbar \partial_t E. \qquad (20.10)$$

With this, while not aiming to be exhaustive, we can recall some of the main ideas leading to canonical quantization. This procedure, originally introduced by Paul Dirac in his 1926 PhD thesis, aims to recast a classical theory into a quantum one by preserving (as much as possible) its symmetries and formal structures. Since the Hamiltonian formalism is one of the key tools used to understand the symmetries of classical mechanical systems, a natural idea in this framework is to try to (at least partially) preserve the Hamiltonian structure in the quantum description of nature.

To this end, a simple observation is that the Hamiltonian formalism relies on conjugated variables q and p, with the physical meanings of position and momentum. The first goal of the canonical quantization is therefore to rephrase position and momentum in a way that is compatible with the quantum description of the observables.

In particular, in quantum mechanics, all significant features of a particle are contained in a certain state, ψ. The observables are represented by operators acting on states. For instance, the eigenvalues of an operator represent the values of the measurements of the corresponding fundamental states of a particle (namely, the eigenfunctions or eigenstates); since any state is represented as a linear combination of eigenstates, the application of an operator to the state ψ corresponds to the determination of a measurable parameter, while the physical act of measuring corresponds to causing the values of the state to "collapse" from a superposition of eigenstates to a single eigenstate due to interaction with the external world.

With these basic principles in mind, we therefore aim to replace the classical position and momentum variables q and p with two operators, say Q and P, while preserving the original structure as much as possible.

The most natural setting is therefore to consider the position operator Q as an operator that corresponds to the position evaluation of a state (simply, applying Q to a state ψ being the evaluation of the function ψ at a given point in space).

As for the corresponding momentum operator, in light of (20.9), a natural choice is to take P as the differential operator $-i\hbar\nabla$, thus obtaining a quantum analogue of (20.9) via the relation

$$P\psi = -i\hbar\nabla\psi. \tag{20.11}$$

Pushing this analogy a bit further, we can also obtain a quantum counterpart $\mathcal{H}$ of the total energy $\mathcal{E}$ (say, the Hamiltonian) in (20.10). In this setting, the operator analog of (20.10) would then be

$$\mathcal{H}\psi = -i\hbar\partial_t\psi. \tag{20.12}$$

While this canonical formalism for quantum mechanics was obtained by analyzing the special case of the wave produced by a photon, we can hypothesize that the same protocol governs the evolution of the wave function of a particle (not necessarily a photon) subject to a given potential V. For instance, for a particle of mass $m > 0$, one can extend the previous construction by considering the quantum analog of the mechanical energy (perhaps neglecting for the moment some relativistic effects, which we will briefly discuss in the forthcoming footnote 3). Indeed, in classical mechanics, the total energy would be given by the sum of the kinetic energy of a particle (equal to $\frac{m|v|^2}{2}$, with v being its velocity) and its potential energy V. That is, recalling the classical momentum definition $p = mv$,

$$\mathcal{E} = \frac{m|v|^2}{2} + V = \frac{|p|^2}{2m} + V.$$

By formally applying the quantization in (20.11) and (20.12), we thus obtain its quantum counterpart as

$$i\hbar\partial_t = \mathcal{H} = \frac{|P|^2}{2m} + V = \frac{P \cdot P}{2m} + V = -\frac{\hbar^2\nabla \cdot \nabla}{2m} + V = -\frac{\hbar^2\Delta}{2m} + V,$$

which is the Schrödinger equation[3] in (20.1).

---

[3]It is interesting to point out that variations of the previous arguments lead to other equations of interest; for instance, by considering the full relativistic energy formula in (20.3) and formally inserting the quantum mechanical operator in (20.9), the quantum analogue of the kinetic energy becomes

$$\sqrt{-\hbar^2 c^2\Delta + (mc^2)^2},$$

which is the "square root of the Laplacian" (see e.g. [AV19] for a basic introduction to this very interesting object). Thus, in the presence of an external potential V, the conservation of the full energy and the quantization in (20.12) lead to the balance

$$-i\hbar\partial_t = \sqrt{-\hbar^2 c^2\Delta + (mc^2)^2} - V,$$

which produces the equation

$$-i\hbar\partial_t\psi = \sqrt{-\hbar^2 c^2\Delta\psi + (mc^2)^2} - V\psi.$$

This equation and its variants are usually called the "relativistic Schrödinger equations", see e.g. equation (1.4) in [CMS90].

Another approach to taking the relativistic effects into account consists of taking the square of the relativistic energy formula in (20.3), thus writing that

$$\mathcal{E}^2 = (|p|\,c)^2 + (mc^2)^2.$$

In the absence of external potentials (i.e. if this represents the square of the total energy of the system), one can proceed with exploiting the quantization procedure in (20.9) and (20.12), formally finding that

$$-\hbar^2\partial_{tt} = -\hbar^2 c^2\Delta + (mc^2)^2.$$

This leads to the equation

$$\partial_{tt}\psi = c^2\Delta\psi - \frac{m^2 c^4\psi}{\hbar^2},$$

which is often referred to by the name "Klein–Gordon equation", see e.g. equation (34) in [Ioa84].

Chapter 21

Bouncing Balls and Whispers

The popular game of billiard, see Figure 21.1, has a mathematical counterpart in the study of dynamical billiards, namely a dynamical system of a material particle confined in a (typically, bounded and sufficiently smooth) region Ω of $\mathbb{R}^n$ (or a more general manifold), which alternates between the free motion in the interior of Ω (given by a straight line traveled at unit speed, or, for general manifolds, by a geodesic path) and specular reflections from the boundary. See Figure 21.2 for a visual animation[1] of a dynamic billiard.

More precisely, the billiard flow in the region Ω corresponds to the motion of a material point which is moving in Ω with a constant (e.g. unit) velocity in the interior of Ω, causing reflections at the boundary of Ω according to the usual law of geometric optics, namely prescribing that the angle of incidence is equal to the angle of reflection.

It is interesting to note that this motion can be obtained, at least in the limit and at least exploiting some formal expansions and approximations, from a smooth Hamiltonian system. Namely, one can consider a large parameter $M > 0$ and a potential $V_M : \mathbb{R}^n \to \mathbb{R}$ defined as

$$V_M(q) := \begin{cases} 0 & \text{if } q \in \Omega, \\ \dfrac{M}{2} \left(\text{dist}(q, \partial\Omega) \right)^2 & \text{if } q \in \mathbb{R}^n \setminus \Omega. \end{cases}$$

---

[1]Actually, Figure 21.2 depicts the motion of a particle in a famous example of billiard introduced in [Bun79].

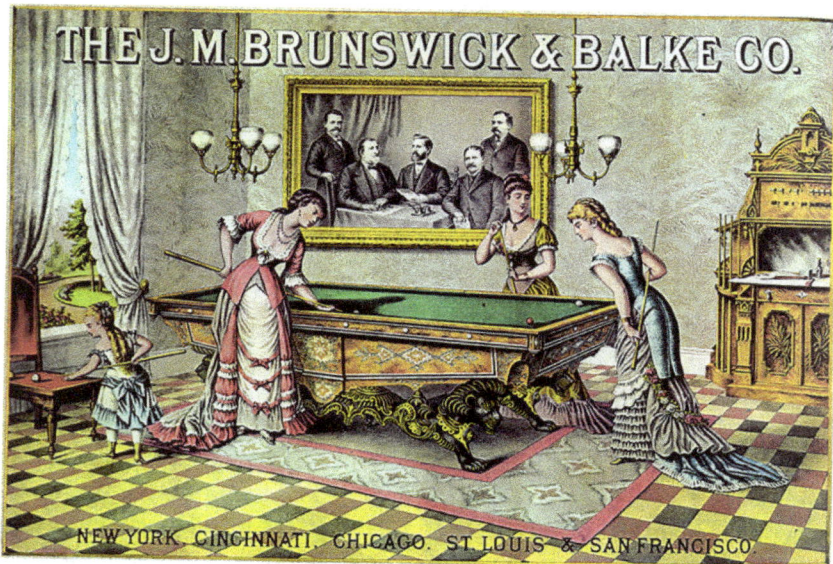

Figure 21.1.   Advertising poster, early 1880s (Public Domain image from Wikipedia).

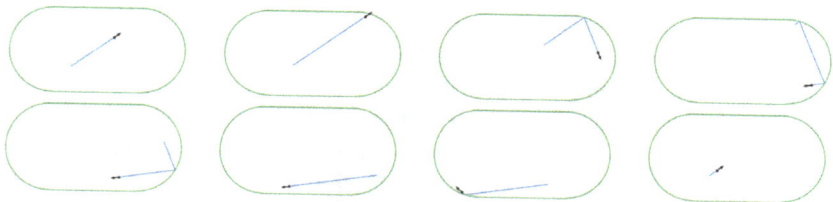

Figure 21.2.   A dynamical billiard (animation by George Datseris, image from Wikipedia, licensed under the Creative Commons Attribution-Share Alike 4.0 International license).

Then, at least at a formal level,

the Hamiltonian $H_M(q,p) := \dfrac{|p|^2}{2} + V_M(q)$

$$(21.1)$$

formally recovers the billiard motion as $M \to +\infty$.

To check this, we observe that the equations of motion induced by H_M are

$$\begin{cases} \dot{q} = p, \\ \dot{p} = -\nabla V_M(q). \end{cases}$$

Consequently, as long as the trajectory lies in Ω, we have that $\dot{p} = 0$, and thus q evolves linearly (precisely as in the billiard motion, and this property is stable as $M \to +\infty$).

We thereby investigate what happens when a particle hits the boundary of Ω, and we show that the motion induced by H_M in this case reduces to the billiard reflection law when $M \to +\infty$. Let us assume that the particle hits the boundary of Ω at some time t_0, which we can assume to be 0, some point $p_0 \in \partial\Omega$, which, up to a translation, can be taken to be the origin. Also, up to a rotation, we can assume that

$$\text{the exterior unit normal of } \Omega \text{ at the origin equals } e_n. \qquad (21.2)$$

We also suppose, for simplicity, that the velocity of the particle upon reaching Ω is not tangential to $\partial\Omega$, and hence $p_n(0) > 0$. In particular, if $p(0) = 0$, we can assume that $p(t) \in \mathbb{R}^n \setminus \Omega$, for all $t \in (0, t_1)$, for some $t_1 > 0$.

Actually, by the conservation of energy, we have that $|\text{dist}(q(t), \partial\Omega)| \leqslant \frac{C}{\sqrt{M}}$ (that is, the particle cannot leave a small neighborhood of Ω) and, therefore,

$$\nabla V_M(q(t)) = M\text{dist}(q(t), \partial\Omega)\nabla\text{dist}(q(t), \partial\Omega)$$
$$= M\text{dist}(q(t), \partial\Omega)\nu(\pi(q(t))),$$

with π being the projection along $\partial\Omega$.

Additionally,

$$\nu(\pi(q(t))) = \nu(\pi(0)) + O(|\pi(q(t)) - \pi(0)|)$$
$$= \nu(0) + O(|q(t)|) = e_n + O\left(\frac{1}{\sqrt{M}}\right)$$

due to (21.2). Hence, for a formal justification of (21.1), we disregard this higher-order correction and simply write

$$\nabla V_M(q(t)) \simeq M\text{dist}(q(t), \partial\Omega)e_n.$$

Again by energy conservation, we know that $|\dot{q}(t)| = |p(t)| \leqslant C$; therefore, we can suppose that $q(t)$ is in a small neighborhood of the origin as long as t_1 is small enough; hence, using (21.2), at a formal level, we can use the approximation

$$\text{dist}(q(t), \partial\Omega) \simeq q_n(t),$$

where we use the notation $q = (q', q_n) \in \mathbb{R}^{n-1} \times \mathbb{R}$.

As a result, for all $t \in (0, t_1)$, if t_1 is small enough,

$$\ddot{q}(t) = \dot{p}(t) = -\nabla V_M(q(t)) \simeq -M \mathrm{dist}(q(t), \partial\Omega)e_n \simeq -M q_n(t)e_n.$$

Therefore,

$$\begin{cases} \ddot{q}'(t) \simeq 0, \\ \ddot{q}_n(t) \simeq -M q_n(t), \end{cases}$$

and we observe that

$$q(0) = 0 \quad \text{and} \quad \dot{q}(0) = p(0).$$

In this way, solving the above ordinary differential equations, we obtain that formally

$$q'(t) \simeq p'(0)t \quad \text{and} \quad q_n(t) \simeq \frac{p_n(0)}{\sqrt{M}} \sin\left(\sqrt{M}t\right). \qquad (21.3)$$

Now, we detect the "reentering" of the particle in the domain Ω. To this end, assuming M to be appropriately large, we claim that

$$q(t) \in \mathbb{R}^n \setminus \Omega \text{ for all } t \in \left(0, \frac{\pi}{\sqrt{M}} - \frac{1}{\sqrt{M}\ln M}\right] \qquad (21.4)$$

and that

$$\text{there exists } t_M \in \left(\frac{\pi}{\sqrt{M}} - \frac{1}{\sqrt{M}\ln M}, \frac{\pi}{\sqrt{M}} + \frac{1}{\sqrt{M}\ln M}\right]$$

$$\text{such that } q(t_M) \in \Omega. \qquad (21.5)$$

To prove this, consistently with (21.2), we parameterize Ω near the origin as a subgraph of a function $f : \mathbb{R}^{n-1} \to \mathbb{R}$, with $f(0) = 0$ and $\partial_k f(0) = 0$ for each $k \in \{1, \ldots, n-1\}$.

In this setting, if $t \in \left(0, \frac{\pi}{\sqrt{M}} - \frac{1}{\sqrt{M}\ln M}\right)$, using the approximation in (21.3),

$$q_n(t) - f(q'(t)) \geqslant q_n(t) - C|q'(t)|^2 \simeq \frac{p_n(0)}{\sqrt{M}} \sin\left(\sqrt{M}t\right) - C|p'(0)|^2 t^2$$

$$= \frac{p_n(0)}{\sqrt{M}} \left(\sin\left(\sqrt{M}t\right) - \frac{C(\sqrt{M}t)^2}{\sqrt{M}}\right),$$

$$(21.6)$$

up to renaming $C > 0$ line after line.

Thus, if $t \in \left(0, \frac{1}{\sqrt{M}}\right)$, we use that $\sin\left(\sqrt{M}t\right) \geqslant \frac{\sqrt{M}t}{2}$, whence, in the formal approximation above,

$$q_n(t) - f(q'(t)) \geqslant \frac{p_n(0)}{2\sqrt{M}} \left(\sqrt{M}t - \frac{C(\sqrt{M}t)^2}{\sqrt{M}}\right)$$

$$= \frac{p_n(0)\,t}{2}\left(1 - \frac{C\sqrt{M}t}{\sqrt{M}}\right) \geqslant \frac{p_n(0)\,t}{2}\left(1 - \frac{C}{\sqrt{M}}\right) > 0.$$

If instead $t \in \left(\frac{1}{\sqrt{M}}, \frac{\pi}{\sqrt{M}} - \frac{1}{\sqrt{M}\ln M}\right]$, we infer from (21.6) that

$$q_n(t) - f(q'(t)) \geqslant \frac{p_n(0)}{\sqrt{M}}\left(\sin\left(\pi - \frac{1}{\ln M}\right) - \frac{C}{\sqrt{M}}\right)$$

$$= \frac{p_n(0)}{\sqrt{M}}\left(\sin\left(\frac{1}{\ln M}\right) - \frac{C}{\sqrt{M}}\right) \geqslant \frac{p_n(0)}{2\sqrt{M}}\left(\frac{1}{\ln M} - \frac{C}{\sqrt{M}}\right) > 0,$$

and so (21.4) follows.

Furthermore,

$$q_n\left(\frac{\pi}{\sqrt{M}} + \frac{1}{\sqrt{M}\ln M}\right) - f\left(q'\left(\frac{\pi}{\sqrt{M}} + \frac{1}{\sqrt{M}\ln M}\right)\right)$$

$$\leqslant q_n\left(\frac{\pi}{\sqrt{M}} + \frac{1}{\sqrt{M}\ln M}\right) + C\left|q'\left(\frac{\pi}{\sqrt{M}} + \frac{1}{\sqrt{M}\ln M}\right)\right|^2$$

$$\simeq \frac{p_n(0)}{\sqrt{M}}\sin\left(\pi + \frac{1}{\ln M}\right) + \frac{C}{M}$$

$$= -\frac{p_n(0)}{\sqrt{M}}\sin\left(\frac{1}{\ln M}\right) + \frac{C}{M}$$

$$< 0,$$

which gives (21.5).

Therefore, the velocity of the particle reentering Ω is of the form

$$p(t_M) = \dot{q}(t_M) \simeq \left(p'(0),\, p_n(0)\cos\left(\sqrt{M}t_M\right)\right)$$

$$= \left(p'(0),\, p_n(0)\cos\left(\pi + \theta_M\right)\right),$$

with $\theta_M \in \left(-\frac{1}{\ln M}, \frac{1}{\ln M}\right]$.

In the limit of $M \to +\infty$, this bouncing velocity becomes $(p'(0), p_n(0) \cos \pi) = (p'(0), -p_n(0))$, which is precisely the elastic reflection of geometric optics due to (21.2).

This completes the formal proof of (21.1).

The formal Hamiltonian presentation of billiards as an infinite potential well in (21.1) is sometimes written as a description of the billiard motion as a Hamiltonian system run by the Hamiltonian

$$H_\infty(q, p) := \frac{|p|^2}{2} + V_\infty(q), \qquad (21.7)$$

with

$$V_\infty(q) := \begin{cases} 0 & \text{if } q \in \Omega, \\ +\infty & \text{if } q \in \mathbb{R}^n \setminus \Omega, \end{cases}$$

whatever this expression means.

One of the advantages of such a (perhaps teetering but quite practical) description is that it readily opens the possibility of studying a quantum analogue of the classical billiards, simply by considering the quantization proposed in (20.11). This method suggests replacing the classical Hamiltonian in (21.7) with an operator of the form

$$\frac{|-i\hbar\nabla|^2}{2} + V_\infty(x) = \frac{\hbar^2}{2} \sum_{k=1}^{n} \partial_k^2 + V_\infty(x) = \frac{\hbar^2}{2}\Delta + V_\infty(x).$$

The corresponding partial differential equation would thereby become

$$Eu(x) = \frac{\hbar^2}{2}\Delta u(x) + V_\infty(x)u(x),$$

for some scalar E.

Now, since $V_\infty = +\infty$ outside Ω, to make sense of the right-hand side above, one typically assumes that u vanishes outside Ω (that is, the condition that the billiard particle is confined to Ω is reflected into a restriction on the support of the corresponding wave function in its quantum analogue). This leads to the problem $Eu = \frac{\hbar^2}{2}\Delta u$, with $u = 0$ outside Ω, or, up to a change of notation,

$$\begin{cases} \Delta u = -\lambda u & \text{in } \Omega, \\ u = 0 & \text{on } \partial\Omega, \end{cases} \tag{21.8}$$

where $\lambda := -\frac{2E}{\hbar^2}$ corresponds now to a large scalar (given the smallness of the reduced Planck constant $\hbar$). Equation (21.8) is sometimes called the Helmholtz equation.

In analogy with the finite-dimensional case of matrices, (21.8) is considered an eigenvalue problem for the Laplace operator (in this case, with homogeneous Dirichlet conditions). The study of quantum billiards is thus often quite related to the analysis of an eigenvalue problem and often specifically focused on the case of large eigenvalues.

Interestingly, problems of this type also surface in the theory of sound[2] and provide an essential ingredient for the understanding of whispering gallery waves. This phenomenon was deeply investigated by John William Strutt, 3rd Baron Rayleigh (most commonly addressed as Lord[3] Rayleigh, see Figure 21.3 for a caricature of him published in the London magazine Vanity Fair in 1899).

To study the whispering phenomenon, Lord Rayleigh took inspiration from the concrete case of the whispering gallery in St. Paul's Cathedral, London, see Figure 21.4, in which whispers can be heard clearly from one side to the other of the gallery.

---

[2] And, of course, similar phenomena can exist for light and other electromagnetic radiations, making whispering phenomena very attractive also for technological reasons.

[3] Here is a funny anecdote about Lord Rayleigh. Given his Anglican faith, at some point, Lord Rayleigh wanted to include a religious quotation from the Bible at the opening of a collection of papers written by him. He was discouraged from doing so by the staff of the Cambridge University Press, as he reported: "When I was bringing out my Scientific Papers I proposed a motto from the Psalms, *The Works of the Lord are great, sought out of all them that have pleasure therein.* The Secretary to the Press suggested with many apologies that the reader might suppose that I was the *Lord*".

The moral of the story, if any, could also be that if one happens to be a member of the aristocracy, it would be advisable not to include the noble title in the author's name.

Figure 21.3. Caricature of Lord Rayleigh (Public Domain image from Wikipedia).

Figure 21.4. (left) Aerial view of Cathedral Church of St. Paul the Apostle, London; (right) its whispering gallery (images from Wikipedia, photo by Mark Fosh, licensed under the Creative Commons Attribution 2.0 Generic license for the first, photo by Femtoquake, licensed under the Creative Commons Attribution-Share Alike 3.0 Unported license for the second).

Actually, Lord Rayleigh proposed several approaches to the question. In [LR10], he related the whispering phenomenon with problem (21.8). Specifically, we can model the whispering gallery in Figure 21.4 as a planar disk, say $\Omega = B_1 \subset \mathbb{R}^2$. We can take the model of the propagation of sound waves put forth in Chapter 7, that is, in light of (7.2), we consider the solutions of

$$\partial_{tt}\rho = c^2 \Delta\rho, \tag{21.9}$$

where $\rho = \rho(x,t)$ represents the density of the air, as perturbed by the propagating sound, and $c > 0$ is the speed of this propagation.

We assume that the density is constant, say ρ_0, along $\partial\Omega$, we consider the eigenfunctions of the Laplacian with zero boundary datum along $\partial\Omega$, see [Eva98, Section 6.5.1], and we look for solutions in the form

$$\rho(x,t) = \rho_0 + \sum_{k=1}^{+\infty} \rho_k(t)\,\eta_k(x), \tag{21.10}$$

with η_k being the eigenfunctions corresponding to the eigenvalue λ_k, with λ_k nondecreasing and $\lambda_k \to +\infty$ as $k \to +\infty$ (and, say $\|\eta_k\|_{L^2(\Omega)} = 1$ as a possible normalization).

By plugging (21.10) into (21.9), we obtain the equation $\ddot{\rho}_k(t) = -c^2\lambda_k\rho_k(t)$, giving e.g. $\rho_k(t) = b_k\sin\left(c\sqrt{\lambda_k}t\right)$ for some $b_k \in \mathbb{R}$, whence

$$\rho(x,t) = \rho_0 + \sum_{k=1}^{+\infty} b_k\sin\left(c\sqrt{\lambda_k}t\right)\eta_k(x).$$

This suggests that the propagation of sound is modeled by the linear superposition of harmonic modes of the form $b_k\sin\left(c\sqrt{\lambda_k}t\right)\eta_k(x)$. Lord Rayleigh's idea is thus to detect modes corresponding to the eigenfunctions η_k, whose mass is mostly concentrated in the vicinity of $\partial\Omega$. From a technical point of view, Lord Rayleigh's analysis consists of using polar coordinates (r, ϑ) to write conveniently $\eta_k(x) = \beta_k(r)\,\alpha_k(\vartheta)$, for suitable functions α_k and β_k, and obtain the radial component β_k in terms of "known" special functions, in this case the so-called Bessel functions of the first kind. Then, he chooses conveniently the parameters to exhibit cases in which the mass of these special functions is mostly concentrated near the boundary.

More explicitly, one can check that, for all $m \in \mathbb{N}$, the function

$$[0, +\infty) \times [0, 2\pi) \ni (r, \vartheta) \longmapsto J_m(\sqrt{\lambda} r) \cos(m\vartheta) \qquad (21.11)$$

satisfies (21.8) provided that[4] the special function J_m is a solution of

$$\begin{cases} s^2 J_m''(s) + s J_m'(s) + (s^2 - m^2) J_m(s) = 0 & \text{for all } s > 0, \\ J_m(\sqrt{\lambda}) = 1, \end{cases} \qquad (21.12)$$

known as the Bessel differential equation (the boundary condition here being related to the boundary datum in (21.8)). See Figures 21.5 and 21.6 for a sketch of these functions.

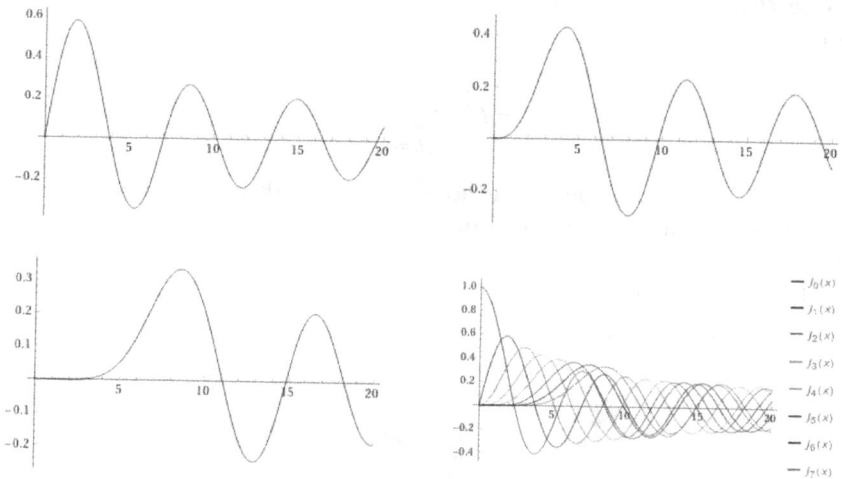

Figure 21.5. Bessel functions of the first kind J_m (corresponding respectively to $m = 1$, $m = 3$, $m = 7$ and $m \in \{0, \ldots, 7\}$).

---

[4]Actually, the Bessel differential equation (21.12) has two families of solutions, say J_m and Y_m. The solutions in the first family are the ones considered here: they are called Bessel functions of the first kind, and they are continuous at the origin. The solutions in the second family are singular at the origin; therefore, they do not provide solutions for eigenvalue problems in a disk (for this reason, they do not appear in the calculations given in these pages); they are called Bessel functions of the second kind, Neumann functions or Weber functions. See e.g. [Wat95] and the references therein for additional information.

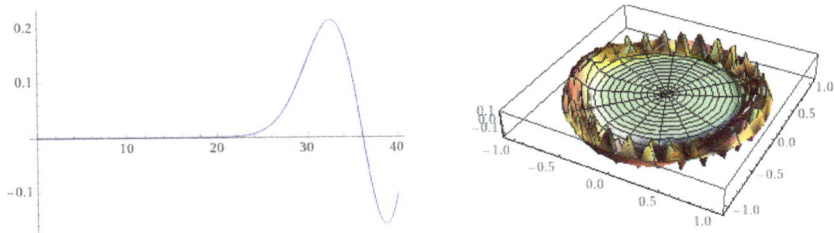

Figure 21.6. Bessel function of the first kind J_{30} and the corresponding eigen-function according to (21.11).

As we can see from Figures 21.5 and 21.6, these functions J_m for large values of m have the tendency to become increasingly flatter at the origin and concentrate proportionally more mass near their first root. This fact can be proven rigorously, see e.g. p. 227 in [Wat95], and it can also be heuristically guessed from the Bessel differential equation since if near 0 we make the *ansatz* that $J_m(s) = s^\ell + o(s^\ell)$ for some $\ell \geqslant 0$, we formally obtain from (21.12) that

$$0 = s^2\big(\ell(\ell-1)s^{\ell-2} + o(s^{\ell-2})\big) + s\big(\ell s^{\ell-1} + o(s^{\ell-1})\big)$$
$$+ \big(s^2 - m^2\big)\big(s^\ell + o(s^\ell)\big)$$
$$= \big(\ell(\ell-1) + \ell - m^2\big)s^\ell + o(s^\ell)$$
$$= \big(\ell^2 - m^2\big)s^\ell + o(s^\ell);$$

hence, $0 = \ell^2 - m^2 + o(1)$ and consequently $\ell = m$.

The possible localization of eigenfunctions in a small region around the boundary of the domain is particularly evident in the second image in Figure 21.6.

The whispering phenomenon described by the concentration of eigenfunctions in the proximity of the boundary can also be considered the counterpart of the caustics in dynamical billiards (namely, curves to which the trajectory remains tangent); in concrete cases, when these caustics are localized near the boundary, they can provide a confinement for the trajectory, see Figure 21.7 (to be compared with the second image in Figure 21.6).

See [NG13] and the references therein for further information about localization of eigenfunctions, Bessel functions, whispering gallery waves and their links to bouncing balls.

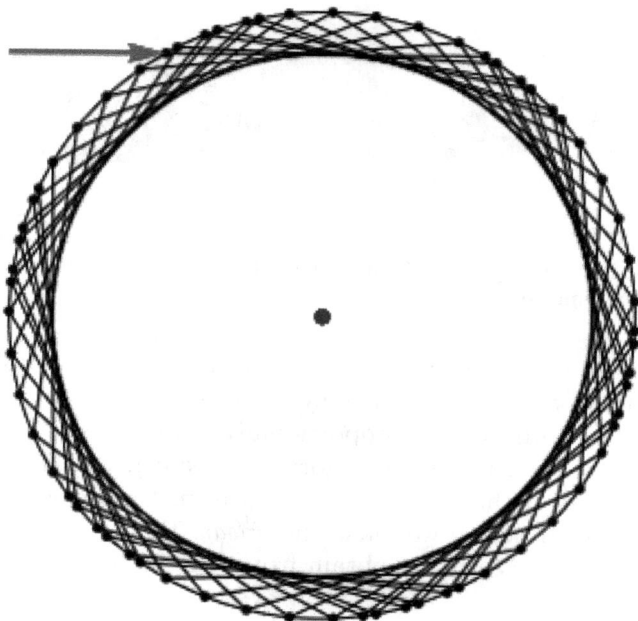

Figure 21.7. Dynamics of circular billiard (obtained from the free resource, https://demonstrations.wolfram.com/DynamicBilliardsInEllipse/).

In general, dynamical billiards have been widely studied from many mathematical viewpoints, also acting as excellent training camps for a number of complex systems; yet, many problems related to billiards remain completely open, and mathematicians often feel behind the eight ball. See [PS17] and the references therein for additional information about billiards and their intimate connections with the Laplace operator.

Chapter 22

Strange Attractions

Another pioneering work by Lord Rayleigh (see Figure 21.3) was his study of thermal convection problems, addressed in [LR16]. The problem can be explained by considering, for simplicity, a two-dimensional incompressible and viscous fluid in a gravity field, with coordinates $(x, z) \in \mathbb{R}^2$. The top and bottom surfaces of the fluid are horizontal, say at levels $z = H$ and $z = 0$, respectively, supposed to be kept at constant temperature. Up to normalization, we suppose that the top surface is at temperature $\overline{T}$ and the bottom surface is at some temperature $\underline{T}$, which we suppose to be larger than $\overline{T}$.

The idea of this model is that gravity tends to push down the molecules of the fluid, but heat from the bottom may favor upward motions, thus producing interesting convective patterns, see e.g. Figure 22.1.

To describe this phenomenon, one can utilize the Navier–Stokes equation in (5.8) complemented with the incompressibility condition in (5.9) and the assumption that the top and bottom surfaces remain unchanged. Writing the velocity of the fluid as $v = (v_x, v_z) \in \mathbb{R}^2$, this leads to

$$\begin{cases} \rho \partial_t v + \rho(v \cdot \nabla)v = \mu \Delta v - \rho g(0, 1) - \nabla p & \text{in } \mathbb{R} \times (0, H), \\ \operatorname{div} v = 0 & \text{in } \mathbb{R} \times (0, H), \quad (22.1) \\ v_z(x, 0) = v_z(x, H) = 0. \end{cases}$$

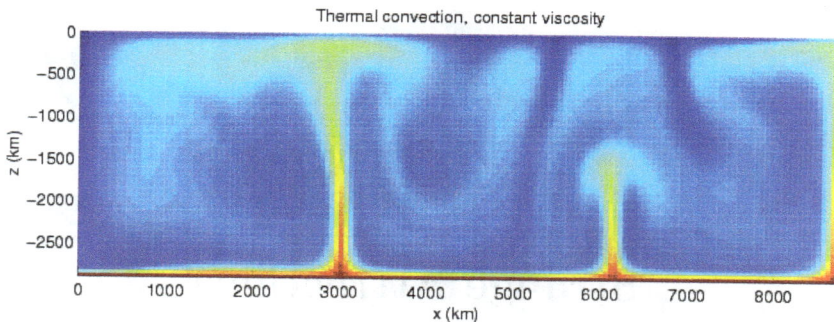

Figure 22.1.   Simulation of thermal convection in the Earth's mantle (by Har-roschmeling; image from Wikipedia, licensed under the Creative Commons Attribution-Share Alike 3.0 Unported license).

We recall that ρ in this framework is the density of the fluid, t is time and p is pressure.

To simplify this setting, one can perform some convenient approximations. First of all, the density is supposed to depend only on the temperature T, which in turn evolves according to the heat equation in (1.6), with no external sources, but with the caveat that the particles are moving with the velocity field. This gives that

$$\frac{d}{dt}T(x(t), z(t), t) = \kappa\Delta T$$

for some diffusion coefficient $\kappa > 0$; therefore, since $(\dot{x}(t), \dot{z}(t)) = v(x(t), z(t), t)$,

$$\partial_t T + v \cdot \nabla T = \kappa\Delta T.$$

We suppose that the higher the temperature, the lower the density of the fluid, which can be understood by considering that when the temperature increases, the molecules move faster and bump into each other more frequently, consequently spreading apart, taking up more space and reducing the density (and this reduction in density at higher temperatures is the reason for which high-temperature parcels of fluid happen to be lighter and have the tendency to move up). The simplest possible model to consider is that in which the density variation is linear with respect to the temperature, say $\rho = \rho(T) = \rho_0\big(1 - \alpha(T - \overline{T})\big)$, for some $\alpha > 0$, which we consider as a small parameter (that is, the density does depend on the temperature, but it is almost constant).

The other simplifying assumption is that the velocity and the velocity variations of the fluid are small. Therefore, we will disregard quadratic terms in v, as well as terms involving the product of the two small quantities α and v (or α and the derivatives of v). With this motivation, we have that

$$\rho\partial_t v + \rho(v \cdot \nabla)v = \rho_0\big(1 - \alpha(T - \overline{T})\big)\partial_t v + \rho_0\big(1 - \alpha(T - \overline{T})\big)(v \cdot \nabla)v$$
$$\simeq \rho_0\partial_t v.$$
$$(22.2)$$

These observations and (22.1) suggest[1] to study the problem

$$\begin{cases} \rho_0\partial_t v = \mu\Delta v - \rho_0\big(1 - \alpha(T - \overline{T})\big)g(0,1) - \nabla p & \text{in } \mathbb{R} \times (0, H), \\ \text{div } v = 0 & \text{in } \mathbb{R} \times (0, H), \\ v_z(x,0) = v_z(x, H) = 0, \\ \partial_t T + v \cdot \nabla T = \kappa\Delta T & \text{in } \mathbb{R} \times (0, H), \\ T(x,0) = \underline{T}, \\ T(x, H) = \overline{T}. \end{cases}$$
$$(22.3)$$

We observe that this system of equations possesses a simple solution in which the fluid is at rest, i.e. $v = 0$, and the temperature increases linearly, i.e. $T(x,z) = \underline{T} + \frac{(\overline{T} - \underline{T})z}{H}$. Indeed, substituting these relations in (22.3), we obtain the pressure prescription

$$\nabla p = -\rho_0\big(1 - \alpha(T - \overline{T})\big)g(0,1)$$
$$= -\rho_0\left[1 - \alpha(\underline{T} - \overline{T})\left(1 - \frac{z}{H}\right)\right]g(0,1),$$

---

[1]One can observe that, when passing from (22.1) to (22.3), essentially we have simply considered the density to be constant except when it appears in terms multiplied by the gravity acceleration g. In a sense, this approximation is based on the *ansatz* that the difference in density is negligible in itself, but the gravity field is supposed to be sufficiently strong to make it appreciable. This method is called the Boussinesq approximation.

which is satisfied by choosing

$$p(x,z) = p_0 - \rho_0 g \left[z - \alpha(\underline{T} - \overline{T}) \left(z - \frac{z^2}{2H} \right) \right],$$

for some $p_0 \in \mathbb{R}$.

It is therefore interesting to seek solutions for (22.3) which bifurcate from these basic states; for this, we look for solutions of (22.3) in which

$$T = \underline{T} + \frac{(\overline{T} - \underline{T})z}{H} + \Theta \quad \text{and}$$

$$p = p_0 - \rho_0 g \left[z - \alpha(\underline{T} - \overline{T}) \left(z - \frac{z^2}{2H} \right) \right] + P, \qquad (22.4)$$

for some unknown $\Theta = \Theta(x,z)$ and $P = P(x,z)$.

We insert (22.4) into (22.3), and we obtain the following system of equations for the new unknown functions:

$$\begin{cases} \rho_0 \partial_t v = \mu \Delta v + \alpha \rho_0 g \Theta\,(0,1) - \nabla P & \text{in } \mathbb{R} \times (0,H), \\ \operatorname{div} v = 0 & \text{in } \mathbb{R} \times (0,H), \\ v_z(x,0) = v_z(x,H) = 0, \\ \partial_t \Theta - \frac{(\underline{T}-\overline{T})v_z}{H} + v \cdot \nabla \Theta = \kappa \Delta \Theta & \text{in } \mathbb{R} \times (0,H), \\ \Theta(x,0) = 0, \\ \Theta(x,H) = 0. \end{cases} \qquad (22.5)$$

One can also consider a scaled version of (22.5) by setting

$$(\bar{x},\bar{z}) := \frac{(x,z)}{H}, \quad \bar{t} := \frac{\kappa t}{H^2}, \quad \bar{v}(\bar{x},\bar{z},\bar{t}) := \frac{Hv(H\bar{x},H\bar{z},H^2\kappa^{-1}\bar{t})}{\kappa}$$

$$\bar{\Theta}(\bar{x},\bar{z},\bar{t}) := \frac{\alpha\rho_0 g H^3 \Theta(H\bar{x},H\bar{z},H^2\kappa^{-1}\bar{t})}{\mu\kappa}$$

$$\text{and} \quad \bar{P}(\bar{x},\bar{z},\bar{t}) := \frac{H^2 P(H\bar{x},H\bar{z},H^2\kappa^{-1}\bar{t})}{\mu\kappa},$$

$$(22.6)$$

thus obtaining

$$
\begin{cases}
\partial_t \bar{v} = \dfrac{\mu}{\rho_0 \kappa}\left(\Delta_{\bar{x}\bar{z}}\bar{v} + \bar{\Theta}\,(0,1) - \nabla \bar{P}\right) & \text{in } \mathbb{R}\times(0,1), \\[2mm]
\operatorname{div}\bar{v} = 0 & \text{in } \mathbb{R}\times(0,1), \\[2mm]
\bar{v}_{\bar{z}}(x,0) = \bar{v}_{\bar{z}}(x,1) = 0, \\[2mm]
\partial_t \bar{\Theta} + \bar{v}\cdot\nabla\bar{\Theta} = \Delta\bar{\Theta} + \dfrac{\alpha \rho_0 g H^3\left(\underline{T}-\overline{T}\right)\bar{v}_{\bar{z}}}{\mu\kappa} & \text{in } \mathbb{R}\times(0,1), \\[2mm]
\bar{\Theta}(x,0) = 0, \\[2mm]
\bar{\Theta}(x,1) = 0.
\end{cases}
\tag{22.7}
$$

The parameter

$$
\mathcal{P} := \frac{\mu}{\rho_0 \kappa}
$$

is called the Prandtl number.

The parameter

$$
\mathcal{R} := \frac{\alpha \rho_0 g H^3\left(\underline{T}-\overline{T}\right)}{\mu\kappa}
$$

is called the Rayleigh number and takes into consideration the thermal gradient (note indeed that $\mathcal{R}$ is large when the temperature gap $\underline{T}-\overline{T}$ between the bottom and top of the fluid is large).

Hereinafter, to ease the notation, we drop the bars in (22.7). Also, as observed in (9.6), since we are considering a two-dimensional setting, the divergence-free condition in (22.7) allows us to write $v = (-\partial_z\psi, \partial_x\psi)$, for some scalar function $\psi : \mathbb{R}\times(0,1)\to\mathbb{R}$.

In this way, we have that

$$
v\cdot\nabla\Theta = (-\partial_z\psi, \partial_x\psi)\cdot(\partial_x\Theta, \partial_z\Theta) = -\partial_z\psi\partial_x\Theta + \partial_x\psi\partial_z\Theta
$$

$$
= \det\begin{pmatrix}\partial_x\psi & \partial_z\psi \\ \partial_x\Theta & \partial_z\Theta\end{pmatrix} =: \mathcal{J}(\psi,\Theta).
$$

Therefore, the temperature equation in (22.7) can be written in the form

$$
\partial_t\Theta + \mathcal{J}(\psi,\Theta) = \Delta\Theta + \mathcal{R}\partial_x\psi.
$$

As a result, (22.7) becomes

$$\begin{cases} \partial_t \partial_z \psi = \mathcal{P} \left(\Delta \partial_z \psi + \partial_x P \right) & \text{in } \mathbb{R} \times (0,1), \\ \partial_t \partial_x \psi = \mathcal{P} \left(\Delta \partial_x \psi + \Theta - \partial_z P \right) & \text{in } \mathbb{R} \times (0,1), \\ \partial_x \psi(x,0) = \partial_x \psi(x,1) = 0, \\ \partial_t \Theta + \mathcal{J}(\psi, \Theta) = \Delta \Theta + \mathcal{R} \partial_x \psi & \text{in } \mathbb{R} \times (0,1), \\ \Theta(x,0) = \Theta(x,1) = 0. \end{cases} \quad (22.8)$$

We now perform an additional approximation, by considering both ψ and Θ as small quantities, together with their derivatives, and thus dropping the quadratic term in (22.8). With this, we reduce the expression to

$$\begin{cases} \partial_t \partial_z \psi = \mathcal{P} \left(\Delta \partial_z \psi + \partial_x P \right) & \text{in } \mathbb{R} \times (0,1), \\ \partial_t \partial_x \psi = \mathcal{P} \left(\Delta \partial_x \psi + \Theta - \partial_z P \right) & \text{in } \mathbb{R} \times (0,1), \\ \partial_x \psi(x,0) = \partial_x \psi(x,1) = 0, \\ \partial_t \Theta = \Delta \Theta + \mathcal{R} \partial_x \psi & \text{in } \mathbb{R} \times (0,1), \\ \Theta(x,0) = \Theta(x,1) = 0. \end{cases} \quad (22.9)$$

We observe that

$$\text{the first two equations in (22.9) are}$$
$$\text{equivalent to } \partial_t \Delta \psi = \mathcal{P}(\Delta^2 \psi + \partial_x \Theta), \quad (22.10)$$

where Δ^2 is the Laplace operator applied twice.

Indeed, if the first two equations in (22.9) hold true, then

$$\partial_t \Delta \psi = \partial_x \left(\partial_t \partial_x \psi \right) + \partial_z \left(\partial_t \partial_z \psi \right)$$
$$= \partial_x \left(\mathcal{P}(\Delta \partial_x \psi + \Theta - \partial_z P) \right) + \partial_z \left(\mathcal{P}(\Delta \partial_z \psi + \partial_x P) \right)$$
$$= \mathcal{P}(\Delta^2 \psi + \partial_x \Theta).$$

Conversely, if the equation in (22.10) is satisfied, we consider the differential form

$$\omega := \left(\frac{1}{\mathcal{P}} \partial_t \partial_z \psi - \Delta \partial_z \psi \right) dx - \left(\frac{1}{\mathcal{P}} \partial_t \partial_x \psi - \Delta \partial_x \psi - \Theta \right) dz$$

and we observe that

$$dw = - \left[\partial_z \left(\frac{1}{\mathcal{P}} \partial_t \partial_z \psi - \Delta \partial_z \psi \right) \right.$$
$$\left. + \partial_x \left(\frac{1}{\mathcal{P}} \partial_t \partial_x \psi - \Delta \partial_x \psi - \Theta \right) \right] dx \wedge dz$$
$$= - \frac{1}{\mathcal{P}} \left[\partial_t \Delta \psi - \mathcal{P} \Delta^2 \psi - \mathcal{P} \partial_x \Theta \right] dx \wedge dz$$
$$= 0.$$

Consequently, using the Poincaré lemma (see e.g. [Con93, Theorem 8.3.8]), we obtain that there exists a scalar function P : $\mathbb{R} \times (0,1) \to \mathbb{R}$ for which $\omega = dP$, and hence

$$\frac{1}{\mathcal{P}} \partial_t \partial_z \psi - \Delta \partial_z \psi = \partial_x P \quad \text{and} \quad - \frac{1}{\mathcal{P}} \partial_t \partial_x \psi + \Delta \partial_x \psi + \Theta = \partial_z P,$$

which are precisely the first two equations in (22.9), thereby completing the proof of (22.10).

As a result, by (22.10), we can rewrite (22.9) in the form

$$\begin{cases} \partial_t \Delta \psi = \mathcal{P}(\Delta^2 \psi + \partial_x \Theta) & \text{in } \mathbb{R} \times (0,1), \\ \partial_x \psi(x,0) = \partial_x \psi(x,1) = 0, \\ \partial_t \Theta = \Delta \Theta + \mathcal{R} \partial_x \psi & \text{in } \mathbb{R} \times (0,1), \\ \Theta(x,0) = \Theta(x,1) = 0. \end{cases} \qquad (22.11)$$

Note that this formulation is convenient not only because it involves one less equation but also because we got rid of the dependence on the pressure, and hence we also have one less unknown.

One can now look for solutions in the form of exponential and trigonometric functions by taking the pressure to be constant. For instance, if we seek solutions for (22.11) in the form

$$\psi = \psi_0 \sin(qx) \sin(k\pi z) e^{\lambda t},$$
$$\Theta = \Theta_0 \cos(qx) \sin(k\pi z) e^{\lambda t}, \qquad (22.12)$$

for some $q \in \mathbb{R}$, $k \in \mathbb{N}$ and $\lambda \in \mathbb{C}$, when we insert these expressions into (22.11), we obtain that

$$\begin{cases} -\lambda(q^2 + k^2\pi^2)\psi_0 = \mathcal{P}((q^2 + k^2\pi^2)^2\psi_0 - q\Theta_0), \\ \lambda\Theta_0 = -(q^2 + k^2\pi^2)\Theta_0 + \mathcal{R}q\psi_0. \end{cases}$$

This system can be conveniently written in a matrix form by

$$MV = \lambda V, \quad \text{where } V := \begin{pmatrix} \psi_0 \\ \Theta_0 \end{pmatrix},$$

$$M := \begin{pmatrix} -\mathcal{P}\gamma^2 & \mathcal{P}q/\gamma^2 \\ \mathcal{R}q & -\gamma^2 \end{pmatrix} \quad \text{and} \quad \gamma := \sqrt{q^2 + k^2\pi^2}.$$

The question is now, *for what choice of physical parameters this system allows solutions corresponding to* $\lambda > 0$. This is extremely relevant in practice because, in such a situation, the basic state in (22.4), corresponding to a steady fluid with linearly increasing temperature, becomes unstable, due to the exponential in the time divergence in (22.12).

Thus, finding a positive eigenvalue λ for M is equivalent to looking for $\lambda > 0$ that solves

$$0 = (-\mathcal{P}\gamma^2 - \lambda)(-\gamma^2 - \lambda) - \frac{\mathcal{P}\mathcal{R}q^2}{\gamma^2} = \lambda^2 + (\mathcal{P}+1)\gamma^2\lambda + \frac{\mathcal{P}}{\gamma^2}(\gamma^6 - \mathcal{R}q^2),$$

which, upon solving in λ, gives

$$2\lambda = -(\mathcal{P}+1)\gamma^2 \pm \sqrt{(\mathcal{P}+1)^2\gamma^4 - \frac{4\mathcal{P}}{\gamma^2}(\gamma^6 - \mathcal{R}q^2)}$$

$$= -(\mathcal{P}+1)\gamma^2 \pm \frac{\sqrt{(\mathcal{P}-1)^2\gamma^6 + 4\mathcal{P}\mathcal{R}q^2}}{\gamma}.$$

Note that the above radicand is nonnegative and hence a real solution does exist; however, to have $\lambda > 0$, we must impose that

$$(\mathcal{P}+1)^2\gamma^6 < (\mathcal{P}-1)^2\gamma^6 + 4\mathcal{P}\mathcal{R}q^2,$$

that is,

$$\mathcal{R} > \frac{\gamma^6}{q^2}. \tag{22.13}$$

This condition thus ensures the Rayleigh number to be sufficiently large, i.e. when the thermal gradient between the bottom and top of the fluid is sufficiently large, the (approximated) convection problem presents new stable solutions which bifurcate from the basic state, as prescribed in (22.12). We also note that (22.13) can be written in the form

$$\mathcal{R} > \frac{(q^2 + k^2\pi^2)^3}{q^2},$$

the minimal threshold of $\mathcal{R} > \frac{27\pi^4}{4}$ corresponding to the choice of $k := 1$ and $q := \frac{\pi}{\sqrt{2}}$.

It is thus interesting to plot the level sets of the function $\cos\left(\frac{\pi x}{\sqrt{2}}\right)\sin(\pi z)$ since they correspond to the long-time dominant asymptotics of the temperature correction in (22.12), see Figure 22.2.

Actually, Figure 22.2 describes a typical regular pattern arising in convection phenomena, with the formation of distinctive convection cells known as Bénard cells, see e.g. Figure 22.3. See also [Get98] and

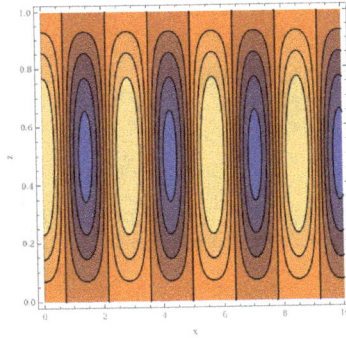

Figure 22.2. Level sets of the function $\cos\left(\frac{\pi x}{\sqrt{2}}\right)\sin(\pi z)$.

Figure 22.3. Altocumulus clouds formed by convective activity as seen from shuttle (Public Domain image from Wikipedia).

the references therein for additional information about convection patterns.

Interestingly, the description of the convection phenomena by Lord Rayleigh is intimately related to one of the paradigmatic models used in the study of chaotic dynamical systems. Indeed, mathematician and meteorologist Edward Norton Lorenz, in [Lor63], proposed to reconsider (22.11) as well as the nonlinear interactions of the form $(v \cdot \nabla)v$ and $\mathcal{J}(\psi, \Theta)$ that were dropped to have a simpler and more manageable expression (recall (22.2) and (22.8)).

This leads[2] to the study of the system

$$
\begin{cases}
\partial_t \Delta \psi + \mathcal{J}(\psi, \Delta \psi) = \mathcal{P}(\Delta^2 \psi + \partial_x \Theta) & \text{in } \mathbb{R} \times (0,1), \\
\partial_x \psi(x,0) = \partial_x \psi(x,1) = 0, \\
\partial_t \Theta + \mathcal{J}(\psi, \Theta) = \Delta \Theta + \mathcal{R}\partial_x \psi & \text{in } \mathbb{R} \times (0,1), \\
\Theta(x,0) = \Theta(x,1) = 0.
\end{cases}
\tag{22.14}
$$

---

[2]We point out that $\mathcal{J}(\psi, \Delta \psi)$ in (22.14) is obtained by retaining the term $\rho_0(v \cdot \nabla)v$ in (22.2). In this way, the first equation in (22.5) becomes

$$\rho_0 \partial_t v + \rho_0 (v \cdot \nabla)v = \mu \Delta v + \alpha \rho_0 g \Theta (0,1) - \nabla P \quad \text{in } \mathbb{R} \times (0, H);$$

therefore, after the scaling in (22.6) (and dropping the bars), the first equation in (22.7) becomes

$$\partial_t \bar{v} + (v \cdot \nabla)v = \mathcal{P}(\Delta v + \Theta(0,1) - \nabla P) \quad \text{in } \mathbb{R} \times (0,1).$$

Hence, using the divergence-free condition in (22.7), we write $v = (-\partial_z \psi, \partial_x \psi)$, for some scalar function $\psi : \mathbb{R} \times (0,1) \to \mathbb{R}$, and we replace the first two equations in (22.8) with

$$
\begin{cases}
\partial_t \partial_z \psi + \partial_x \psi \Delta \psi = \mathcal{P}(\Delta \partial_z \psi + \partial_x P) & \text{in } \mathbb{R} \times (0,1), \\
\partial_t \partial_x \psi - \partial_z \psi \Delta \psi = \mathcal{P}(\Delta \partial_x \psi + \Theta - \partial_z P) & \text{in } \mathbb{R} \times (0,1).
\end{cases}
$$

Assuming that both ψ and Θ are small quantities, we obtain that the first two equations in (22.9) become

$$
\begin{cases}
\partial_t \partial_z \psi + \partial_x \psi \Delta \psi = \mathcal{P}(\Delta \partial_z \psi + \partial_x P) & \text{in } \mathbb{R} \times (0,1), \\
\partial_t \partial_x \psi - \partial_z \psi \Delta \psi = \mathcal{P}(\Delta \partial_x \psi + \Theta - \partial_z P) & \text{in } \mathbb{R} \times (0,1).
\end{cases}
$$

These equations are equivalent to

$$\partial_t \Delta \psi + \mathcal{J}(\psi, \Delta \psi) = \mathcal{P}(\Delta^2 \psi + \partial_x \Theta),$$

which is the first equation in (22.14).

Inspired by (22.12), Lorenz proposed to look for approximate solutions in the form

$$\psi = \mathcal{A}(t) \sin(qx) \sin(\pi z),$$
$$\Theta = \mathcal{B}(t) \cos(qx) \sin(\pi z) + \mathcal{C}(t) \sin(2\pi z), \tag{22.15}$$

for some $q \in \mathbb{R}$ and functions $\mathcal{A}$, $\mathcal{B}$ and $\mathcal{C}$ to be determined.

Note that, with this choice, the nonlinear term $\mathcal{J}(\psi, \Delta\psi) = 0$.

By plugging (22.15) into (22.14), one obtains

$$\begin{cases} -(q^2 + \pi^2)\dot{\mathcal{A}} = \mathcal{P}((q^2 + \pi^2)^2\mathcal{A} - q\mathcal{B}), \\ \left[\dot{\mathcal{B}} + (q^2 + \pi^2)\mathcal{B} - \mathcal{R}q\mathcal{A}\right] \cos(qx) \sin(\pi z) \\ \quad + \left[\dfrac{\pi q\mathcal{A}\mathcal{B}}{2} + \dot{\mathcal{C}} + 4\pi^2\mathcal{C}\right] \sin(2\pi z) \\ \quad + 2\pi q\mathcal{A}\mathcal{C} \cos(qx) \sin(\pi z) \cos(2\pi z) = 0. \end{cases} \tag{22.16}$$

It is also useful to apply the identity

$$2 \sin(\pi z) \cos(2\pi z) = \frac{(e^{\pi z} - e^{-\pi z})(e^{2\pi z} + e^{-2\pi z})}{2i}$$

$$= \frac{e^{3\pi z} - e^{-3\pi z} - e^{\pi z} + e^{-\pi z}}{2i} = \sin(3\pi z) - \sin(\pi z)$$

to manipulate the last term in (22.16) and reduce the expression to

$$\begin{cases} -(q^2 + \pi^2)\dot{\mathcal{A}} = \mathcal{P}((q^2 + \pi^2)^2\mathcal{A} - q\mathcal{B}), \\ \left[\dot{\mathcal{B}} + (q^2 + \pi^2)\mathcal{B} - \mathcal{R}q\mathcal{A} - \pi q\mathcal{A}\mathcal{C}\right] \cos(qx) \sin(\pi z) \\ \quad + \left[\dfrac{\pi q\mathcal{A}\mathcal{B}}{2} + \dot{\mathcal{C}} + 4\pi^2\mathcal{C}\right] \sin(2\pi z) \\ \quad + \pi q\mathcal{A}\mathcal{C} \cos(qx) \sin(3\pi z) = 0. \end{cases} \tag{22.17}$$

Now, comparing with the approximation put forth in (22.15), the term containing $\sin(3\pi z)$ has a spatial dependence of higher frequency than what the previous assumption was accounting for.

Therefore, the suggestion by Lorenz was to *drop this term* and simplify (22.17) into

$$\begin{cases} -(q^2 + \pi^2)\dot{\mathcal{A}} = \mathcal{P}((q^2 + \pi^2)^2 \mathcal{A} - q\mathcal{B}), \\[2mm] \left[\dot{\mathcal{B}} + (q^2 + \pi^2)\mathcal{B} - \mathcal{R}q\mathcal{A} - \pi q \mathcal{A} \mathcal{C}\right] \cos(qx)\,\sin(\pi z) \\[2mm] \qquad + \left[\dfrac{\pi q \mathcal{A}\mathcal{B}}{2} + \dot{\mathcal{C}} + 4\pi^2 \mathcal{C}\right] \sin(2\pi z) = 0, \end{cases}$$

leading to

$$\begin{cases} -(q^2 + \pi^2)\dot{\mathcal{A}} = \mathcal{P}((q^2 + \pi^2)^2 \mathcal{A} - q\mathcal{B}), \\[2mm] \dot{\mathcal{B}} = \mathcal{R}q\mathcal{A} - (q^2 + \pi^2)\mathcal{B} + \pi q \mathcal{A}\mathcal{C}, \\[2mm] -\dot{\mathcal{C}} = \dfrac{\pi q \mathcal{A}\mathcal{B}}{2} + 4\pi^2 \mathcal{C}. \end{cases} \qquad (22.18)$$

To make these equations look more attractive, it is customary to perform some cosmetic modifications by defining

$$\omega := \frac{q^2}{(q^2 + \pi^2)^3}, \quad \sigma := \mathcal{P}, \quad \beta := \frac{4\pi^2}{q^2 + \pi^2}, \quad \rho := \omega\mathcal{R},$$

$$X(t) := \frac{\pi q}{\sqrt{2}(q^2 + \pi^2)} \mathcal{A}\left(\frac{t}{q^2 + \pi^2}\right), \quad Y(t) := \frac{\pi \omega}{\sqrt{2}} \mathcal{B}\left(\frac{t}{q^2 + \pi^2}\right)$$

and $\quad Z(t) := -\pi\omega\mathcal{C}\left(\dfrac{t}{q^2 + \pi^2}\right).$

With this, using the notation $\tau := \frac{t}{q^2 + \pi^2}$, one deduces from (22.18) that

$$\dot{X}(t) = \frac{\pi q}{\sqrt{2}(q^2 + \pi^2)^2} \dot{\mathcal{A}}(\tau)$$

$$= -\frac{\pi \mathcal{P} q}{\sqrt{2}(q^2 + \pi^2)^3} \left((q^2 + \pi^2)^2 \mathcal{A}(\tau) - q\mathcal{B}(\tau)\right)$$

$$= \frac{\pi \mathcal{P} q^2}{\sqrt{2}(q^2 + \pi^2)^3} \mathcal{B}(\tau) - \frac{\pi \mathcal{P} q}{\sqrt{2}(q^2 + \pi^2)} \mathcal{A}(\tau)$$

$$= \frac{\mathcal{P} q^2}{\omega(q^2 + \pi^2)^3} Y(t) - \mathcal{P}X(t)$$

$$= \sigma(Y(t) - X(t)),$$

that

$$\dot{Y}(t) = \frac{\pi \omega}{\sqrt{2}(q^2 + \pi^2)} \dot{\mathcal{B}}(\tau)$$

$$= \frac{\pi \omega}{\sqrt{2}(q^2 + \pi^2)} \left(\mathcal{R} q \mathcal{A}(\tau) - (q^2 + \pi^2)\mathcal{B}(\tau) + \pi q \mathcal{A}(\tau)\mathcal{C}(\tau) \right)$$

$$= \frac{\pi \omega}{\sqrt{2}(q^2 + \pi^2)} \left(\frac{\sqrt{2}\mathcal{R}(q^2 + \pi^2)}{\pi} X(t) \right.$$

$$\left. - \frac{\sqrt{2}(q^2 + \pi^2)}{\pi \omega} Y(t) - \frac{\sqrt{2}(q^2 + \pi^2)}{\pi \omega} X(t) Z(t) \right)$$

$$= \rho X(t) - Y(t) - X(t) Z(t)$$

and that

$$\dot{Z}(t) = -\frac{\pi \omega}{q^2 + \pi^2} \dot{\mathcal{C}}(\tau)$$

$$= \frac{\pi \omega}{q^2 + \pi^2} \left(\frac{\pi q \mathcal{A}(\tau)\mathcal{B}(\tau)}{2} + 4\pi^2 \mathcal{C}(\tau) \right)$$

$$= \frac{\pi \omega}{q^2 + \pi^2} \left(\frac{q^2 + \pi^2}{\pi \omega} X(t) Y(t) - \frac{4\pi}{\omega} Z(t) \right)$$

$$= X(t) Y(t) - \beta Z(t).$$

Hence, we can write (22.18) in the simpler form

$$\begin{cases} \dot{X} &= \sigma(Y - X), \\ \dot{Y} &= X(\rho - Z) - Y, \\ \dot{Z} &= XY - \beta Z. \end{cases} \qquad (22.19)$$

The set of these three ordinary differential equations is known as the Lorenz system. Note that ρ is proportional to the Rayleigh number $\mathcal{R}$; hence, recalling (22.13), we may suspect that the specific value of ρ may play an important role in a possible bifurcation diagram for this system.

This is indeed the case, though a full understanding of the complexity of the Lorenz system overcomes the present knowledge on the subject: as pointed out in [HSD13, Section 15.1], "we are decades (if not centuries) away from rigorously understanding all of the fascinating dynamical phenomena that occur as the parameters change".

A first glimpse of the importance of the values of the parameters in (22.19) can be understood by studying the equilibria of the system. For example (see e.g. [HSD13, Section 14.2]), one can show that (22.19) always presents the origin as an equilibrium and two additional equilibria

$$Q_\pm := \left(\pm\sqrt{\beta(\rho - 1)}, \pm\sqrt{\beta(\rho - 1)}, \rho - 1 \right)$$

when $\rho > 1$. When $\rho < 1$, the origin is a sink, but it becomes a saddle when $\rho > 1$. Assuming that $\sigma > 1 + \beta$, the equilibria in $Q_\pm$ are sinks when $\rho \in (1, \rho_\star)$, with

$$\rho_\star := \frac{\sigma(\sigma + \beta + 3)}{\sigma - \beta - 1}, \qquad (22.20)$$

but when $\rho > \rho_\star$, the eigenvalues of the linearized system at $Q_\pm$ become purely imaginary, and in very broad terms, the local dynamics tend to circulate around these points, with opposite velocities.

Also, the system shrinks volumes exponentially fast, and trajectories are confined within a bounded region (see e.g. [HSD13, p. 309]), thus suggesting that the system presents an "attractor" (or in rough terms, a set to which trajectories asymptotically approach in the course of dynamic evolution). Actually, before the introduction of the Lorenz system, the only types of stable attractors known in differential equations were equilibria and closed orbits, while the Lorenz system exhibited a very different structure: namely, for certain values of the parameters, the Lorenz system possesses what has come to be known as a "strange attractor" (see e.g. [Tuc02, Vis04]), that is, roughly, an object of fractal dimension which is strongly sensitive to initial conditions.

Without going into the difficult details of this matter, let us mention that, as a model case, Lorenz considered $\sigma := 10$ and $\beta := \frac{8}{3}$, for which (22.20) produces $\rho_\star = \frac{470}{19} = 24.7368421\ldots$; see Figure 22.4 for a plot of the relatively regular dynamics occurring[3]

---

[3]See also https://itp.uni-frankfurt.de/~gros/Vorlesungen/SO/simulation_example/; http://www.malinc.se/m/Lorenz.php; https://fusion809.github.io/Lorenz/ for interactive animated simulations of the Lorenz system.

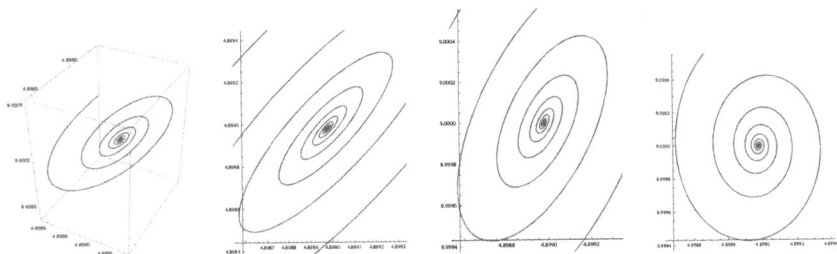

Figure 22.4. Mathematica simulation of the trajectory of the Lorenz system with parameters $\sigma := 10$, $\rho := 10$, $\beta := \frac{8}{3}$, starting at the point $(10, 10, 10)$ (three-dimensional plot and projections onto the coordinate planes).

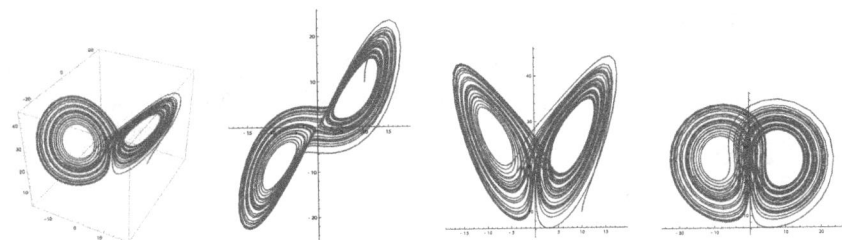

Figure 22.5. Mathematica simulation of the trajectory of the Lorenz system with parameters $\sigma := 10$, $\rho := 28$, $\beta := \frac{8}{3}$, starting at the point $(10, 10, 10)$ (three-dimensional plot and projections onto the coordinate planes).

when $\rho := 10 < \rho_\star$. Instead, for larger values of ρ, a number of complex phenomena arise, see Figure 22.5 for a sketch of the case $\rho := 28 > \rho_\star$, which was the one originally presented[4] by Lorenz:

---

[4]Concerning the traditional choice of $\sigma := 10$, $\rho := 28$ and $\beta := \frac{8}{3}$, as stressed in [BSF13], "there is nothing magical about these values; indeed, any other values within a fairly wide range would produce qualitatively similar behavior": in [BSF13], it is also stressed that some authors have however tried to infer from these rather arbitrary values a "positivity ratio" which, with apparently "no theoretical or empirical justification for the use of differential equations drawn from fluid dynamics", should predict "an individual's degree of flourishing". The authors introducing this questionable positivity ratio accomplished however some kind of academic prestige since this ratio "has had an extensive influence on the field of positive psychology", their work "has been frequently cited", and popular books are "devoted to expounding this *huge discovery*, which has also been enthusiastically brought to a wider audience" in dissemination books.

Well, sometimes producing an easy and user-friendly recipe to cope with complex and very topical problems does pay off (but sometimes it doesn't).

note in this picture the very complicated structure of the trajectory, which seems repeatedly to approach and then spiral away from one of the two equilibria $Q_\pm$ to move close to the other, then escaping again toward the previous one, jumping back and forward on and on, producing dense leaves which are interwoven in a rather tangled way.

Additionally, the resulting attractor seems to be highly sensitive[5] to initial conditions, showing a great deal of unpredictability which is typically considered one of the main ingredients of "chaos". To understand this sensitivity, we can look at Figure 22.6, which plots the X-trajectories of the Lorenz system starting at the points $(10, 10, 10)$ and $\left(10 + \frac{1}{100}, 10, 10\right)$. While these two trajectories are almost indistinguishable for small times (e.g. about $t = 18$, in the left images), they become completely different and essentially unrelated as time flows (e.g. for $t \in [18, 50]$ in the images on the right).

Note, however, that while a tiny change in the initial position may result in drastic changes in the eventual behavior of the orbits, the trajectories end up approaching the same attractor, just traveled at different instants of time: this is shown for instance in Figure 22.7, in which the global geometric pattern described by the journey of the two trajectories happen to be quite similar.

Moreover, the Lorenz system seems to show alternating patterns of chaos and periodic motion, depending on the values of ρ (see e.g. [Spa82, Chapter 4]), with "windows" of the values of ρ allowing for periodic behaviors, see Figure 22.8 for an example of this situation.

Furthermore, for very large values of ρ, perhaps quite surprisingly, the dynamics of the Lorenz system simplifies again, with globally attracting limit cycles (see [MS06, p. 109] and the references therein). This phenomenon is showcased here in Figures 22.9 and 22.10, in which the periodic limit trajectory appears quite distinctly.

---

[5]Lorenz himself was astonished by the discovery of the high sensitivity on the initial data. In his own words (see https://eapsweb.mit.edu/sites/default/files/Scientist_by_Choice.pdf): "At first I suspected trouble with the computer, which occurred fairly often, but, when I compared the new solution step by step with the older one, I found that at first the solutions were the same, and then they would differ by one unit in the last decimal place, and then the differences would become larger and larger".

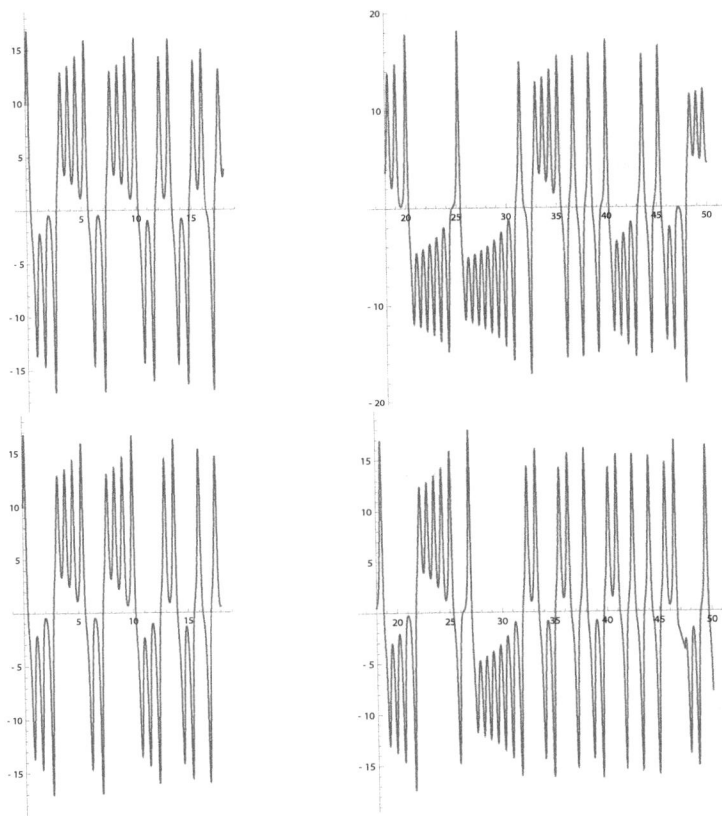

Figure 22.6. Mathematica simulation of the X-trajectories of the Lorenz system with parameters $\sigma := 10$, $\rho := 28$, $\beta := \frac{8}{3}$, starting at the points $(10, 10, 10)$ (above) and $\left(10 + \frac{1}{100}, 10, 10\right)$ (below).

We stress, however, that these claims are at the moment mostly based only on numerical experiments, and we are still in need of complete analytic proofs.

Interestingly, there is also a simple and practical mechanical model of a particular case of the Lorenz system, which was developed by Willem Malkus in the 1960s and consists of a chaotic waterwheel. The model consists indeed of a toy waterwheel with cups. A water source at the top pours a constant flow in at the top bucket, making the wheel turn. At low flow rates, the wheel turns in one direction or the other, but at high flow rates, the wheel spins one way, and

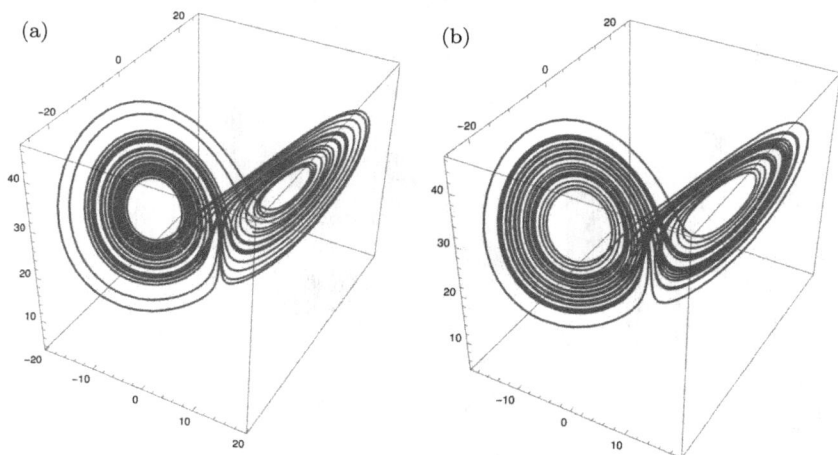

Figure 22.7. Mathematica simulation of the trajectories of the Lorenz system with parameters $\sigma := 10$, $\rho := 28$, $\beta := \frac{8}{3}$, starting at the points $(10, 10, 10)$ (a) and $\left(10 + \frac{1}{100}, 10, 10\right)$ (b) for $t \in [18, 50]$.

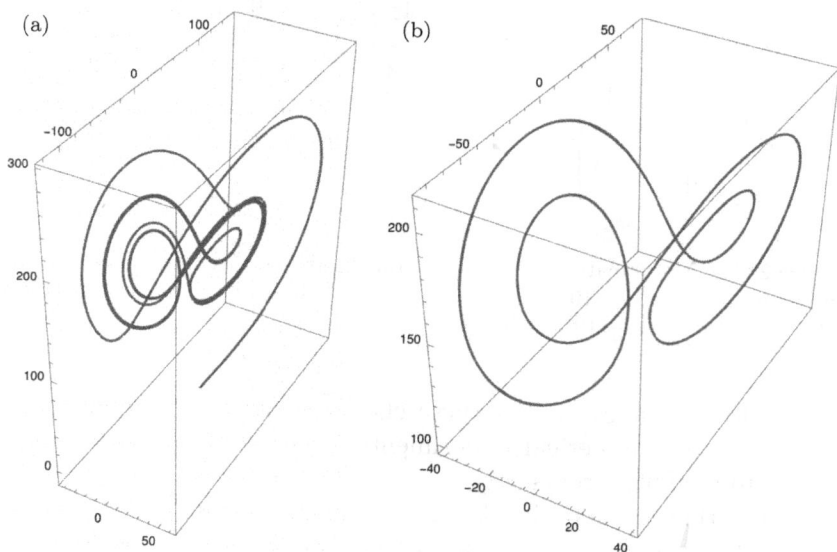

Figure 22.8. Mathematica simulations of the trajectories of the Lorenz system with parameters $\sigma := 10$, $\rho := 155$, $\beta := \frac{8}{3}$, starting at the point $(10, 10, 10)$ with $t \in [0, 100]$ (a) and $t \in [90, 100]$ (b).

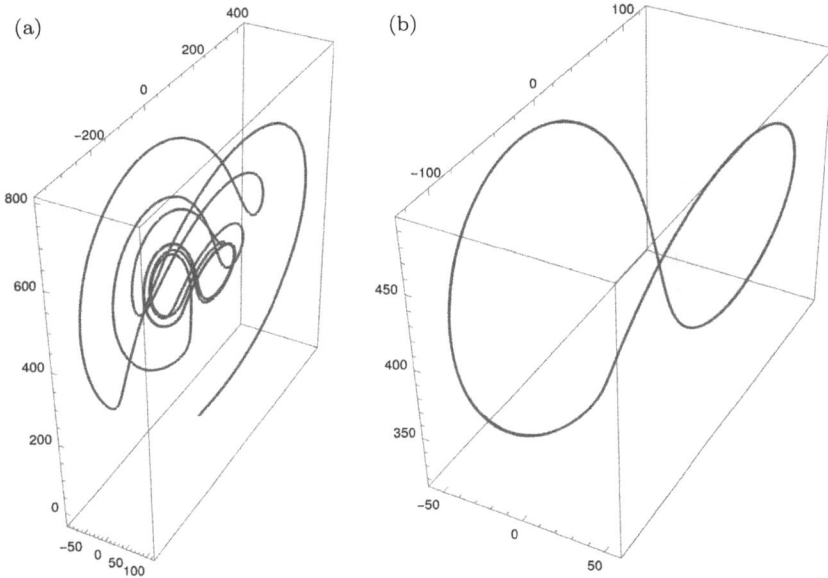

Figure 22.9.   Mathematica simulations of the trajectories of the Lorenz system with parameters $\sigma := 10$, $\rho := 400$, $\beta := \frac{8}{3}$, starting at the point $(10, 10, 10)$ with $t \in [0, 100]$ (a) and $t \in [90, 100]$ (b).

then the other, since water-filled cups are heavier, oppose the spinning of the wheel and make it turn the other way. Furthermore, the waterwheel exhibits a sensitive dependence on the initial data, see e.g. Figure 22.11, in which the initial angles of the wheels differ by only 1 degree, but the evolution of the plotted centers of mass, as well as the disposition of water in the buckets, diverge significantly in time. The unpredictable manner in which spins reverse orientation in this model is precisely a manifestation of the chaotic patterns exhibited by the Lorenz system, see [Mat07], [Str15, Section 9.1] and the references therein for a detailed mathematical discussion of the chaotic waterwheel and its link to the Lorenz system.

All in all, the analysis pioneered by Lord Rayleigh and Edward Lorenz was not only important towards gaining a partial understanding of a rather simplified model for convection, it also truly opened up new horizons, showcasing a number of novel and breathtaking

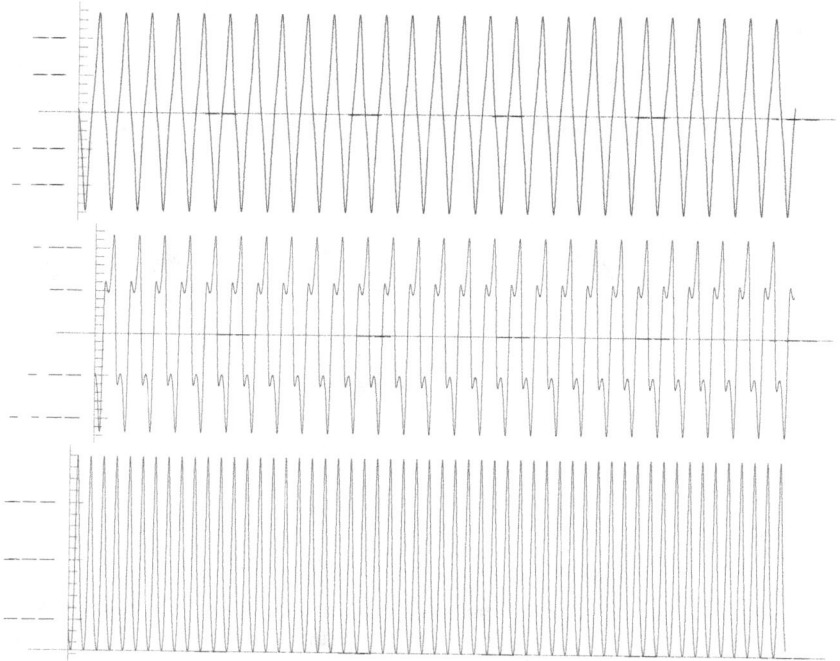

Figure 22.10. Mathematica simulations of the trajectories X, Y and Z of the Lorenz system with parameters $\sigma := 10$, $\rho := 400$, $\beta := \frac{8}{3}$, starting at the point $(10, 10, 10)$ with $t \in [90, 100]$.

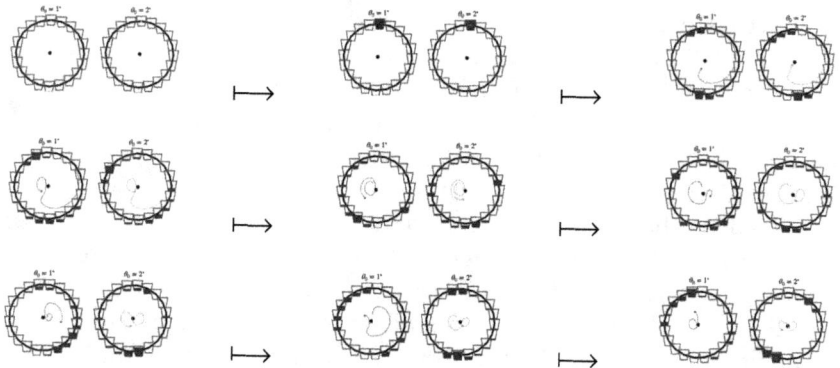

Figure 22.11. Simulation of two waterwheels with initial angles differing by 1 degree (extract from a video by Aiyopasta from Wikipedia, licensed under the Creative Commons Attribution-Share Alike 4.0 International license).

phenomena which were later discovered virtually in every area of science.

See [Spa82, HSD13, Str15] for a thorough introduction to the dynamics of the Lorenz system. For additional information about the close relationship between convection models in fluids and the Lorenz system, see e.g. [BPV86, Hil94, CP06, PMSB21] and the references therein.

Chapter 23

Who Wants to Be a Millionaire?

One way to become a millionaire is to solve a Millennium Prize Problem (see p. 62 here and Section 3.3 of the companion book [DV23]), but this is probably by far the most difficult way to get rich. Also, one should take into consideration the possibility that mathematics may not make us rich, but it makes us happy, which is invaluable.

There are however truly outstanding mathematicians, such as Jim Simons (see Figure 23.1), who, after having revolutionized the theory of minimal surfaces and the topological quantum field theory and also after having broken codes for the NSA during the Cold War, decided to improve his portfolio: allegedly, Simons' net worth is estimated to be US$25.2 billion, making him the 66th-richest person in the world and likely the richest[1] among the mathematicians. His financial

---

[1]Simons has not forgotten his love for mathematics, not only because he owns a motor yacht named Archimedes, but also because he co-founded the Simons Foundation, which supports projects related to scientific research, education and health, and he also sponsored the Simons Institute for the Theory of Computing and the Mathematical Sciences Research Institute at Berkeley.

Here is an interesting story by the way. While Simons was working at the Institute for Defense Analyses by cracking Russian codes, his boss General Maxwell Taylor wrote an article in *The New York Times* magazine strongly supporting the Vietnam War. Then, Simons published a counter-editorial in *The Times*, claiming that not everyone who worked for Taylor subscribed to his views. This resulted in an interview with Simons by *Newsweek*. When Simons told his superiors about this interview, they fired him right away.

Allegedly, Simons said, "Getting fired once can be a good experience. You just don't want to make a habit of it".

Figure 23.1.   James Harris Simons in 2007 (image from Wikipedia, licensed under the Creative Commons Attribution-Share Alike 2.0 Germany license).

success is related to the foundation of a quantitative hedge fund (see the following for the notion of hedging) which trades using quantitative mathematical models, so mathematics can be sometimes useful after all!

One of the chief uses of partial differential equations in mathematical finance is related to the Black–Scholes equation

$$\frac{\partial V}{\partial t} + \frac{1}{2}\sigma^2 S^2 \frac{\partial^2 V}{\partial S^2} = rV - rS\frac{\partial V}{\partial S}, \qquad (23.1)$$

named after Fischer Black and Myron Scholes.

The meaning of this equation is the following. Suppose that we have a stock (that is, a security that represents the ownership of a fraction of a corporation) whose price is denoted by S. The price of a stock fluctuates based on supply and demand (if more people want to buy than sell it, the price will rise), hence we take $S = S(t)$ to be a function of time.

To model the fluctuations of S in the simplest possible way, we can make two assumptions.

On the one hand, we can assume that the stock refers to some company with revenues, earnings, dividends, etc., and we can estimate the growth rate of the company and of the corresponding stocks; for this, we assume that the variation in S in the unit of time is proportional, by some factor μ, to the value of S (the more stocks one possesses, and the higher the coefficient μ, the more one receives from the growth of the company's revenues; actually positive values of μ correspond to growth and negative ones to degrowth).

On the other hand, the market presents a number of uncertainties, which are almost impossible to fully take into account, such as economic crises, political developments, real estate bubbles, technological innovations, pandemics, and landing of aliens from outer space, etc. If we have no specific information on the matter, we can simply assume that this volatility is modeled by a Brownian motion (pretty much like the random walk presented on p. 19). This is accounted for by a process $W(t)$ that "randomly wiggles up and down" and that quantifies the source of uncertainty in the price. This term is modulated by a volatility coefficient $\sigma \in [0, +\infty)$ (the higher this coefficient, the bigger the impact of the uncertainties in the price oscillation).

Assuming that these two effects contribute somewhat consistently to the variation in time of the stock, we therefore write that[2]

$$dS = \mu S\, dt + \sigma S\, dW. \qquad (23.2)$$

Strictly, this is not really a differential equation since we cannot "divide both terms by dt", as Brownian motions are "not differentiable"; see e.g. [Eva13, Section 3.4.2] for a precise statement and for an elegant introduction to stochastic calculus. Indeed, (23.2) is a stochastic differential equation, which is a type of equation that we do not treat here, considering it only at an intuitive level. Basically,

---

[2]Equation (23.2) is sometimes called the "geometric Brownian motion" equation. It is a very popular tool in quantitative finance since it relies on the simple, but often reasonable, assumption that the relative variation $\frac{dS}{S}$ of a value follows a diffusive process.

the only bit of information that we need from stochastic calculus is the so-called[3] Itô's chain rule (see e.g. Sections 1.3, 4.3 and 4.4 as well as, in particular, the theorem on p. 80 of [Eva13]), according to which if $dX = F\,dt + G\,dW$, then the random variable $H(X(t), t)$ satisfies

$$dH(X, t) = \left(\partial_t H + F \partial_X H + \frac{G^2}{2} \partial_X^2 H \right) dt + G \partial_X H\, dW. \quad (23.3)$$

Let us now return to our financial setting. Suppose that we perceive the buying of a given stock as too risky and we prefer instead to buy only "the possibility of buying the stock in the future", if this turns out to be convenient. This type of contract is called an "option" and conveys its owner the right, but not the obligation, to buy the

---

[3]Kiyosi Itô was the pioneer of the stochastic calculus, who lends his name to this equation. See Figure 23.2 for a picture of Kiyosi Itô at age 22 (Kiyosi is the second from the left; the first from the left is his brother Seizô, who also became a mathematician).

The heuristic explanation of Itô's chain rule (23.3) is that, formally, "$(dW)^2 = dt$", in agreement with the fact that, in the random walk, the unit of time depends "quadratically" on the unit of space, according to (2.2). This (by keeping linear orders in dt and up to quadratic in dW) leads to

$$H(X + dX, t + dt) - H(X, t)$$

$$= \partial_X H\, dX + \partial_t H\, dt + \frac{1}{2} \partial_X^2 H\, (dX)^2 + \frac{1}{2} \partial_t^2 H\, (dt)^2$$
$$\quad + \partial_{Xt}^2 H\, dX\, dt + o((dX)^2 + (dt)^2)$$

$$= \partial_X H\, (F\,dt + G\,dW) + \partial_t H\, dt + \frac{1}{2} \partial_X^2 H\, (F\,dt + G\,dW)^2 + \frac{1}{2} \partial_t^2 H\, (dt)^2$$
$$\quad + \partial_{Xt}^2 H\, (F\,dt + G\,dW)\, dt + o(dt\,dW + (dW)^2 + (dt)^2)$$

$$= \partial_X H\, (F\,dt + G\,dW) + \partial_t H\, dt + \frac{G^2}{2} \partial_X^2 H\, (dW)^2 + o(dt + (dW)^2)$$

$$= \left(\partial_t H + F \partial_X H + \frac{G^2}{2} \partial_X^2 H \right) dt + G \partial_X H\, dW + o(dt),$$

that gives Itô's Chain Rule (23.3).

See e.g. [Eva13] and the references therein for a more exhaustive treatment of Itô's chain rule and a more accurate approach to its proof.

Figure 23.2. The Itô family (Public Domain image from Wikipedia).

stock at a specified strike price either prior to or on a specified date. For simplicity, let us consider the case in which the option allows, but not obliges, the holder to buy the stock for a specified price K (called "strike price" in jargon) at a specific time T, which is on the date of option maturity: these options are called in jargon "European options"; the options which instead can be exercised any time up to and including the date of expiration are called "American options" (perhaps the name originated from the alleged fact that one type of option was commonly traded in London and the other in New York).

We denote by V the price of this option, and we suppose that V depends on time and the value S of the underlying stock, that is, $V = V(t, S)$, for t less than the maturity time T. For instance, at the maturity date, the value of this option is either $S(T) - K$, namely the difference between the actual value of the stock S at time T and the strike price K that was agreed upon at the beginning, if $S(T) - K$ is positive (in which case the option turned out to be a convenient investment), or null, if $S(T) - K$ is negative (in this case, the value of the stock is lower than the strike price, so no need to use the option to pay more, and if one is interested in the stock, they can just forget about the option and buy the stock at the market price). More specifically, these observations yield that[4]

$$V(T, S) = \max\{S - K, 0\}. \tag{23.4}$$

Our objective is thus to understand the time evolution of the value V of the option; this is important since if one possesses an option and may decide to sell it at some time t, knowing just the strike price K agreed upon at the beginning and the present value of the stock $S(t)$, hence it is relevant to establish a fair price $V(t, S(t))$ for the option under this information (since one would like to answer this question for all possible values of the stocks, this reduces to understanding $V(t, S)$ only in knowing the strike price K, with this one would substitute $S = S(t)$ to obtain the fair price of the option).

To model the time evolution of V and thus provide a convincing derivation of the Black–Scholes equation in (23.1), we use a method called in finance "delta hedging"; namely, we adopt a trading strategy that reduces, or "hedges", the risk related to the volatility of the process. For this, we suppose that we buy a quantity $\alpha = \alpha(t, S(t))$ of options and some quantity $\beta = \beta(t, S(t))$ of the stocks (here, α and β are real, negative numbers, formally corresponding to options

---

[4]For American options, one also has that

$$V(t, S(t)) \geqslant \max\{S(t) - K, 0\}$$

for all $t \in [0, T]$, the inequality coming from the fact that one can, but is not obliged to, utilize the option at time t, and this provides an "obstacle problem" for equation (23.1) since it forces solutions of (23.1) to stay above a constraint (or to drop the equation when they meet the constraint; this situation is significantly more complicated than that of a single equation, and this is the reason for which we limited ourselves here to the case of European options).

With respect to (23.4), we may consider it as a "terminal condition" for equation (23.1).

Interestingly, with respect to the classification discussed in footnote 13 on p. 9, we have that equation (23.1) is of parabolic type, but the sign in front of the time derivative is opposite to the case of the heat equation (compare with (1.6); a closer relation between the Black–Scholes equation and the heat equation will be discussed on p. 249). In this sense, equation (23.1) shares some similarities with the heat-type equations but going backward in time. This makes sense since, while the heat equation smooths out differences as time flows, the value of an option becomes quite rigid close to its maturity date.

or stock that are actually sold). The goal is to choose α and β to fully cover the risk (or, at least, to cover the risk to the best of our possibilities). More precisely, we consider a portfolio $P := \alpha V + \beta S$, and we compare this investment with another one that carries zero risk, that is, some operation which ensure a certain (or, more realistically, almost certain) future return and (virtually) no possibility of loss: for instance, a treasury bill, or treasury bond, namely a bond in which the face value is repaid at the time of maturity with some interest. We denote by $r \geqslant 0$ the risk-free interest rate (which will be precisely the parameter r appearing in (23.1)). The value of this zero-risk portfolio P_0 is thus described by $dP_0 = r P_0 \, dt$.

Since, via a suitable choice of α and β, the portfolio P is set up to have zero risk, it must be equivalent to the portfolio P_0; otherwise, there would be a better choice to gain money at zero risk! This also states that $dP = r P \, dt$ and, therefore,

$$\left(r\alpha V + r\beta S \right) dt = r P \, dt = dP. \tag{23.5}$$

We stress that the portfolio P is assumed to be self-financing, namely there are no inflows or outflows of money. More specifically, at every instant of time, the purchase of a new option or stock must be financed by the sale of an old one; hence, since there is no external infusion or withdrawal of money, the variation in the value of the portfolio corresponds precisely to the variation in the values of the options and stocks possessed, that is,

$$dP = \alpha \, dV + \beta \, dS. \tag{23.6}$$

We stress that this does not mean necessarily that α and β are constants, only that the market value of the portfolio P at a given time equals the purchase value of the new portfolio. Namely, if in the infinitesimal interval of time dt, the value of the options has increased by dV and the value of the stocks by dS, then the gain obtained would correspond to the number of options possessed times the options' increment value (that is, α times dV) plus the number of stocks possessed times the stocks' increment value (that is, β times dS). This gives that the total gain in an infinitesimal interval of time dt is $\alpha \, dV + \beta \, dS$, which is precisely the right-hand side of (23.6). Accordingly, the balance prescribed by (23.6) states that this gain is used precisely to reinvest to enlarge the portfolio itself (this, in the optimistic scenario that there is a gain; if there is a loss, then the right-hand side of (23.6) is negative and the portfolio needs to be correspondingly shrunk).

Now, we let ϖ to be the ratio[5] between β and α; namely, we take $\varpi = \varpi(t, S(t))$ such that $\beta = \alpha\varpi$. Then, (23.5) can be stated in the form

$$\left(r\alpha V + r\alpha\varpi S\right) dt = \left(r\alpha V + r\beta S\right) dt$$
$$= dP = \alpha\, dV + \beta\, dS = \alpha\, dV + \alpha\varpi\, dS$$

and, therefore (assuming for simplicity $\alpha \neq 0$),

$$\left(rV + r\varpi S\right) dt = dV + \varpi\, dS. \tag{23.7}$$

We also point out that

$$dV = \left(\partial_t V + \mu S \partial_S V + \frac{\sigma^2 S^2}{2} \partial_S^2 V\right) dt + \sigma S \partial_S V\, dW,$$

thanks to (23.3).

From this, (23.2) and (23.7), we infer that

$$0 = dV + \varpi\, dS - \left(rV + r\varpi S\right) dt$$
$$= \left(\partial_t V + \mu S \partial_S V + \frac{\sigma^2 S^2}{2} \partial_S^2 V\right) dt$$
$$\quad + \sigma S \partial_S V\, dW + \varpi\, dS - \left(rV + r\varpi S\right) dt$$
$$= \left(\partial_t V + \mu S \partial_S V + \frac{\sigma^2 S^2}{2} \partial_S^2 V - rV - r\varpi S\right) dt + \sigma S \partial_S V\, dW$$
$$\quad + \varpi\left(\mu S\, dt + \sigma S\, dW\right)$$
$$= \left(\partial_t V + \mu S(\partial_S V + \varpi) + \frac{\sigma^2 S^2}{2} \partial_S^2 V - rV - r\varpi S\right) dt$$
$$\quad + \sigma S\left(\partial_S V + \varpi\right) dW.$$

$$\tag{23.8}$$

---

[5]In most literature, this ratio (actually, the negative this ratio) is denoted by either Δ or δ, hence the name delta hedging. Here, for typographical convenience and to avoid confusion with the Laplace operator or with increments and derivatives, we preferred to call it ϖ.

Consequently, since ϖ is chosen to make the portfolio free from the risk caused by the stochastic term, this gives that one must choose ϖ such that $\partial_S V + \varpi = 0$ (note that this cancels in (23.8) the term involving W and the linear term in σ). That is, we take

$$\varpi = -\partial_S V, \tag{23.9}$$

and we thus plug[6] this information into (23.8), concluding that

$$0 = \left(\partial_t V + \frac{\sigma^2 S^2}{2} \partial_S^2 V - rV + rS\partial_S V \right) dt,$$

which leads to the Black–Scholes equation in (23.1), as desired.

It is also interesting to revisit the Black–Scholes equation in (23.1) in view of the financial model that it describes. We note in particular that the right-hand side of (23.1) is described only in terms of the risk-free interest rate r; this is therefore the risk-free part of the investment, and it consists of the superposition of a long-term strategy dictated by the option V (which will become effective only at its maturity time) and a short-term one embodied by the stock S (which fluctuates in a market subject to randomness).

On the left-hand side of (23.1), we instead see two terms. The first is the time derivative of the option. The second term on the left-hand side of (23.1) is more "geometric" since it reflects the convexity properties of the dependence of the option on the underlying stock.

It is also useful to relate the Black–Scholes equation in (23.1) to, the classical heat equation. As already observed on p. 246, to make this connection, one needs to revert the arrow of time; hence,

---

[6]We observe that in financial models, it is customary to assume that $\partial_S V \in [0, 1]$. Indeed, if the price of a stock increases, we may expect that the price of the corresponding option also increases (since it would allow us the possibility of buying a valuable stock at a prescribed price), which implies that $\partial_S V \geqslant 0$. Moreover, one expects the price of an option to increase slower than the price of a stock since the change in value of the option is somehow in consequence to the change in value of the stock, which suggests that $\partial_S V \leqslant 1$.

In this spirit, the prescription in (23.9) reads $\frac{\beta}{\alpha} = \varpi \leqslant 0$, giving that, in our notation, α and β have opposite signs. This is in agreement with the intuition that, say, the purchase of stocks (corresponding to a positive β) corresponds to a sale of options (i.e. corresponding to a negative α).

it is convenient to look at a new time $\tau := T - t$ which transforms the terminal condition (23.4) into an initial condition. Furthermore, it comes in handy to introduce the nonlinear transformation $x = \ln\left(\frac{S}{K}\right) + \left(r - \frac{\sigma^2}{2}\right)\tau$ in order to simplify the right-hand side of (23.1). More specifically, one sets

$$u(x, \tau) := e^{r\tau}\, V\left(T - \tau, K e^{x - \left(r - \frac{\sigma^2}{2}\right)\tau}\right)$$

and observes that if V solves the Black–Scholes equation in (23.1), then, using the notation $S := K e^{x - \left(r - \frac{\sigma^2}{2}\right)\tau}$ and $t := T - \tau$, it follows that

$$
\begin{aligned}
e^{-r\tau} &\left[\frac{\sigma^2}{2}\partial_{xx}u(x, \tau) - \partial_\tau u(x, \tau)\right] \\
&= \frac{\sigma^2}{2}\partial_x\left[K e^{x - \left(r - \frac{\sigma^2}{2}\right)\tau}\partial_S V\left(T - \tau, K e^{x - \left(r - \frac{\sigma^2}{2}\right)\tau}\right)\right] \\
&\quad - rV\left(T - \tau, K e^{x - \left(r - \frac{\sigma^2}{2}\right)\tau}\right) \\
&\quad + K e^{x - \left(r - \frac{\sigma^2}{2}\right)\tau}\left(r - \frac{\sigma^2}{2}\right)\partial_S V\left(T - \tau, K e^{x - \left(r - \frac{\sigma^2}{2}\right)\tau}\right) \\
&\quad + \partial_t V\left(T - \tau, K e^{x - \left(r - \frac{\sigma^2}{2}\right)\tau}\right) \\
&= \frac{\sigma^2}{2}\left[K e^{x - \left(r - \frac{\sigma^2}{2}\right)\tau}\partial_S V\left(T - \tau, K e^{x - \left(r - \frac{\sigma^2}{2}\right)\tau}\right)\right. \\
&\quad \left. + \left(K e^{x - \left(r - \frac{\sigma^2}{2}\right)\tau}\right)^2\partial_{SS}V\left(T - \tau, K e^{x - \left(r - \frac{\sigma^2}{2}\right)\tau}\right)\right] \\
&\quad - rV\left(T - \tau, K e^{x - \left(r - \frac{\sigma^2}{2}\right)\tau}\right)
\end{aligned}
$$

$$+ K e^{x - \left(r - \frac{\sigma^2}{2} \right) \tau} \left(r - \frac{\sigma^2}{2} \right) \partial_S V \left(T - \tau, K e^{x - \left(r - \frac{\sigma^2}{2} \right) \tau} \right)$$

$$+ \partial_t V \left(T - \tau, K e^{x - \left(r - \frac{\sigma^2}{2} \right) \tau} \right)$$

$$= \frac{\sigma^2}{2} \left[S \partial_S V(t, S) + S^2 \partial_{SS} V(t, S) \right] - r V(t, S)$$

$$+ S \left(r - \frac{\sigma^2}{2} \right) \partial_S V(t, S) + \partial_t V(t, S)$$

$$= \frac{\sigma^2 S^2}{2} \partial_{SS} V(t, S) - r V(t, S) + r S \partial_S V(t, S) + \partial_t V(t, S)$$

$$= 0.$$

Also,

$$u(x, 0) = V(T, K e^x).$$

Consequently, the Black–Scholes equation in (23.1) and the terminal condition (23.4) produce the following heat diffusion problem for time $\tau > 0$ with an initial datum:

$$\begin{cases} \partial_\tau u(x, \tau) = \dfrac{\sigma^2}{2} \partial_{xx} u(x, \tau), \\ u(x, 0) = K(e^x - 1) \chi_{(0, +\infty)}(x), \end{cases} \tag{23.10}$$

where, as usual, we used the notation for the characteristic function of a set, namely

$$\chi_A(x) := \begin{cases} 1 & \text{if } x \in A, \\ 0 & \text{otherwise.} \end{cases}$$

The plot of a solution of (23.10) is sketched in Figure 23.3. The advantage of this formulation is that one can focus on the solution of the classical heat equation in (23.10) and deduce useful information about the solution of the Black–Scholes equation in (23.1) by transforming it back in terms of the original variables.

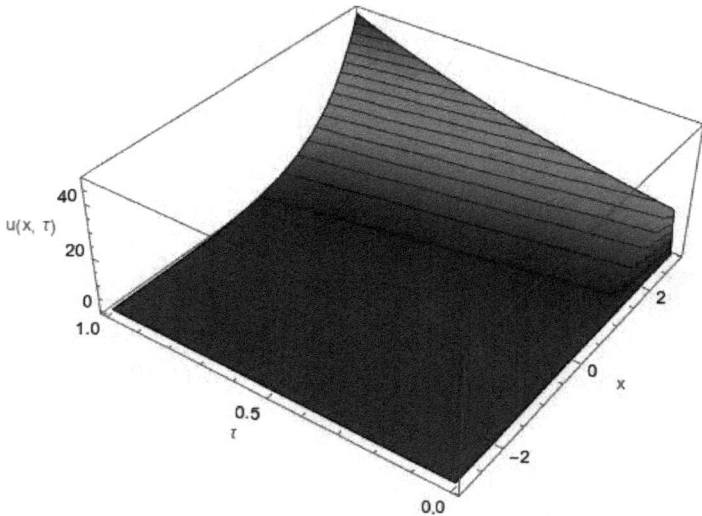

Figure 23.3.   Plot of a solution of (23.10) (with $K := 1$ and $\sigma := \sqrt{2}$).

See e.g. [DS06, Bly14] for additional readings on the Black–Scholes equation[7] and related models.

---

[7]Of course, like every equation, the Black–Scholes equation must be taken with a pinch of salt. For instance, according to Ian Stewart's article, which appeared in the British newspaper *The Observer* https://www.theguardian.com/science/2012/feb/12/black-scholes-equation-credit-crunch, the Black–Scholes equation was "one ingredient in a rich stew of financial irresponsibility, political ineptitude, perverse incentives and lax regulation" that ultimately led to the 2007 financial crisis. "The equation itself wasn't the real problem. It was useful, it was precise, and its limitations were clearly stated [...]. The trouble was its potential for abuse". In particular, we stress that the equation relies on the knowledge of the market volatility σ. "This is a measure of how erratically its market value changes. The equation assumes that the asset's volatility remains the same for the lifetime of the option, which needs not be correct. Volatility can be estimated by statistical analysis of price movements but it can't be measured in a precise, foolproof way, and estimates may not match reality [...]. The equation also assumes that there are no transaction costs, no limits on short-selling and that money can always be lent and borrowed at a known, fixed, risk-free interest rate. Again, reality is often very different".

Chapter 24

Diffusion of Transition Probabilities

In a stochastic model, one is often interested in the so-called transition probability, namely the likelihood of transitioning from one state to another.

In particular, Markov processes are stochastic models "without memory", i.e. in which predictions about future outcomes can be done based solely on the knowledge of its present state. In this setting, if $P(x_0, t_0 | x, t)$ denotes the probability of going from the point x_0 at time t_0 to the point x at time t (say, with $t > t_0$), we have that such a probability can be expressed as the superposition of all the probabilities of going first to a point y at some time $s \in (t_0, t)$ and from there reaching x at time t; more explicitly, given $s \in (t_0, t)$, it holds that

$$P(x_0, t_0 | x, t) = \int_{\mathbb{R}^n} P(x_0, t_0 | y, s)\, P(y, s | x, t)\, dy. \qquad (24.1)$$

This relation is often called Chapman–Kolmogorov equation, after[1] Sydney Chapman and Andrey Nikolaevich Kolmogorov.

---

[1]Chapman was a mathematician and geophysicist. At age 16, he entered the University of Manchester with a scholarship (he was the last student selected) and graduated with an engineering degree, but his passion for mathematics drove him to study for one further year to take a mathematics degree.

Besides his eminent contributions to stochastic processes, Chapman contributed to the theory of geomagnetism, investigating the beautiful phenomenon of aurorae

Figure 24.1.   Collection of pictures of aurorae from around the world (photos by Mila Zinkova, Samuel Blanc, Joshua Strang, Varjisakka and Jerry Magnum Porsbjer; images from Wikipedia, licensed under the Creative Commons Attribution-Share Alike 1.0 Generic license).

(see Figure 24.1) in relation to the Earth's magnetic field and the solar wind, contributing to the understanding of the photochemical mechanisms that produce the ozone layer, and predicting the existence of the magnetosphere (confirmed experimentally 30 years later).

Kolmogorov's scientific contributions are paramount and comprise basically all fields of mathematics, including probability, topology, logic, mathematical physics, harmonic analysis, mathematical biology and numerical analysis. His mathematical talent appeared quite early in his life: at the age of five, he wrote

Figure 24.2.   Getting ready for a big presentation (photo by Terrence L. Fine; image from Wikipedia, licensed under the Creative Commons Attribution-Share Alike 3.0 Unported license).

his first mathematical paper, published in his school journal. The content of the paper was the formula

$$\sum_{k=0}^{N-1}(2k+1) = N^2,$$

which may seems simple for all of us, but not quite a piece of cake for a five-year-old boy (by the way, the little kid also became an editor of the mathematical section of the school journal).

His former student and prominent mathematician Vladimir Igorevich Arnol′d used to place Kolmogorov in his top five mathematicians list (with Poincaré, Gauß, Euler and Newton, see Figure 24.2 for a picture of Kolmogorov preparing his talk during a conference in Tallin in 1973).

A mark on Kolmogorov's reputation is however produced by his involvement in the so-called "Luzin Affair" during the Great Purge in 1936 (see also footnote 1 on p. 274; on this occasion, Nikolai Nikolaevich Luzin, prominent mathematical analyst, point-set topologist and Kolmogorov's former doctoral advisor (and author

It is clearly desirable, given an initial point x_0 at an initial time t_0, to know more about the likelihood of being somewhere else at some future time, which involves having more information on the function $p(x,t) := P(x_0,t_0|x,t)$.

This is obtained through a partial differential equation of the type[2]

$$\partial_t p(x,t)\, dx = -\operatorname{div}\big(\alpha(x,t)\, p(x,t)\big) + \sum_{i,j=1}^{n} \partial_{ij}\big(\beta_{ij}(x,t)\, p(x,t)\big),$$

$$(24.2)$$

---

of the classical Luzin's theorem, see e.g. [WZ15, Theorem 4.20]) became a target of Stalin's regime. Luzin was accused of plagiarism and nepotism and of being an enemy of the Soviet people and a servant to "fascistoid science" (whatever that means), as confirmed by the fact that Luzin published some of his results in foreign journals.

In view of this allegation, Luzin lost his academic position (not too bad after all, his advisor Dimitri Fyodorovich Egorov, author of Egorov's theorem, see [WZ15, Theorem 4.17], was arrested during a previous purge and died after a hunger strike initiated in prison).

Regrettably, Kolmogorov took part in the hearing at the Commission of the Academy of Sciences of the USSR, where the allegations against Luzin were formalized. The question of whether Kolmogorov was coerced by the police into testifying against Luzin, possibly using as a threat an alleged long-lasting homosexual relationship, remains a topic of speculation, see e.g. [GK09]. In any case, it is a sad story of envy, violence, indifference, discrimination and brutality, which is nonetheless good to know since those who fail to learn from the mistakes of the past are doomed to repeat them.

[2]Equation (24.2) is usually called the Fokker–Planck equation. The name comes after Adriaan Daniël Fokker and Max Karl Ernst Ludwig Planck.

Fokker was a physicist and a musician, inventor of a 31-tone equal-tempered organ (we recall that the 12-tone equal temperament is the most common musical system today; recall the musical digression on p. 150). A picture of Fokker playing his organ is available at https://upload.wikimedia.org/wikipedia/en/a/ae/AdriaanFokker.jpg

Planck is famous for his prioneering work on quantum theory; fortunately, he did not follow the advice of his professor Philipp von Jolly at the Ludwig-Maximilians-Universität München, according to whom it was not worth to go into physics since "in this field, almost everything is already discovered, and all that remains is to fill a few holes".

The Planck family lived for many years in a villa in Berlin-Grunewald, located in Wangenheimstraße 21, which became a gathering place for other professors, such as Albert Einstein and theologian Adolf von Harnack (brother of the mathematician portrayed in Section 2.15 of the companion book [DV23]).

Figure 24.3. Nobel Laureates only (Public Domain image from Wikipedia).

The Planck family suffered hard times due to the deplorable political situation at the time. At the onset of World War I, Planck was one of the signatories of the so-called "Manifesto of the Ninety-Three", stating an unequivocal support for German military actions; among the other 92 prominent signatories, Fritz Haber, Adolf von Harnack, Philipp Lenard (Nobel Prize for cathode rays research, future supporter of the Nazi ideology and future crusader against Einstein's relativity theory) and Felix Klein (the outstanding mathematician who devised the "the Klein bottle"). During World War I, Planck's oldest son Karl was killed in action at the Battle of Verdun, and his second son Erwin was taken prisoner by the French.

When the Nazis came to power, Planck was 74, and when asked to gather distinguished professors to issue a public proclamation against the expulsion of Jewish academics, Planck replied, "If you are able to gather today 30 such gentlemen, then tomorrow 150 others will come and speak against it, because they are eager to take over the positions of the others". However, Planck tried to avoid the expulsion of Fritz Haber (who had pioneered chemical warfare by introducing the use of poisonous gases during World War I and personally overseeing the first use of chlorine gas during the Second Battle of Ypres and who had received the Nobel Prize in Chemistry in 1918 for the invention of a process to synthesize ammonia from nitrogen and hydrogen gases, with broad applications in the synthesis of where α and β_{ij} are called drift vector and diffusion tensor, respectively.

We now give a motivation for equation (24.2). Let φ be a smooth and compactly supported function in $\mathbb{R}^n$. Using the Chapman–Kolmogorov equation in (24.1), we know that, given $s \in (t_0, t)$,

$$\int_{\mathbb{R}^n} \varphi(x)\, p(x, t)\, dx = \int_{\mathbb{R}^n} \varphi(x)\, P(x_0, t_0 | x, t)\, dx$$

$$= \iint_{\mathbb{R}^n \times \mathbb{R}^n} P(x_0, t_0 | y, s)\, P(y, s | x, t)\, \varphi(x)\, dx\, dy$$

and, as a result, differentiating in t,

$$\int_{\mathbb{R}^n} \varphi(x)\, \partial_t p(x, t)\, dx$$

$$= \iint_{\mathbb{R}^n \times \mathbb{R}^n} P(x_0, t_0 | y, s)\, \partial_t P(y, s | x, t)\, \varphi(x)\, dx\, dy. \qquad (24.3)$$

It is also useful to recall that the total probability is normalized to 1, namely

$$\int_{\mathbb{R}^n} P(y, s | x, t)\, dx = 1,$$

whence, taking a derivative in t,

$$\int_{\mathbb{R}^n} \partial_t P(y, s | x, t)\, dx = 0. \qquad (24.4)$$

Given $y \in \mathbb{R}^n$, we also use a Taylor expansion to write that, when $|x - y|$ is small,

$$\varphi(x) = \varphi(y) + \nabla\varphi(y) \cdot (x - y) + \frac{1}{2} D^2\varphi(y)(x - y) \cdot (x - y) + O(|x - y|^3).$$

---

fertilizers and explosives). Planck's support in this circumstance was in vain, and Haber died in exile the following year.

Planck's son Erwin was later arrested by the Gestapo, following an attempted assassination of Hitler on July 20, 1944, sentenced to death and executed by hanging.

On a positive note, see Figure 24.3 for a social dinner among Nobel Laureates: from left to right: Walther Nernst (1920 Nobel Prize for Chemistry), Albert Einstein (1921 Nobel Prize for Physics), Max Planck (1918 Nobel Prize for Physics), Robert A. Millikan (1923 Nobel Prize for Physics) and Max von Laue (1914 Nobel Prize for Physics).

This and (24.4) give that

$$\int_{\mathbb{R}^n} \partial_t P(y, s|x, t)\, \varphi(x)\, dx$$

$$= \int_{\mathbb{R}^n} \partial_t P(y, s|x, t) \left(\varphi(y) + \nabla\varphi(y) \cdot (x - y) \right.$$

$$+ \frac{1}{2} D^2 \varphi(y)(x - y) \cdot (x - y) + O(|x - y|^3) \Bigg)\, dx$$

$$= \int_{\mathbb{R}^n} \partial_t P(y, s|x, t) \left(\nabla\varphi(y) \cdot (x - y) \right. \tag{24.5}$$

$$+ \frac{1}{2} D^2 \varphi(y)(x - y) \cdot (x - y) + O(|x - y|^3) \Bigg)\, dx$$

$$= \alpha(y, s, t) \cdot \nabla\varphi(y) + \frac{1}{2} \sum_{i,j=1}^{n} \beta_{ij}(y, t) \partial_{ij} \varphi(y)$$

$$+ \int_{\mathbb{R}^n} \partial_t P(y, s|x, t)\, O(|x - y|^3)\, dx,$$

where[3]

$$\alpha(y, s, t) := \int_{\mathbb{R}^n} \partial_t P(y, s|x, t)\, (x - y)\, dx$$

$$\text{and} \quad \beta_{ij}(y, s, t) := \frac{1}{2} \int_{\mathbb{R}^n} \partial_t P(y, s|x, t)\, (x_i - y_i)(x_j - y_j)\, dx.$$

Combining this and (24.3), we find that

$$\int_{\mathbb{R}^n} \varphi(x)\, \partial_t p(x, t)\, dx$$

$$= \int_{\mathbb{R}^n} \Bigg[\alpha(y, s, t) \cdot \nabla\varphi(y) + \sum_{i,j=1}^{n} \beta_{ij}(y, s, t) \partial_{ij} \varphi(y)$$

$$+ \int_{\mathbb{R}^n} \partial_t P(y, s|x, t)\, O(|x - y|^3)\, dx \Bigg] p(y, s)\, dy$$

---

[3]To be precise, we note that the latter integral in (24.5) occurs only in the support of φ, which is a bounded set.

$$= \int_{\mathbb{R}^n} \alpha(y, s, t) \cdot \nabla\varphi(y) \, p(y, s) \, dy$$

$$+ \int_{\mathbb{R}^n} \sum_{i,j=1}^{n} \beta_{ij}(y, s, t)\partial_{ij}\varphi(y) \, p(y, s) \, dy + \mathcal{R}(s, t), \qquad (24.6)$$

where

$$\mathcal{R}(s, t) := \iint_{\mathbb{R}^n \times \mathbb{R}^n} \partial_t P(y, s|x, t) \, O(|x - y|^3) \, dx \, dy.$$

We suppose additionally that the stochastic process is homogeneous in time, namely the transition probability depends only on the points x_0 and x and on the elapsed time $t - t_0$, that is,

$$P(x_0, t_0 + \tau|x, t + \tau) = P(x_0, t_0|x, t)$$

for all $\tau \in \mathbb{R}$.

Differentiating in τ, this gives that

$$\partial_{t_0} P(x_0, t_0|x, t) + \partial_t P(x_0, t_0|x, t) = 0.$$

As a consequence, observing that $P(y, t|x, t) = 0$ when $x \neq y$ (unless one is provided with the "gift of ubiquity"),

$$\alpha(y, t) := \lim_{\varepsilon \searrow 0} \frac{1}{\varepsilon} \int_{t-\varepsilon}^{t} \alpha(y, s, t) \, ds$$

$$= \lim_{\varepsilon \searrow 0} \frac{1}{\varepsilon} \iint_{\mathbb{R}^n \times (t-\varepsilon, t)} \partial_t P(y, s|x, t) \, (x - y) \, dx \, ds$$

$$= -\lim_{\varepsilon \searrow 0} \frac{1}{\varepsilon} \iint_{\mathbb{R}^n \times (t-\varepsilon, t)} \partial_s P(y, s|x, t) \, (x - y) \, dx \, ds$$

$$= \lim_{\varepsilon \searrow 0} \frac{1}{\varepsilon} \int_{\mathbb{R}^n} \Big(P(y, t - \varepsilon|x, t) - P(y, t|x, t)\Big) (x - y) \, dx$$

$$= \lim_{\varepsilon \searrow 0} \frac{1}{\varepsilon} \int_{\mathbb{R}^n} P(y, t - \varepsilon|x, t) \, (x - y) \, dx$$

$$= \lim_{\varepsilon \searrow 0} \int_{\mathbb{R}^n} \frac{P(y, 0|x, \varepsilon) \, (x - y)}{\varepsilon} \, dx.$$

Similarly,

$$\beta_{ij}(y,t) := \lim_{\varepsilon \searrow 0} \frac{1}{\varepsilon} \int_{t-\varepsilon}^{t} \beta_{ij}(y,s,t)\, ds$$

$$= \lim_{\varepsilon \searrow 0} \frac{1}{2} \int_{\mathbb{R}^n} \frac{P(y,0|x,\varepsilon)}{\varepsilon} (x_i - y_i)(x_j - y_j)\, dx$$

and

$$\mathcal{R}(t) := \lim_{\varepsilon \searrow 0} \frac{1}{\varepsilon} \int_{t-\varepsilon}^{t} \mathcal{R}(s,t)\, ds$$

$$= \lim_{\varepsilon \searrow 0} \iint_{\mathbb{R}^n \times \mathbb{R}^n} \frac{P(y,0|x,\varepsilon)}{\varepsilon} O(|x-y|^3)\, dx\, dy. \qquad (24.7)$$

Formally, we also have that

$$\lim_{\varepsilon \searrow 0} \frac{1}{\varepsilon} \iint_{\mathbb{R}^n \times (t-\varepsilon,t)} \alpha(y,s,t) \cdot \nabla \varphi(y)\, p(y,s)\, dy\, ds$$

$$= \lim_{\varepsilon \searrow 0} \frac{1}{\varepsilon} \left(\iint_{\mathbb{R}^n \times (t-\varepsilon,t)} \alpha(y,s,t) \cdot \nabla \varphi(y)\, p(y,t)\, dy\, ds \right.$$

$$\left. + \iint_{\mathbb{R}^n \times (t-\varepsilon,t)} \alpha(y,s,t) \cdot \nabla \varphi(y)\, (p(y,s) - p(y,t))\, dy\, ds \right)$$

$$= \int_{\mathbb{R}^n} \alpha(y,t) \cdot \nabla \varphi(y)\, p(y,t)\, dy$$

$$+ \lim_{\varepsilon \searrow 0} \frac{1}{\varepsilon} \iint_{\mathbb{R}^n \times (t-\varepsilon,t)} \alpha(y,s,t) \cdot \nabla \varphi(y)\, (p(y,s) - p(y,t))\, dy\, ds$$

$$= \int_{\mathbb{R}^n} \alpha(y,t) \cdot \nabla \varphi(y)\, p(y,t)\, dy$$

$$+ \lim_{\varepsilon \searrow 0} \frac{1}{\varepsilon} \iint_{\mathbb{R}^n \times (t-\varepsilon,t)} |\nabla \varphi(y)|\, O(\varepsilon)\, dy\, ds$$

$$= \int_{\mathbb{R}^n} \alpha(y,t) \cdot \nabla \varphi(y)\, p(y,t)\, dy$$

and, in a similar way,

$$\lim_{\varepsilon \searrow 0} \frac{1}{\varepsilon} \iint_{\mathbb{R}^n \times (t-\varepsilon, t)} \beta_{ij}(y, s, t) \partial_{ij} \varphi(y) \, p(y, s) \, dy$$

$$= \int_{\mathbb{R}^n} \beta_{ij}(y, t) \partial_{ij} \varphi(y) \, p(y, t) \, dy.$$

Therefore, integrating (24.6) in $s \in (t - \varepsilon, \varepsilon)$, dividing by ε and sending $\varepsilon \searrow 0$, we see that

$$\int_{\mathbb{R}^n} \varphi(x) \, \partial_t p(x, t) \, dx = \int_{\mathbb{R}^n} \alpha(y, t) \cdot \nabla \varphi(y) \, p(y, t) \, dy$$

$$+ \int_{\mathbb{R}^n} \sum_{i,j=1}^n \beta_{ij}(y, t) \partial_{ij} \varphi(y) \, p(y, t) \, dy + \mathcal{R}(t).$$

$$(24.8)$$

Now, for a sufficiently small time ε, if the stochastic process is continuous, we expect the probability of going from y to x in the elapsed time ε to be "rather small" except when $|x - y|$ is small. It is therefore customary to consider the quantity $\mathcal{R}(t)$ in (24.7) as a "negligible term" (one may also want to recall the observation in footnote 3 to get rid of the contribution in $\mathcal{R}(t)$ coming from infinity). Hence, taking the liberty of formally dropping it from the equation, we reduce (24.8) to

$$\int_{\mathbb{R}^n} \varphi(x) \, \partial_t p(x, t) \, dx = \int_{\mathbb{R}^n} \alpha(y, t) \cdot \nabla \varphi(y) \, p(y, t) \, dy$$

$$+ \int_{\mathbb{R}^n} \sum_{i,j=1}^n \beta_{ij}(y, t) \partial_{ij} \varphi(y) \, p(y, t) \, dy.$$

Integrating by parts (and renaming y as x in the variable of integration), we obtain

$$\int_{\mathbb{R}^n} \varphi(x) \, \partial_t p(x, t) \, dx$$

$$= - \int_{\mathbb{R}^n} \varphi(x) \, \mathrm{div} \left(\alpha(x, t) \, p(x, t) \right) \, dx$$

$$+ \int_{\mathbb{R}^n} \sum_{i,j=1}^n \varphi(x) \, \partial_{ij} \left(\beta_{ij}(x, t) \, p(x, t) \right) \, dx,$$

which, by the arbitrariness of φ, produces the Fokker–Planck equation in (24.2), as desired.

Another possible derivation of the Fokker–Planck equation in (24.2) comes from stochastic differential equations (we present this argument in one dimension for the sake of simplicity, but the same ideas would carry over to higher dimensions as well). Namely, if $dX = F\,dt + G\,dW$, we can use Itô's chain rule in (23.3) to every random variable $H(X(t),t)$, integrate and find that

$$H(X(T),T) - H(X(0),0) = \int_0^T dH$$

$$= \int_0^T \left[\left(\partial_t H + F\partial_X H + \frac{G^2}{2}\partial_X^2 H \right) dt + G\partial_X H\,dW \right].$$

In particular, if $H = H(X(t))$,

$$H(X(T)) - H(X(0)) = \int_0^T \left[\left(F\partial_X H + \frac{G^2}{2}\partial_X^2 H \right) dt + G\partial_X H\,dW \right].$$
$$(24.9)$$

Now, we take the expected value $\mathbb{E}$ of this identity (say, at a given time T). For this, it is useful to recall that, for a "reasonable" ϕ, we have that

$$\mathbb{E}\left(\int_0^T \phi\,dW \right) = 0, \qquad (24.10)$$

see e.g. [Eva13, equation (ii) on p. 68] (the bottom line of this formula being that W presents independent increments, and hence the expectation of a function ϕ that is adapted to the stochastic process will vanish).

Thus, it follows from (24.9) and (24.10) that

$$\mathbb{E}\Big(H(X(T)) - H(X(0)) \Big) = \mathbb{E}\left(\int_0^T \left(F\partial_X H + \frac{G^2}{2}\partial_X^2 H \right) dt \right).$$
$$(24.11)$$

Now, we recall the definition of expected value of a random variable Y (see e.g. [Eva13, equation (4) on p. 14]), according to which

$$\mathbb{E}(Y) = \int_{\mathbb{R}^n} Y(x)\,p(x,T)\,dx.$$

In this way, equation (24.11) becomes

$$\int_{\mathbb{R}^n} \Big(H(x) - H(X(0)) \Big) p(x,T)\, dx$$

$$= \iint_{\mathbb{R}^n \times (0,T)} \left(F(x,t)\partial_x H(x) + \frac{G^2(x,t)}{2} \partial_x^2 H(x) \right) p(x,T)\, dx\, dt.$$

We now write this identity for $T+\varepsilon$, subtract the one for T, divide by ε and formally send $\varepsilon \searrow 0$: namely, recalling (24.4),

$$\int_{\mathbb{R}^n} H(x)\, \partial_t p(x,T)\, dx$$

$$= \int_{\mathbb{R}^n} \Big(H(x) - H(X(0)) \Big) \partial_t p(x,T)\, dx$$

$$= \lim_{\varepsilon \searrow 0} \int_{\mathbb{R}^n} \Big(H(x) - H(X(0)) \Big) \frac{p(x,T+\varepsilon) - p(x,T)}{\varepsilon}\, dx$$

$$= \lim_{\varepsilon \searrow 0} \frac{1}{\varepsilon} \left[\int_{\mathbb{R}^n} \Big(H(x) - H(X(0)) \Big) p(x,T+\varepsilon)\, dx \right.$$

$$\left. - \int_{\mathbb{R}^n} \Big(H(x) - H(X(0)) \Big) p(x,T)\, dx \right]$$

$$= \lim_{\varepsilon \searrow 0} \frac{1}{\varepsilon} \left[\iint_{\mathbb{R}^n \times (0,T+\varepsilon)} \left(F(x,t)\partial_x H(x) + \frac{G^2(x,t)}{2} \partial_x^2 H(x) \right) \right.$$

$$\times p(x,T+\varepsilon)\, dx\, dt$$

$$\left. - \iint_{\mathbb{R}^n \times (0,T)} \left(F(x,t)\partial_x H(x) + \frac{G^2(x,t)}{2} \partial_x^2 H(x) \right) p(x,T)\, dx\, dt \right]$$

$$= \lim_{\varepsilon \searrow 0} \left[\iint_{\mathbb{R}^n \times (0,T+\varepsilon)} \left(F(x,t)\partial_x H(x) + \frac{G^2(x,t)}{2} \partial_x^2 H(x) \right) \right.$$

$$\times \frac{p(x,T+\varepsilon) - p(x,T)}{\varepsilon}\, dx\, dt + \frac{1}{\varepsilon} \iint_{\mathbb{R}^n \times (T,T+\varepsilon)}$$

$$\left. \times \left(F(x,t)\partial_x H(x) + \frac{G^2(x,t)}{2} \partial_x^2 H(x) \right) p(x,T)\, dx\, dt \right]$$

$$= \mathcal{S}(T) + \int_{\mathbb{R}^n} \left(F(x,T)\partial_x H(x) + \frac{G^2(x,T)}{2} \partial_x^2 H(x) \right) p(x,T)\, dx,$$

where

$$S(T) := \iint_{\mathbb{R}^n \times (0,T)} \left(F(x,t)\partial_x H(x) + \frac{G^2(x,t)}{2}\partial_x^2 H(x) \right)$$
$$\times \partial_t p(x,T)\, dx\, dt.$$

Since the term $S(T)$ is small when T is small, if we neglect it from the previous computation, we reduce the problem to

$$\int_{\mathbb{R}^n} H(x)\, \partial_t p(x,T)\, dx$$
$$= \int_{\mathbb{R}^n} \left(F(x,T)\partial_x H(x) + \frac{G^2(x,T)}{2}\partial_x^2 H(x) \right) p(x,T)\, dx.$$

Integrating by parts in x, we thus conclude that

$$\int_{\mathbb{R}^n} H(x)\, \partial_t p(x,T)\, dx$$
$$= \int_{\mathbb{R}^n} \left(-\partial_x\Big(F(x,T)\, p(x,T) \Big) + \frac{1}{2}\partial_x^2\Big(G^2(x,T)\, p(x,T) \Big) \right) H(x)\, dx.$$

From the arbitrariness of H, we thus obtain

$$\partial_t p = -\partial_x(Fp) + \frac{1}{2}\partial_x^2(G^2 p), \qquad (24.12)$$

which is the Fokker–Planck equation (24.2) in this setting; interestingly, in the expression (24.12), one can directly relate the drift coefficient of the Fokker–Planck equation to the "deterministic" term of the stochastic equation and the diffusion coefficient of the Fokker–Planck equation to the "random" term of the stochastic equation.

A classical application of the Fokker–Planck equation in (24.12) is given by the so-called[4] Ornstein–Uhlenbeck stochastic differential equation. In this case, one considers in the background a stochastic differential equation of the form

$$dX = -\vartheta X\, dt + \sigma\, dW, \qquad (24.13)$$

for suitable positive constant parameters ϑ and σ.

---

[4]The process in (24.13) is named after Leonard Salomon Ornstein and George Eugene Uhlenbeck, see Figure 24.4.

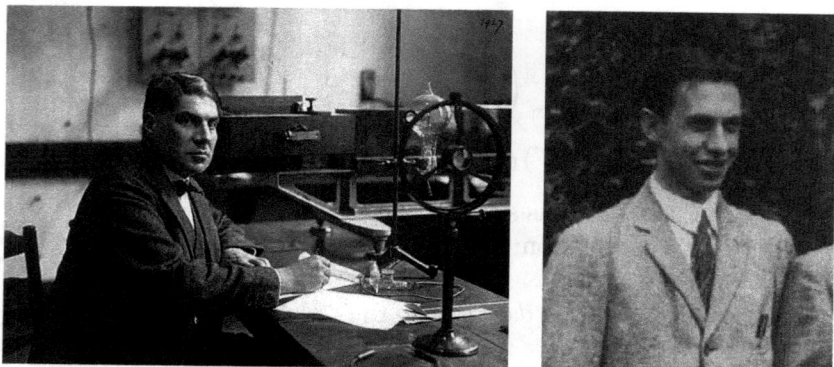

Figure 24.4. Leonard Ornstein and George Uhlenbeck (Public Domain images from Wikipedia).

To develop an intuition of the process in (24.13), we can imagine an elastic spring in the presence of a large dumping and thermal fluctuations. In a nutshell, Hooke's law would prescribe an equation of motion of the type $\ddot{X} = -\kappa X$, with $\kappa > 0$ being the elastic constant of the spring. In the presence of friction, the previous equation becomes $\ddot{X} = -\kappa X - \gamma \dot{X}$, with γ being the friction coefficient. Rewriting this equation in the form $\frac{\ddot{X}}{\gamma} = -\vartheta X - \dot{X}$, with $\vartheta := \kappa/\gamma$, for large values of elastic and dumping coefficients we can reduce the expression to $0 = -\vartheta X - \dot{X}$, which we can formally write as $dX = -\vartheta X \, dt$. In this sense, the Ornstein–Uhlenbeck equation in (24.13) consists of simply adding to this model a random fluctuation (e.g. due to thermal deviations); therefore it can be considered describing a noisy relaxation process of an oscillator.

Compare also the photo of Ornstein in his lab with Figure 24.5, showing a very elegant mural in Utrecht, the Netherlands, depicting Ornstein, the random walk (embodied by a drunkard, check the bottle of booze in his left hand) and the Ornstein–Uhlenbeck process (note the formula next to Ornstein's head, as enlarged in Figure 24.6).

The mural has been painted by the Dutch painting collective De Strakke Hand, see Figure 24.7. See also http://www.destrakkehand.nl/ for other beautiful pieces of art in an urban landscape.

Sometimes, models similar to (24.13) are also referred to by the name Langevin equation, after Paul Langevin.

Figure 24.5. Leonard Ornstein mural (image by Hansmuller from Wikipedia, licensed under the Creative Commons Attribution-Share Alike 4.0 International license).

Note that the setting in (24.13) corresponds to that in (24.12) with the choices $F := -\vartheta X$ and $G := \sigma$, whence the corresponding Fokker–Planck equation for the Ornstein–Uhlenbeck process takes the form

$$\partial_t p = \vartheta \partial_x (xp) + \frac{\sigma^2}{2} \partial_x^2 p. \qquad (24.14)$$

It is interesting to observe that the drift term modulated by ϑ in (24.14) corresponds to an attraction toward the center $x = 0$ for the transition probability p. To convince ourselves of this fact, one can simply compare (24.14) with (2.10), which was used to describe the evolution of a biological population following a chemical attractant. Note that w in (2.10) would correspond, up to constants, to $-\vartheta x^2$

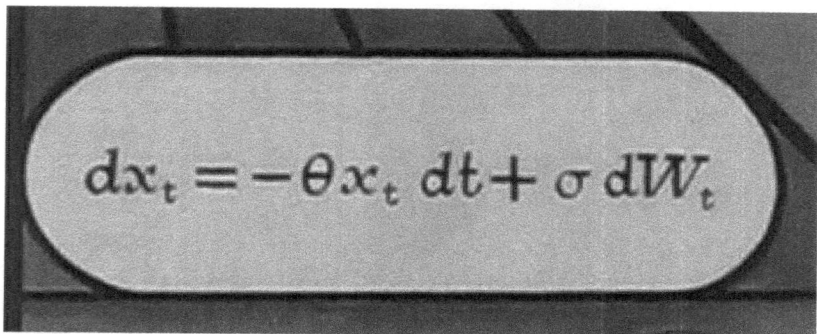

Figure 24.6.   The Ornstein–Uhlenbeck process, detail from the Leonard Ornstein mural (image by Hansmuller from Wikipedia, licensed under the Creative Commons Attribution-Share Alike 4.0 International license).

Figure 24.7.   De Strakke Hand Team (image from De Strakke Hand website http://www.destrakkehand.nl/about).

in (24.14), and hence p is "attracted" toward the higher values of $-\vartheta x^2$ (namely, $x = 0$) in the same way as a biological population is attracted toward the higher-density regions of the chemotactic factor.

A plot of the solution p of (24.14), for a suitable choice of the parameters ϑ and σ, is sketched in Figure 24.8; the initial datum at $t = 0$ corresponds to the probability of finding the random particle located at $x = 1$ (and for this reason, the picture exhibits a blowup at $(x, t) = (1, 0)$), and we can appreciate that, as time flows, the solution has the tendency to move its mass toward $x = 0$.

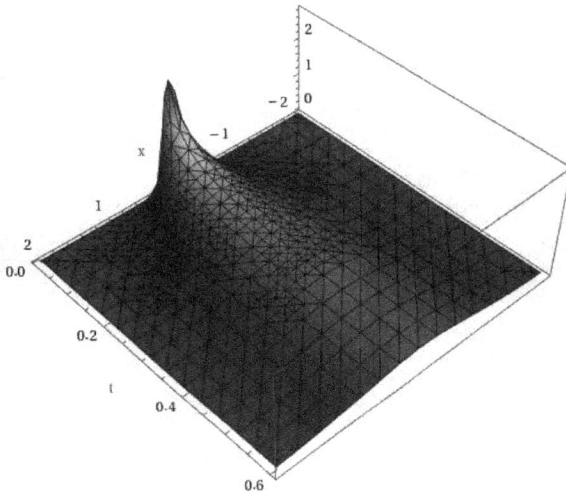

Figure 24.8. Plotting the evolution of the transition probability for the Ornstein–Uhlenbeck process according to the Fokker–Planck equation in (24.14).

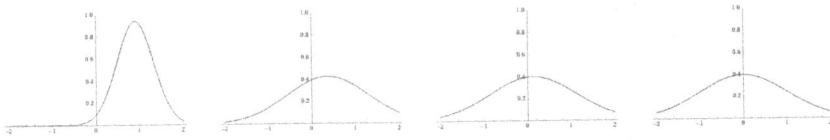

Figure 24.9. Evolution of the transition probability for the Ornstein–Uhlenbeck process at subsequent times.

See also Figure 24.9 for frames of this solution at times $t = 0.1$, 1, 2 and 10.

The situation in which the general tendency of reaching an objective (such as the origin in Figures 24.8 and 24.9) is weighted against the possibility of diffusing around and missing it is actually quite common in our everyday experience; hence, the plots in Figures 24.8 and 24.9 should be somewhat close to our intuition. One can think, for instance, of the case of rogaining, in which, in principle compass bearings point you straight to the next control, but in reality one has to account for some fluctuations around the goal (due to some miscalculations of angles and distances for which the teammate has to be blamed, see e.g. Figure 24.10, the roughness of the terrain, possible modifications of the landscape, possible inaccuracies of the

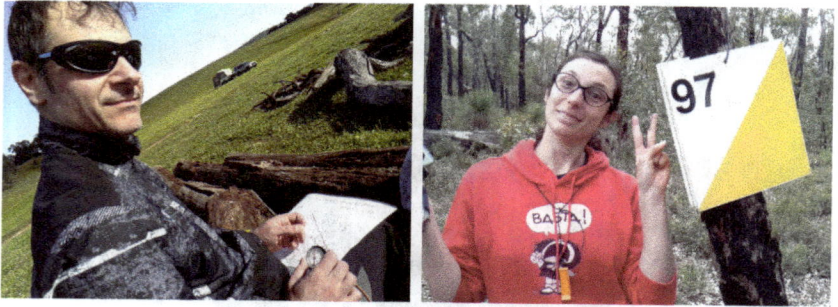

Figure 24.10.   The difference between planning routes on a map and getting there.

map, etc.). Accordingly, the evolution of an initial position is realistically not given by a single point (that is, a Dirac delta function located at a moving point) but rather by a probability distribution describing how likely one is exactly at that point. In this scenario, the origin in Figures 24.8 and 24.9 represents the ideal destination corresponding to the maximal value of the probability distribution, but the tails of this distribution represent the possibilities of missing the control during navigation due to whatever fluctuations (at that point, the rogainer will have to seek the control patiently and accurately; knowing the probability density would be perhaps a good indication on how far one should search).

For additional information about Brownian motion, random disturbances and the Fokker–Planck equation, see [Kra40, Moy49, Ris89, Gar09, Per15] and the references therein.

Chapter 25

Phase Coexistence Models

A topical problem in material sciences is the understanding of the separation patterns between different phases of a given substance. The two phases could be related to molecule orientation, magnetization, multi-component alloy systems or the state of the matter (including superconductive or superfluid states). In this context, the description of a physical system often relies on the understanding of suitable "order parameters", which are suitable quantities capable of distinguishing different phases of a given system.

For instance, the magnetization vector M expresses the density of magnetic dipole moment in a magnetic material. In broad terms, at each point x of a given magnetic material, the direction of $M(x)$ reproduces the direction of the magnetic field near x, and the magnitude $|M(x)|$ represents the amount of magnetization.

Another classical example of order parameter arises in nematic crystals. These materials consist of long, thin molecules that prefer to align with one another. In this situation, the two ends of the elongated molecules are essentially indistinguishable; therefore, an efficient order parameter for these crystals is given by a vector $N(x)$ modulo a sign (that is, an element of the projective plane, or equivalently a vector on a hemisphere with opposing points along the equator identified). That is, while magnetic materials tend to displace the magnetic vectors in a parallel fashion, the nematic vectors can be equally displayed either in a parallel or antiparallel form (the technological advantage is thus that these crystals can be aligned by an external magnetic or electric field, as it happens for instance in liquid crystal displays for video games).

While the above-mentioned M and N are vectorial order parameters, there are also ideal cases in which a scalar order parameter is sufficiently efficient to describe some features of a system. For example, nematic crystals may exhibit at some sufficiently large scale a "preferred direction" along which most of its elements are aligned, which is called in jargon the "local nematic director". In this context, scalar state parameters which are often adopted for practical purposes are the functions of the angle ϑ between the liquid-crystal molecular axis and the local nematic director.

The simplest of this function may be constructed as follows. We take, up to a rotation, the local nematic director to be the first vector e_1 of the Euclidean basis, and we consider a state parameter $S(\vartheta)$ such that:

(i) S is 2π-periodic (consistently with the fact that ϑ is an angle),

(ii) $S(\vartheta) = S(\vartheta + \pi)$ (consistently with the fact that the two endings of the nematic crystals are indistinguishable),

(iii) $S(\vartheta) = S(-\vartheta)$ (consistently with the fact that this order parameter should account for the "angular distance" of a molecule from the local nematic director),

(iv) $S(\vartheta) \leqslant 1 = S(0)$ (normalizing the maximal of the state parameter with 1, corresponding to the full alignment case),

(v) $\int_{\partial B_1} S(\arccos(\omega \cdot e_1)) \, d\mathcal{H}_\omega^{n-1} = 0$ (normalizing at 0 the case of random directions, using the notation $\omega \cdot e_1 = \cos \vartheta$),

see Figure 25.1.

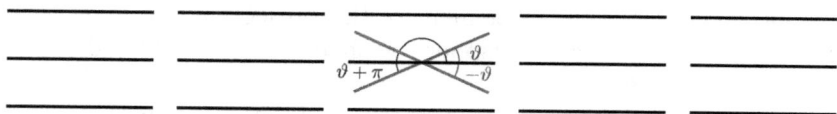

Figure 25.1.   A scalar order parameter using the local nematic director.

By (i), one can seek S in the form of a Fourier Series, say

$$S(\vartheta) = \frac{a_0}{2} + \sum_{j=1}^{+\infty} (a_j \cos(j\vartheta) + b_j \sin(j\vartheta)).$$

By (iii), we have that $b_j = 0$ for all $j \in \{1, 2, \dots\}$. Hence, using (ii),

$$0 = S(\vartheta + \pi) - S(\vartheta)$$

$$= \sum_{j=1}^{+\infty} a_j \left(\cos(j(\vartheta + \pi)) - \cos(j\vartheta)\right) = -2 \sum_{\substack{j \geqslant 1 \\ j \text{ odd}}} a_j \cos(j\vartheta)$$

and, therefore, $a_j = 0$ for all j odd.

Thus, neglecting higher orders, one can take

$$S(\vartheta) = \frac{a_0}{2} + a_2 \cos(2\vartheta).$$

By (iv), we have that $\frac{a_0}{2} = 1 - a_2$, whence, in view of (v),

$$0 = \fint_{\partial B_1} S(\arccos(\omega \cdot e_1)) \, d\mathcal{H}_\omega^{n-1}$$

$$= 1 - a_2 + a_2 \fint_{\partial B_1} \cos(2 \arccos \omega_1) \, d\mathcal{H}_\omega^{n-1}$$

$$= 1 - a_2 + a_2 \fint_{\partial B_1} \left(2 \cos^2(\arccos \omega_1) - 1\right) d\mathcal{H}_\omega^{n-1}$$

$$= 1 - 2a_2 + 2a_2 \fint_{\partial B_1} \omega_1^2 \, d\mathcal{H}_\omega^{n-1}$$

$$= 1 - 2a_2 + \frac{2a_2}{n},$$

which, in the concrete case of $n = 3$, yields that $a_2 = 3/4$.

As a result,

$$S(\vartheta) = \frac{1}{4} + \frac{3}{4} \cos(2\vartheta) = \frac{1}{2}(3 \cos^2 \vartheta - 1).$$

Interestingly, we will discover that this coincides with the second Legendre polynomial (see the companion book [DV23]).

Returning to the study of phase transitions, Landau theory[1] attempts to understand the phase transitions related to continuous order parameters. Its main *ansatz* is that the equilibria of the system should come from a "free energy" produced by the order parameter under consideration. This energy may also be sensitive to the temperature of the system. For instance, one can consider an order parameter η for a given system at temperature T and describe the energy E of the system as a function of η and T (for the sake of simplicity, we take here η to be a scalar order parameter, but the case of vector-valued order parameters can be treated in a similar manner).

Up to a normalization, one can suppose that the energy corresponding to $\eta = 0$ is zero. Moreover, in many concrete cases, the energy of the physical system is invariant if we exchange η with $-\eta$ (this is the case, for instance, with magnetic materials since exchanging the north pole with the south pole should not alter the energy state of the system, and it is also the case with nematic crystals since the ending of their molecules is symmetric, recall e.g. point (iii) here above).

---

[1]This theory is named after Lev Davidovich Landau, who was 1962 Nobel Prize in Physics for his mathematical theory of superfluidity. In 1938, Landau wrote a leaflet, jointly with Mosey Korets, condemning Stalin and the People's Commissariat for Internal Affairs (the forerunner of MVD and KGB) on the occasion of the Great Purge in the Soviet Union (see also footnote 1 on p. 253). According to the leaflet, Stalin and his acolytes "can only beat defenseless prisoners, catch unsuspecting innocent people, plunder national wealth and invent ridiculous trials against nonexistent conspiracies". The authors of the leaflet were sure that "Proletariat [...] will throw off the fascist dictator and his clique".

Things didn't work out as expected, and Landau was arrested and held in the Lubyanka prison for more than a year, see Figure 25.3.

There were however other features which possibly compensated for Landau's dissident leaflet in the eyes of Stalin. Landau led a team of mathematicians supporting the Soviet atomic and hydrogen bomb development, thus receiving for his work the Stalin Prize in 1949 and 1953 and the title "Hero of Socialist Labor" in 1954. We should be always suspicious about ourselves when we receive too many prizes and awards because, as all Spiderman fans know, "with great power comes great responsibility".

By the way, according to Wikipedia, "Landau believed in free love rather than monogamy and encouraged his wife and his students to practice free love. However, his wife was not enthusiastic".

These considerations lead to an energy $E(\eta, T)$ which is even in η and such that $E(0, T) = 0$. The corresponding Taylor expansion must therefore contain only even terms and takes the form

$$E(\eta, T) = a(T)\eta^2 + b(T)\eta^4 + \cdots,$$

and we will indeed neglect higher-order terms, thus reducing the above to the following simple case:

$$E(\eta, T) = a(T)\eta^2 + b(T)\eta^4. \tag{25.1}$$

The energy coefficients $a(T)$ and $b(T)$ depend on the temperature T, and their signs play a decisive role in the formation and separation of phases. More specifically, one usually considers stable solutions in correspondence with minimal energy levels; therefore, to avoid minimizers corresponding to an energy equal to $-\infty$, a natural structural assumption is to suppose that $b(T) > 0$ for every temperature T. Instead, to allow for possible phase changes, one can assume that $a(T)$ changes sign above and below some critical temperature T_c, for instance taking $a(T) > 0$ when $T < T_c$ and $a(T) < 0$ when $T > T_c$.

In this setting, below the critical temperature, the free energy exhibits the null value of the order parameter η as its only minimizer, but above the critical temperature, the stable phase corresponds to $\pm\eta_0$, where η_0 is the minimizer for E, given by

$$\eta_0 := \sqrt{-\frac{a}{2b}},$$

see Figure 25.2.

In particular, if given $T > T_c$, one wishes to normalize the stable phases at the levels ± 1 (as done for instance in point (iv) above), up to normalizing factors, one can choose $a(T) = -\frac{1}{2}$ and $b(T) = \frac{1}{4}$, so that (25.1) reduces to

$$E(\eta) = -\frac{1}{2}\eta^2 + \frac{1}{4}\eta^4 = \frac{(\eta^2 - 1)^2 - 1}{4} \tag{25.2}$$

and the stable phases are the zeros of

$$E'(\eta) = \eta^3 - \eta. \tag{25.3}$$

It is interesting to remark that this analysis of free energy is useful for detecting the stable phases, namely ± 1 in (25.3), but it does not

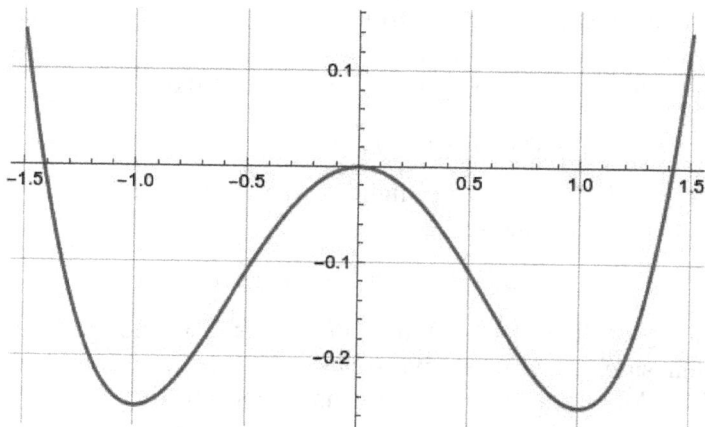

Figure 25.2.   Plot of the function $\eta \mapsto \frac{(\eta^2-1)^2-1}{4}$.

give any information about the separation between the two possible phases; indeed, all configurations only attaining the stable phases ± 1 would indeed be zeros of (25.3), as well as minimizers of (25.2). However, in many concrete situations, one expects the separation between phases to be somewhat minimal as well; one often experiences situations in which the two phases occupy two separate "bulks" of the material, which are separated by an interface, thus showing a distinctive coarsening and phase separation. This phenomenon is possibly the outcome of a "ferromagnetic" effect, which tends to align the direction of the molecules of the material, hence avoiding oscillations of the order parameter.

To include this phenomenon into the phase separation model, one can modify the energy in (25.2) by adding a small penalization term which charges the formation of interfaces. The simplest version of this procedure typically involves adding to (25.2) a small "gradient term" (a gradient indeed detects the "local oscillation" of phases), i.e. replace (25.2) by

$$\mathcal{G}(\eta) = \frac{\varepsilon}{2} \int_\Omega |\nabla \eta(x)|^2 \, dx + \int_\Omega \frac{(\eta^2(x) - 1)^2 - 1}{4} \, dx, \qquad (25.4)$$

for a small parameter $\varepsilon > 0$ (where Ω is the region occupied by the material).

Looking for the corresponding minimal points of the full energy $\mathcal{G}$ leads to the equation

$$\varepsilon \Delta \eta = \eta^3 - \eta, \tag{25.5}$$

which is often referred[2] to by the name Allen–Cahn equation.

See e.g. [KL03] and the references therein for additional details on phase separation models.

---

[2] Equation (25.5) is named after John Cahn and Sam Allen, who presented a theoretical treatment of metal alloys [AC75].

With respect to the classification presented in footnote 13 on p. 9, we see that equation (25.5) is of elliptic type.

We believe that it is interesting to relate the "democratic" features of the Laplace operator, according to the discussions on pp. 24 and 62, to the ferromagnetic tendency of some materials, which avoids oscillations in spins and molecule alignments.

Furthermore, equation (25.5) provides one of the chief examples of "semilinear equations", within a structure that we will investigate further in the companion book [DV23]. See also footnote 2 on p. 202.

One can also appreciate the link between minimizers of the energy functional in (25.4) and interfaces of minimal area, by using the Cauchy–Schwarz inequality and the Coarea formula (see e.g. [EG15]), namely setting $W(\eta) := \frac{(\eta^2 - 1)^2}{4}$ and noting that

$$\mathcal{G}(\eta) + \frac{|\Omega|}{4} = \frac{\varepsilon}{2} \int_{\Omega} |\nabla \eta(x)|^2 \, dx + \int_{\Omega} W(\eta(x)) \, dx \geqslant 2 \int_{\Omega} \sqrt{\frac{\varepsilon}{2} |\nabla \eta(x)|^2 W(\eta(x))} \, dx$$

$$= \sqrt{2\varepsilon} \int_{\Omega} |\nabla \eta(x)| \sqrt{W(\eta(x))} \, dx = \sqrt{2\varepsilon} \int_{\mathbb{R}} \left[\int_{\{\eta = \tau\}} \sqrt{W(\eta(x))} \, d\mathcal{H}_x^{n-1} \right] d\tau.$$

Hence, heuristically, it should be convenient for a minimizer η_ε to be located "as much as possible" near the minima of W (which correspond to the stable phases ± 1). If we assume that the level sets of η_ε are surfaces "more or less parallel" to an interface $\mathcal{S}$ and η_ε approaches "quite fast" the stable phases ± 1, the previous computation suggests that the minimization of $\mathcal{G}$ reduces, as a first approximation, to the minimization of the area of the interface $\mathcal{S}$.

Of course, it is not easy to transform the above heuristic argument into a rigorous proof since the setting in (25.5) is that of a singular perturbation, namely the small parameter ε affects precisely the most significant term in equation (25.5). As a matter of fact, the coherent development of arguments of this sort required the introduction of a novel notion of asymptotics called Γ-convergence, see [DGF75].

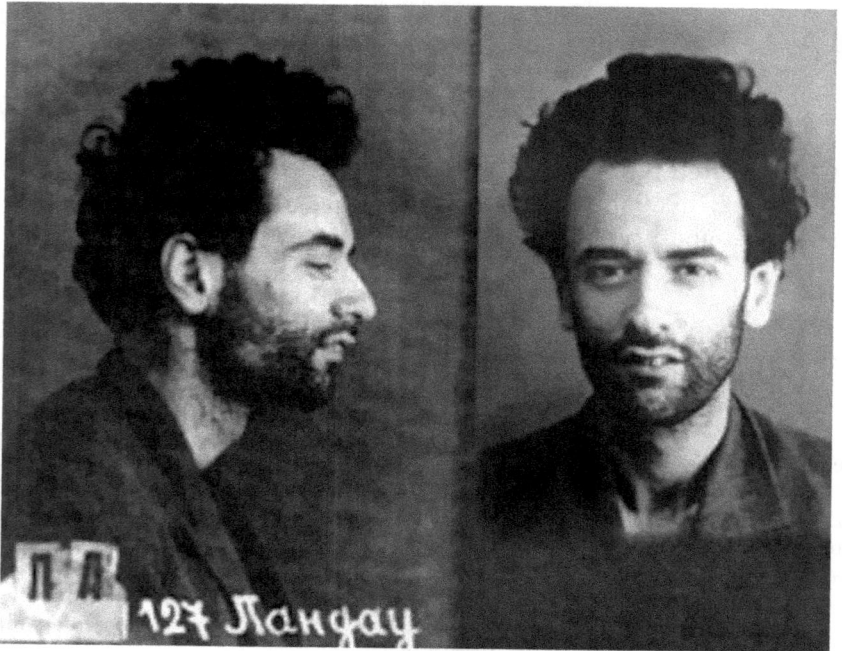

Figure 25.3.   Lev Landau in prison, photo by the People's Commissariat for Internal Affairs (Public Domain image from Wikipedia).

Chapter 26

Growth of Interfaces

A number of scientific problems of interest are associated with the growth of the profile of suitable surfaces which describe, for instance, tumors, flame fronts, and clusters.

A typical model describing this phenomenon is given by the so-called[1] Kardar–Parisi–Zhang equation

$$\partial_t h = \mu \Delta h + \lambda |\nabla h|^2 + f, \qquad (26.1)$$

where $\mu > 0$ is a diffusion parameter, λ is a parameter related to surface growth and f is a forcing term (which could be either a deterministic function or a stochastic term, such as white noise). Of course, when $\lambda = 0$, the setting in (1.6) essentially boils down to that of the heat equation in (1.6), but when $\lambda \neq 0$, the equation presents a nonlinear term of geometric importance.

Indeed, the motivation underpinning (26.1) is as follows. Suppose that we have an aggregate with an active zone of growth on its surface, which is described, at a given time t, by the graph of a certain function, $y = g(x, t)$. We consider an "infinitesimal" time step τ during which the surface growth occurs, and suppose that such growth is regulated by three ingredients: first, in time $\left(t, t + \frac{\tau}{3}\right)$, some particle

---

[1]Equation (26.1) is named after Mehran Kardar, Giorgio Parisi and Yi-Cheng Zhang, who introduced this model in [KPZ86]. Parisi is also known for his contributions to quantum fluctuations, spin glasses and whirling flocks of birds. See Figure 26.1 in which the Italian President Sergio Mattarella (on the right) congratulates Parisi (on the left) for having received the 2021 Nobel Prize in Physics (they are wearing face masks due to COVID-19 regulations).

Figure 26.1. Opening ceremony at the Italian Parliament (October 10, 2021; attribution: Presidenza della Repubblica: https://www.quirinale.it/elementi/60152#&gid=1&pid=10).

Figure 26.2. Time evolution of a growing interface.

is added to the surface; then, in time $\left(t + \frac{\tau}{3}, t + \frac{2\tau}{3}\right)$, a forcing term kicks in; finally, in time $\left(t + \frac{2\tau}{3}, t + \tau\right)$, some diffusion takes place (the order in which these phenomena occur is not really important, but we fix this order only to keep a concrete case in mind).

To describe the first step, we assume that spherical particles of diameter τ are added uniformly along the surface of the cluster, as shown in Figure 26.2. In this setting, if ϑ denotes the angle between the normal of the spherical particle and the vertical direction, we have that the infinitesimal vertical surface growth corresponds to $\frac{\tau}{\cos \vartheta}$. Hence, we write that

$$g\left(x, t + \frac{\tau}{3}\right) = g(x, t) + \frac{\tau}{\cos \vartheta}. \tag{26.2}$$

Actually, the normal of the spherical particle agrees with the normal of the original surface $\frac{(-\nabla g(x,t),1)}{\sqrt{1+|\nabla g(x,t)|^2}}$ and, accordingly,

$$\cos \vartheta = \frac{(-\nabla g(x, t), 1)}{\sqrt{1 + |\nabla g(x, t)|^2}} \cdot e_n = \frac{1}{\sqrt{1 + |\nabla g(x, t)|^2}}.$$

We substitute this information into (26.2), and we find that

$$g\left(x, t + \frac{\tau}{3}\right) = g(x, t) + \tau\sqrt{1 + |\nabla g(x, t)|^2}. \tag{26.3}$$

Now, we describe the second step, namely the action of a forcing term: this effect is encoded in the equation

$$g\left(x, t + \frac{2\tau}{3}\right) = g\left(x, t + \frac{\tau}{3}\right) + \tau f\left(x, t + \frac{\tau}{3}\right). \tag{26.4}$$

As for the third step, we assume that the diffusion takes place in the form

$$g(x, t + \tau) = g\left(x, t + \frac{2\tau}{3}\right) + \tau \Delta g\left(x, t + \frac{2\tau}{3}\right). \tag{26.5}$$

Thus, by collecting the observations in (26.3), (26.4) and (26.5),

$$\frac{g(x, t + \tau) - g(x, t)}{\tau}$$

$$= \frac{g(x, t + \tau) - g\left(x, t + \frac{2\tau}{3}\right)}{\tau} + \frac{g\left(x, t + \frac{2\tau}{3}\right) - g\left(x, t + \frac{\tau}{3}\right)}{\tau}$$

$$+ \frac{g\left(x, t + \frac{\tau}{3}\right) - g(x, t)}{\tau}$$

$$= \Delta g\left(x, t + \frac{2\tau}{3}\right) + f\left(x, t + \frac{\tau}{3}\right) + \sqrt{1 + |\nabla g(x, t)|^2},$$

whence, formally taking the limit as $\tau \searrow 0$,

$$\partial_t g(x,t) = \Delta g(x,t) + f(x,t) + \sqrt{1 + |\nabla g(x,t)|^2}. \tag{26.6}$$

When the slope of the growing interface is small, one can consider $|\nabla g|$ as a small perturbation and employ the approximation $\sqrt{1 + |\nabla g|^2} \simeq 1 + \frac{|\nabla g|^2}{2}$. In this framework, (26.6) reduces to

$$\partial_t g = \Delta g + f + 1 + \frac{1}{2}|\nabla g|^2.$$

Considering the velocity shift $h(x,t) := g(x,t) - t$, we thus conclude that

$$\partial_t h = \partial_t g - 1 = \Delta g + f + \frac{1}{2}|\nabla g|^2 = \Delta h + f + \frac{1}{2}|\nabla h|^2,$$

providing a motivation for the Kardar–Parisi–Zhang equation in (26.1).

We mention that, despite its apparent simplicity, equations such as the one in (26.1) do capture complex phenomena and lead to the study of important universality classes sharing the same characteristic exponents and scale-invariant limits, see e.g. [Cor12].

Furthermore, equation (26.1) is also directly related to other classical equations. In particular, defining $v := \nabla h$, we have that

$$\partial_t v + v \cdot \nabla v - \mu \Delta v - \nabla f = \nabla\left(\partial_t h - \mu \Delta h - f\right) + \sum_{j=1}^{n} \partial_j h \, \partial_j \nabla h$$

$$= \nabla\left(\partial_t h - \mu \Delta h - f\right) + \frac{1}{2}\nabla\left(\sum_{j=1}^{n}|\partial_j h|^2\right)$$

$$= \nabla\left(\partial_t h - \mu \Delta h - f + \frac{1}{2}|\nabla h|^2\right),$$

from which it follows that the Kardar–Parisi–Zhang equation in (26.1), with $\lambda := \frac{1}{2}$, is equivalent to the vectorial equation

$$\partial_t v + v \cdot \nabla v = \mu \Delta v + \nabla f, \tag{26.7}$$

Figure 26.3. Sketch of Harry Bateman (Public Domain image from Wikipedia).

which is a version of the Navier–Stokes equation in (5.8) in the absence of gravity (in this setting, v corresponds to the speed of a given fluid, μ is its viscosity coefficient and f is the negative of the pressure).

For constant pressure, in dimension $n = 1$, equation (26.7) reduces to the scalar equation

$$\partial_t v + v \partial_x v = \mu \partial_{xx} v, \qquad (26.8)$$

which is known in jargon[2] as the Bateman–Burgers equation, which models the speed of a fluid in a thin ideal pipe.

---

[2]Equation (26.8) is named after Harry Bateman and Johannes Martinus Burgers. See Figure 26.3 for a nice drawing of Harry Bateman from the 1931 yearbook of California Institute of Technology.

In this context, it is interesting to point out that in the absence of viscosity, equation (26.8) boils down to

$$\partial_t v + v \partial_x v = 0, \qquad (26.9)$$

known as[3] the inviscid Bateman–Burgers equation. A feature of the solutions of (26.9) is that they may develop "shock waves", i.e. singularities at a wave breaking time. For instance, it is readily seen that the relation $v(x,t) = \arctan(tv(x,t) - x)$ provides, via the implicit function theorem, a solution to (26.9) for small times, with $v(x,0) = -\arctan x$. This solution exhibits a shock wave since, otherwise, for $t > 1$,

$$|\partial_x v(0,t)| = \left| \frac{t \partial_x v(0,t)}{\cos^2\left(tv(0,t)\right)} \right| \geqslant |t \partial_x v(0,t)| > |\partial_x v(0,t)|,$$

which is a contradiction. See Figure 26.4 for a sketch of the formation of such a shock wave.

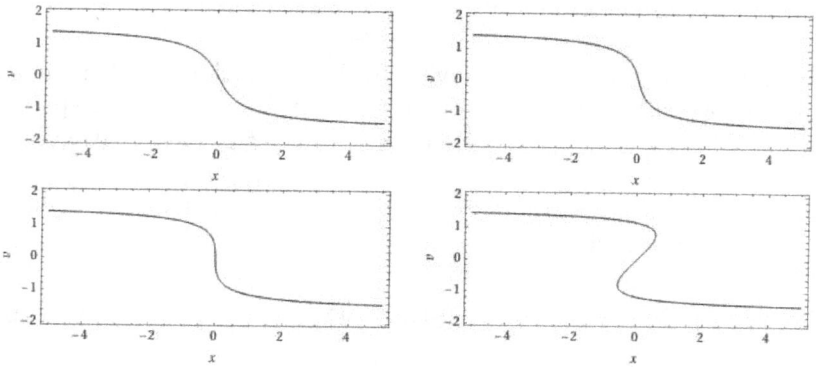

Figure 26.4.   Implicit plot of $v = \arctan(tv - x)$ for $t \in \{0.5, 0.75, 0.99, 2\}$.

---

[3]Note that the inviscid Bateman–Burgers equation in (26.9) can be viewed as a particular case of the Korteweg–de Vries equation in (10.1), with $a := 1$, $b := 0$ and $c := 1$.

Figure 26.5. The battleship USS Iowa of the United States Navy during a training exercise in Puerto Rico (Public Domain image from Wikipedia).

See e.g. [Eva98, pp. 140, 195, 204] for more information on shock waves in the context of the Bateman–Burgers equation. See also Figure 26.5 for a rather dramatic view of expanding spherical atmospheric shock waves resulting from a gun fire on a surface of water.

Chapter 27

What to Do When You Are Stuck in a Traffic Jam

Traffic jams (see Figure 27.1) are maddening, so let's try to understand a bit about traffic flow problems and what to do in the unfortunate circumstance of being stuck in a blockage.

For this, we consider a highway modeled by the real line, with material points (the cars) moving on it. We denote by $\rho = \rho(x, t)$ the density of cars at a given point x at time t. Since (in the absence of accidents) the total number of cars on the highway is preserved, the continuity equation (5.4) prescribes that

$$\partial_t \rho + \partial_x(\rho v) = 0, \qquad (27.1)$$

where $v = v(x, t)$ denotes the corresponding velocity of the traffic flow at the point x at time t.

To model in a somewhat realistic way how drivers react to the traffic, one may suppose that the velocity depends on the density of cars. Typically, if the highway is empty (i.e. if $\rho = 0$), the drivers go as fast as the speed limit allows (i.e. reaching the maximal speed allowed, denoted by V); however, in heavy traffic situations, the drivers slow down and possibly stop in a tailback, where the cars are bumper to bumper (i.e. the velocity is zero when the density reaches a maximal threshold, which we denote by R).

As an example of a velocity function with this property, one can consider the linear function

$$v(x, t) := \frac{(R - \rho(x, t))V}{R}.$$

Figure 27.1.   Traffic in Rome in 1939 and in Berkeley in 2005 ((a) Public Domain image from Wikipedia; (b) photo by Minesweeper, image from Wikipedia, licensed under the GNU Free Documentation License).

By inserting this into (27.1), one finds

$$\partial_t \rho + \partial_x \left(\frac{(R - \rho)V\rho}{R} \right) = 0.$$

Substituting for

$$u(x,t) := \frac{R - 2\rho \left(x, \frac{t}{2V} \right)}{2R},$$

which corresponds to

$$\rho(x,t) = \frac{R}{2} \left(1 - 2u\left(x, 2Vt \right) \right),$$

we find

$$\partial_t u + \partial_x \left(\frac{u^2}{2} \right) = 0,$$

which corresponds to the inviscid Bateman–Burgers equation in (26.9).

The fact that this equation produces shock waves (as depicted in Figure 26.4) is certainly not good news for the drivers (but likely very good news for auto body repairs and insurance companies!).

The bottom line is that it's always your choice what you do in the moment: if you're stuck in a traffic jam, you can either get angry and honk your horn or work on your mathematics!

For additional information on the broad subject of crowd and traffic flow models see e.g. [GP06, BD11] and the references therein.

Chapter 28

Reaching Equilibrium.
Or Maybe Not?

One can believe that, over long periods of time, a system tends to evolve naturally toward a disordered state in which energy is shared equally among all of its various forms. This was certainly the belief of Enrico Fermi (according to [Fer23] and [Bru16, p. 128]), see Figure 28.1.

But life is full of surprises. Fermi left his lab located in Via Panisperna, Rome, in 1938 to escape racial laws that affected his Jewish wife, Laura Capon, and moved to the United States. After some years, Fermi joined the secret laboratory in Los Alamos to design and build the first atomic bomb, to which he contributed[1] very substantially.

---

[1] In particular, Fermi constructed the world's first artificial nuclear reactor, and on the first detonation of a nuclear weapon, he calculated the energy that was going to be released as blast (with that, he allegedly ended up taking bets on whether or not the whole atmosphere would ignite due to the bomb's explosion and if so whether it would incinerate only the state or the entire planet).

Before moving to Los Alamos, Fermi was already one of the world leaders in both theoretical physics and experimental physics. He had been awarded the 1938 Nobel Prize in Physics for his discovery of induced radioactivity by neutron bombardment and of chemical elements with atomic numbers greater than that of uranium. Fermi's prominent contributions range from statistical mechanics to particle physics, pioneering the theory of weak interaction.

Figure 28.1.   Enrico Fermi at the blackboard (image from the Smithsonian Institution Archives).

Fermi remained in close contact with the lab in Los Alamos even after the end of the war and was very interested in the development of the vacuum-tube computer[2] MANIAC I (Mathematical Analyzer Numerical Integrator and Automatic Computer Model I), which

---

In [Fer22], he introduced a moving system of reference adapted to lines and surfaces, which is nowadays called "Fermi Coordinates" and broadly used in geometry, analysis and mathematical physics (we will also utilize them in Section 9.1 of the companion book [DV23]).

[2]In retrospect, the computational powers of these computers were perhaps quite limited. In 1956, MANIAC I would have become the first computer to defeat a human being in chess (the human player was a beginner, whose name is unknown to us, who had been taught the rules of chess only in the preceding week, the final position after the 23rd move Ne5# being depicted in Figure 28.2). Well, not precisely chess but a chess variant on a 6×6 chessboard in which bishops were removed due to the limited amount of memory and computing power. This variant of chess is nowadays called "Los Alamos chess" or, due to the missing bishops, "anti-clerical chess". The rules are as in chess except that there is no

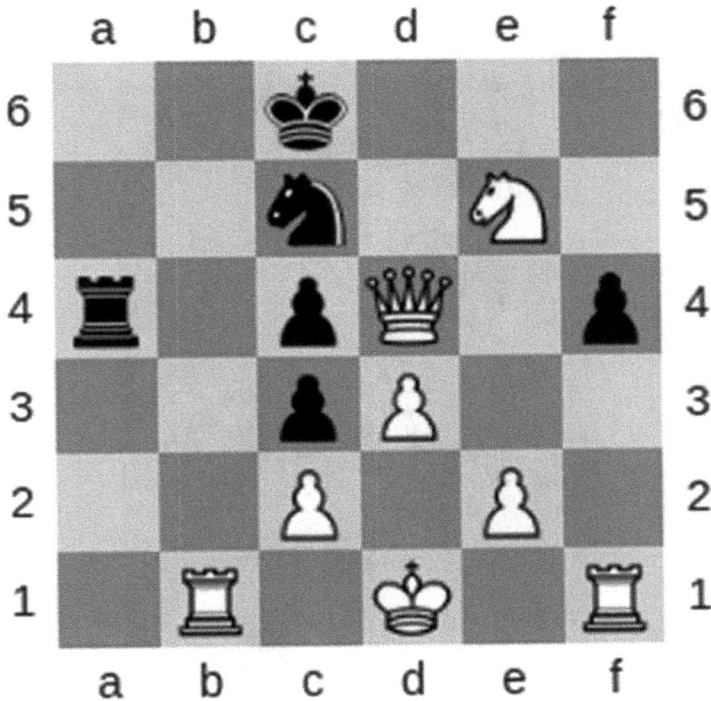

Figure 28.2.   MANIAC I beating a human player at Los Alamos chess.

Fermi regarded not merely as a tool for calculating but rather as a useful device to explore and better understand the complicated dynamics of physical systems. For this, Fermi held extensive discussions with John Pasta and[3] Stanisław Ulam (see Figure 28.3) to

---

pawn initial double-step move, no *en passant* capture and no castling (and, of course, pawns may not promote to bishops).

[3]Ulam was also a prominent scientific figure. Among the other accomplishments of his career, he invented cellular automaton and the Monte Carlo method.

The concept of the Teller–Ulam design refers to a specific configuration of the hydrogen bomb (the terminology for this design joins the name of Ulam with that of theoretical physicist Edward Teller, who, by the way, is often rumored to be one of the inspirations for the character of Dr. Strangelove, the mad scientist in Stanley Kubrick's 1964 satirical comedy). Given the high classification of the documents related to thermonuclear weapons, the details of the design are military secrets, presumably accessible to only a handful of nations.

Figure 28.3.   Stanisław Ulam (image from the Los Alamos National Laboratory).

formulate a physical problem which was simple to state but whose solution would require a computation lengthy enough to be unfeasible by pencil and paper (even[4] for Fermi's remarkable computing skills). The problem finally chosen by Fermi, Pasta and Ulam was that of some masses connected by elastic springs. This problem was easy to visualize and relate to everyday experience (e.g. a string instrument, such as a guitar). Fermi's expectation was that, after a short time, a nonlinear perturbation of this elastic system would have reached an energy equipartition. As Ulam reported (see [Bru16, p. 256]) "the

---

See [Ula76] for Ulam's autobiography.

[4]Fermi was famous for his ability to make exceptionally good approximate calculations. His renowned reputation on this topic is the reason for which nowadays problems that require brilliant applications of dimensional analysis and approximation are named "Fermi quizzes".

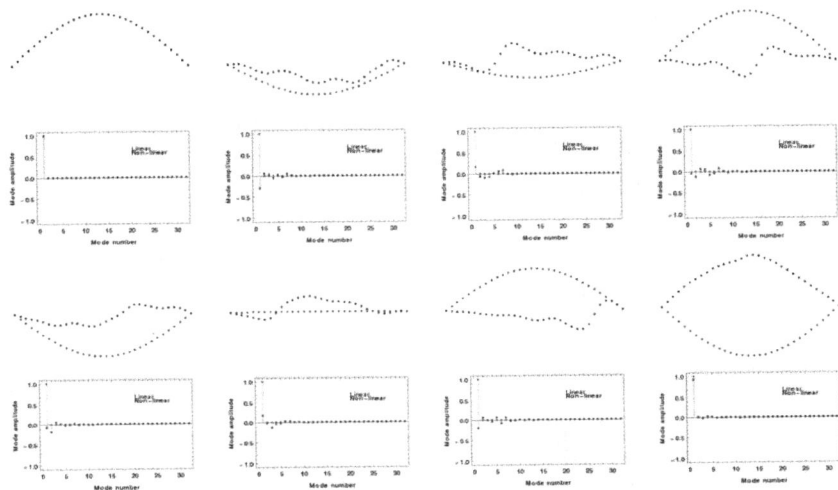

Figure 28.4. An elastic chain with a small perturbation that initially distributes the energy among its modes but later on shifts back almost all the energy to the original mode (extract from a simulation by Jacopo Bertolotti, image from Wikipedia, licensed under the Creative Commons CC0 1.0 Universal Public Domain Dedication).

original objective had been to see at what rate the energy of the string, initially put into a single sine wave (the note was struck as one tone), would gradually develop higher tones with the harmonics, and how the shape would finally become a mess both in the form of the string and in the way the energy was distributed among higher and higher modes. Nothing of the sort happened. To our surprise[5] the string started playing a game of musical chairs, only between several low notes, and perhaps even more amazingly, after what would have been several hundred ordinary up and down vibrations, it came back almost exactly to its original sinusoidal shape". See Figure 28.4 for a sketch of such a dumbfounding phenomenon.

---

[5]The fact that Fermi's original belief about systems (almost always) evolving naturally to energy equipartition was incorrect does not diminish his genius; on the contrary, the experiment run with MANIAC I was one of the cornerstone of the modern theory of dynamical systems, in light of ergodic theory, KAM theory, chaos, etc.

In a further detail, the mathematical framework chosen by Fermi, Pasta and Ulam was used to describe the solution of the chain of oscillators in Fourier Series and to consider the "harmonic energy" associated with each mode of such a decomposition. Their original expectation was to see the system exploring the available energy surface and relax toward an equipartition state in which, for instance, all the harmonic energies (or at least their time average) would approach the same value. Instead, the harmonic energies seemed to exhibit a recurrent behavior, almost returning to the initial configuration after some excursions, and the time average of the harmonic energies seemed to relax not toward a constant distribution but to a distribution that looked exponentially decreasing with the modes (in particular, the system seemed to exhibit the formation of a "packet of modes" with geometrically decaying energies). These behaviors were perceived by Fermi, Pasta and Ulam as paradoxical with respect to the equilibrium theory of statistical mechanics.

As a historical remark, Fermi, Pasta and Ulam were experts in mathematics as well as theoretical, nuclear and computational physics, but they were not specialized programmers; therefore, they contacted one of the programmers of the MANIAC I, Mary Tsingou (see Figure 28.5) to write the code and implement the simulation. According to Tsingou, interestingly, Fermi, Pasta and Ulam "didn't know anything[6] about programming. They set up the equations, and I did all the programming". Even the action of programming was at the time quite different from what programmers do today. According to Pasta (as reported on [Pal97, p. 359]) "the program was of course punched on cards. A DO loop was executed by the operator feeding in the deck of cards over and over".

Shortly after this numerical experiment took place, Fermi was invited to give a honorary lecture at the annual American

---

[6]But perhaps Tsingou's statement was also a bit ungenerous: according to Ulam (see the preface to [FPU55]), "during one summer Fermi learned very rapidly how to program problems for the electronic computers and he not only could plan the general outline and construct the so-called flow diagram but would work out himself the actual coding of the whole problem in detail".

Figure 28.5.   Mary Tsingou (image from the Los Alamos National Laboratory).

Mathematical Society meeting, where he intended to talk about it. Unfortunately, he became ill before the meeting, so his lecture never took place. Fermi died of inoperable stomach cancer, and the results of this work were published in [FPU55] after his death (Enrico Fermi, John Pasta and Stanisław Ulam being the authors of the article; the footnote on p. 979 of [FPU55] reporting the acknowledgment, "We thank Miss Mary Tsingou for efficient coding of the problems and for running the computations on the Los Alamos MANIAC machine"). The problem has been established in the scientific literature as the Fermi–Pasta–Ulam problem but was renamed in 2008 as the Fermi–Pasta–Ulam–Tsingou problem to grant attribution not only to the ideators of the mathematical physics problem but to the computer programmer as well.

Let us now see more specifically what this problem was and how (perhaps quite surprisingly) it relates to the theory of partial

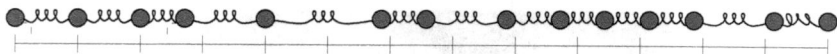

Figure 28.6.   A chain of oscillators.

differential equations (see also [Pal97] for further details). Given $h > 0$ and $N \in \mathbb{N}$, let us put some material points of equal mass m along a line at some rest positions located at jh, with $j \in \{0, \ldots, N\}$. Let us connect[7] these material points by identical elastic springs.

This system mimics a string with restoring forces (for more on that, see Chapter 14). If we move horizontally some of the material points from their rest positions, the springs exert an elastic force producing oscillations. We denote by $X_j(t)$ the position of the jth material point at time t. The force acting on the jth material point comes from the springs to its left- and right-hand sides. Assuming perfect elasticity, since the left-hand spring produces a force proportional to $-(X_j - X_{j-1})$ and the right-hand spring a force proportional to $(X_{j+1} - X_j)$, the total force acting on the jth material point has the form

$$f_j := -\kappa(X_j - X_{j-1}) + \kappa(X_{j+1} - X_j), \tag{28.1}$$

for some elastic coefficient $\kappa > 0$, see Figure 28.6.

Thus, we can also introduce a potential energy,

$$P_j(X) := \frac{\kappa}{2} \left((X_j - X_{j-1})^2 + (X_{j+1} - X_j)^2 \right), \tag{28.2}$$

where $X := (X_0, \ldots, X_N)$, and rewrite the force in (28.1) as

$$f_j = -\partial_{X_j} P_j(X). \tag{28.3}$$

The idea proposed by Fermi is to consider a small anelastic perturbation of this ideal problem. For simplicity, one can simply add a

---

[7]In the original numerical experiment [FPU55], the model focused on a case in which the first and last material points were constrained at their rest position to have a situation with fixed endpoints, but later on periodic arrays of such a configuration were taken into account as well. So, to keep things as simple as possible, we will skate over the details of the precise boundary conditions chosen (a general perception in the literature is that the specific boundary condition taken is possibly not the most relevant feature in this particular setting, but see [BMP08] for a precise discussion of similarities and differences).

small additional potential energy term of cubic (instead of quadratic) type, thus replacing (28.2) with

$$\widetilde{P}_j(X) := \frac{\kappa}{2}\left((X_j - X_{j-1})^2 + (X_{j+1} - X_j)^2\right)$$
$$+ \frac{\varepsilon}{3}\left((X_j - X_{j-1})^3 + (X_{j+1} - X_j)^3\right),$$

where $\varepsilon > 0$ is a small parameter.

Correspondingly, using the notation $Y_j := X_j - X_{j-1}$, the force in (28.3) would become

$$\begin{aligned}
f_j &= -\partial_{X_j}\widetilde{P}_j(X)\\
&= -\kappa\left((X_j - X_{j-1}) - (X_{j+1} - X_j)\right)\\
&\quad -\varepsilon\left((X_j - X_{j-1})^2 - (X_{j+1} - X_j)^2\right)\\
&= -\kappa\left(Y_j - Y_{j+1}\right) - \varepsilon\left(Y_j^2 - Y_{j+1}^2\right)\\
&= -\kappa\left(Y_j - Y_{j+1}\right)\left(1 + \frac{\varepsilon}{\kappa}\left(Y_j + Y_{j+1}\right)\right)\\
&= \kappa\left(X_{j+1} + X_{j-1} - 2X_j\right)\left(1 + \frac{\varepsilon}{\kappa}\left(X_{j+1} - X_{j-1}\right)\right).
\end{aligned}$$

By Newton's law, this gives that

$$m\ddot{X}_j = \kappa\left(X_{j+1} + X_{j-1} - 2X_j\right)\left(1 + \frac{\varepsilon}{\kappa}\left(X_{j+1} - X_{j-1}\right)\right).$$

To understand the link between these equations of motion and the theory of partial differential equations, it is useful to choose $\varepsilon := \kappa h$ and define $c := \sqrt{\frac{\kappa}{m}}\, h$. In this way, we have that

$$\ddot{X}_j(t) = \frac{c^2}{h^2}\left(X_{j+1}(t) + X_{j-1}(t) - 2X_j(t)\right)$$
$$\left(1 + h\left(X_{j+1}(t) - X_{j-1}(t)\right)\right). \qquad (28.4)$$

It is also useful to introduce a smooth function $u(x,t)$ to try to measure (in some appropriate limit sense) the displacement at time t of the particle corresponding to the equilibrium position x.

In principle, the equilibria are discrete and of the form $x = jh$, with $j \in \{0, \ldots, N\}$, but the continuous limit would correspond to the asymptotics as $h \searrow 0$. Hence, the above function u is modeled in such a way that $u(jh, t) = X_j(t)$ (and, say, interpolated to be defined as $u(x, t)$ for a continuum of x).

We thereby consider the formal Taylor expansions as $h \searrow 0$:

$$\frac{X_{j+1}(t) + X_{j-1}(t) - 2X_j(t)}{h^2}$$

$$= \frac{u(jh + h, t) + u(jh - h, t) - 2u(jh, t)}{h^2}$$

$$= \frac{1}{h^2} \left(\left(u(jh, t) + h\partial_x u(jh, t) + \frac{h^2}{2} \partial_x^2 u(jh, t) \right.\right.$$

$$\left. + \frac{h^3}{6} \partial_x^3 u(jh, t) + \frac{h^4}{24} \partial_x^4 u(jh, t) \right)$$

$$+ \left(u(jh, t) - h\partial_x u(jh, t) + \frac{h^2}{2} \partial_x^2 u(jh, t) \right.$$

$$\left. - \frac{h^3}{6} \partial_x^3 u(jh, t) + \frac{h^4}{24} \partial_x^4 u(jh, t) \right)$$

$$\left. -2u(jh, t) + o(h^4) \right)$$

$$= \partial_x^2 u(jh, t) + \frac{h^2}{12} \partial_x^4 u(jh, t) + o(h^2)$$

and

$$X_{j+1}(t) - X_{j-1}(t)$$

$$= u(jh + h, t) - u(jh - h, t)$$

$$= \left(u(jh, t) + h\partial_x u(jh, t) \right) - \left(u(jh, t) - h\partial_x u(jh, t) \right) + o(h)$$

$$= 2h\partial_x u(jh, t) + o(h).$$

Hence, since also $\ddot{X}_j(t) = \partial_t^2 u(jh, t)$, we formally rewrite (28.4) as

$$\partial_t^2 u(jh, t) = c^2 \left(\partial_x^2 u(jh, t) + \frac{h^2}{12} \partial_x^4 u(jh, t) + o(h^2) \right)$$
$$\times \left(1 + h \left(2h \partial_x u(jh, t) + o(h) \right) \right)$$
$$= c^2 \left(\partial_x^2 u(jh, t) + 2h^2 \partial_x u(jh, t) \partial_x^2 u(jh, t) \right.$$
$$\left. + \frac{h^2}{12} \partial_x^4 u(jh, t) + o(h^2) \right).$$

Formally writing x in the place of jh, we thus obtain[8]

$$\frac{1}{c^2} \partial_t^2 u(x, t) - \partial_x^2 u(x, t) = 2h^2 \partial_x u(x, t) \partial_x^2 u(x, t)$$
$$+ \frac{h^2}{12} \partial_x^4 u(x, t) + o(h^2). \qquad (28.5)$$

Defining $U(\xi, \tau) := u \left(\xi + \frac{\tau}{h^2}, \frac{\tau}{ch^2} \right)$, it follows that[9]

$$u(x, t) = U(\xi, \tau), \qquad \text{where } \xi = x - ct \text{ and } \tau = ch^2 t;$$

therefore, omitting the variables for convenience, for each $m \in \mathbb{N}$,

$$\partial_x^m u = \partial_\xi^m U,$$
$$\partial_t u = -c \partial_\xi U + ch^2 \partial_\tau u$$
$$\text{and} \quad \partial_t^2 u = c^2 \partial_\xi^2 U + c^2 h^4 \partial_\tau^2 U - 2c^2 h^2 \partial_{\xi\tau} U.$$

---

[8]Interestingly, if we had sent $h \searrow 0$ in (28.5), we would instead obtain the vibrating string equation $\partial_t^2 u = c^2 \partial_x^2 u$, to be compared with (14.5). This is, in a sense, not too surprising since sending $h \searrow 0$ "too early" would correspond to "neglecting the anelastic perturbation" since ε is proportional to h. The finer analysis that we perform to get (28.6) is therefore necessary to account for the anelastic term in the continuous limit.

[9]One can recall that this substitution is similar to the one utilized to "surf the wave" on p. 112.

In particular,

$$\frac{1}{c^2}\partial_t^2 u - \partial_x^2 u = h^4 \partial_\tau^2 U - 2h^2 \partial_{\xi\tau} U,$$

and (28.5) becomes

$$h^4 \partial_\tau^2 U - 2h^2 \partial_{\xi\tau} U = 2h^2 \partial_\xi U \partial_\xi^2 U + \frac{h^2}{12} \partial_\xi^4 U + o(h^2);$$

thus, dividing by $2h^2$,

$$-\partial_{\xi\tau} U = \partial_\xi U \partial_\xi^2 U + \frac{1}{24} \partial_\xi^4 U + o(1). \tag{28.6}$$

Setting $v := \partial_\xi U$ and formally taking the limit as $h \searrow 0$, we find

$$-\partial_\tau v = v \partial_\xi v + \frac{1}{24} \partial_\xi^3 v, \tag{28.7}$$

which corresponds to the Korteweg–de Vries equation in (10.1) (say, with $a := 1$, $b := 1/24$ and $c := 1$, the numbers being unimportant here).

The link between the Fermi–Pasta–Ulam–Tsingou problem and the Korteweg–de Vries equation is not a mere curiosity based on formal (though a bit sophisticated) Taylor expansions; rather, it has a deep impact on the understanding of the surprising phenomenon highlighted by the oscillators ending up relocating almost all the energy onto the initial mode, as discovered[10] in [ZK65]. Roughly, the existence of solitons in the Korteweg–de Vries equation, which essentially recover their initial shape after interactions (as observed in [ZK65]), is considered the counterpart in the framework of partial differential equations of the near recurrence to the initial state in discretized weakly nonlinear strings. The topic is quite complex and remains a subject of intense scientific investigations; hence, we

---

[10]The name "soliton" was indeed coined in [ZK65] to describe the "solitary waves", which persist after nondestructive interactions. Likely, the name "soliton" is meant to stress the similarities between the behavior of such waves and that of interacting "particles" (since elementary particles have names which end in "-on", such as "electron", "proton", "neutron", "lepton", and "boson", so why not give the name "solitons" to wave packets that maintain their shape while propagating and interacting?).

do not aim at being exhaustive here. Yet, in a sense, the description of the solutions of the Korteweg–de Vries equation (28.7) explored in [ZK65] goes as follows. At the beginning, the term $\partial_\tau v + v \partial_\xi v$ in (28.7) "dominates", and the solution steepens in regions where it has a negative slope (this is indeed the typical behavior of solutions of the Bateman–Burgers equation (26.9), recall Figure 26.4). But, as time flows, the third derivative term $\partial_\xi^3 v$ in equation (28.7) takes over[11] and helps prevent the formation of discontinuities, producing instead some oscillations on the left of the front. Later on, the amplitudes of these oscillations grow until each oscillation achieves an almost steady amplitude (increasing from left to right). These bumps begin to move with a speed proportional to their amplitude (hence the bumps on the right move faster than the ones on the left, causing these bumps to spread apart). Since the domain is bounded[12] these bumps will eventually meet again and interact. But quite surprisingly, shortly after the interaction, the bumps reappear essentially with the same shape.

This is a remarkable property of these solitons, which are able to "pass through each other" by maintaining their silhouettes. The analogy between the behavior of these solitons and the almost return to initial conditions detected in the Fermi–Pasta–Ulam–Tsingou setting is that, from time to time, the solitons return to the positions they had initially, hence almost restoring the initial conditions.

The dynamics of the solitons is clearly explained in [ZK65] with the following wording: "When solitons of very different amplitude approach, their trajectories deviate from straight lines (accelerate) as they 'pass through' one another. During the overlap time interval the joint amplitude of the interacting solitons decreases (in contradistinction to what would happen if two pulses overlapped linearly). [...] When the amplitudes of approaching solitons are comparable they seem to exchange amplitudes and therefore velocities. [...] At [a certain] time T_R all the solitons arrive *almost* in the same phase and almost reconstruct the initial state through the nonlinear interaction.

---

[11]The importance of the third-order term in the Korteweg–de Vries equation and its smoothing effect also appeared in the discussion on p. 123.

[12]Strictly, periodic boundary conditions were considered in [ZK65] rather than fixed endpoints in [FPU55], but we are glossing over this detail, see footnote 7.

This process proceeds onwards, and at $2T_R$ one again has a 'near recurrence' which is not as good as the first recurrence. [...] Because the solitons are remarkably stable entities, preserving their identity through numerous interactions, one would expect this system to exhibit thermalization (complete energy sharing among the corresponding linear normal modes) only after extremely long times, if ever".

For a clear picture of these solitons, see Figure 1 in [ZK65].

The investigation of the Fermi–Pasta–Ulam–Tsingou problem, as well as its continuous counterparts such as the Korteweg–de Vries equation, remains very active, and despite remarkable progress, a definite conclusion has not been reached. A mostly unanimous belief among experts is that the Fermi–Pasta–Ulam–Tsingou problem develops chaotic features for essentially all initial data, but the time required to observe these phenomena may overcome the lifetime of a physical system and significantly challenge the reliability of the numerical simulations. In any case, it was experimentally detected in [IC66] that the recurrent phenomena that appeared in the original Fermi–Pasta–Ulam–Tsingou problem somewhat disappear if the initial data were considered with a sufficiently high energy; in this case, it seems that the system quickly relaxes to energy equipartition, and the time averages of all the harmonic energies approach the same value (compare e.g. Figures 3 and 4 on [BCMM15, p. 240] to appreciate the effect of increasing the initial energy per oscillator). Also, some numerical data seem to indicate that, even when energy is initially given to a small subset of modes of low frequency, for very long times, the system does tend toward a statistical equilibrium, identified by energy equipartition, see e.g. [GGMV92, BG04, BP11, BP21].

In this perspective, the state detected in the original experiment [FPU55], in which, at low energy values, a packet of modes, in addition to the initially excited one, enters the game, possibly returning close to the initial conditions, is only apparently stationary and should instead be considered only a "metastable state" since, on a much longer time scale,[13] the system should evolve toward statistical

---

[13]For completeness, we also mention that the features of the chains of oscillators in dimensions higher than one are quite different from those of the Fermi–Pasta–Ulam–Tsingou problem; for instance, even in two dimensions, the times necessary

equilibrium and energy equipartition. Yet, complete mathematical proofs for these phenomena seem to be missing.

For perturbative approaches to the Fermi–Pasta–Ulam–Tsingou problem (with periodic conditions and a small energy assumption depending on the number of oscillators), see [Rin01, HK08]; in this setting, for small energies, one rigorously obtains many recurrent solutions of the problem of quasi-periodic type.

Another important question related to the Fermi–Pasta–Ulam–Tsingou problem deals with the "thermodynamic limit", in which the number of oscillators N in the chain tends to infinity while keeping the total energy proportional to N. This would be perhaps the most significant question pertaining to the topic in relation to statistical mechanics, whose foundations heavily depend on the understanding of systems with an arbitrarily large number of interacting particles. This problem is also largely open: for questions of this kind, as stated in [BCMM15, Section 3], "of course numerics can just give some indications, while a definitive result can only come from a theoretical result, which is the only one able to attain the limit $N = \infty$". In this context, the results related to the formation of packets and their persistence in chains of N oscillators are typically confined to a regime in which the total energy is of order N^{-3}. Though this limitation is unsatisfactory for physical applications (confining the study to that of large oscillators with extremely low energy), it is somewhat remarkable that different theoretical approaches lead to an assumption about the same energy regime (on the other hand, "numerics do not provide any evidence of changes in the dynamics when energy is increased beyond this limit", see [BCMM15, p. 251]).

The rigorous link between the Fermi–Pasta–Ulam–Tsingou problem and the Korteweg–de Vries equation is also a delicate issue. One of the structural problems in this approximation is that the Fermi–Pasta–Ulam–Tsingou system turns out to be a "singular perturbation" of the Korteweg–de Vries equation, in the sense that the approximation error contains higher-order derivatives.

---

to numerically observe states of equilibrium are substantially shorter than in one dimension, see e.g. [Ben05].

The dependence of chaotic behaviors upon the number of dimensions is also a general feature of the modern theory of dynamical systems since in low dimensions, some systems possess a "topological obstruction" to chaos, see e.g. [FNT06].

See [BCMM15, Section 4.1] for rigorous results about this asymptotic problem.

The influence of the theory of solitons on the rigorous study of the Fermi–Pasta–Ulam–Tsingou problem is showcased in a series of papers: [FP99, FP02, FP04a, FP04b]. See also [Gal08] for an account of the various facets of the Fermi–Pasta–Ulam–Tsingou problem and [PB05, GPR21] for rigorous links between this problem and the Korteweg–de Vries equation.

But let us also emphasize a significant difference between the dynamics of the Fermi–Pasta–Ulam–Tsingou system and that of the Korteweg–de Vries equation: while the recurrent states are nowadays expected to be a rare or transient phenomenon in chain of oscillators, the Korteweg–de Vries equation is a "completely integrable" system (see e.g. [Fad16, pp. 277–284]) and all its solutions (under appropriate boundary conditions) are periodic, quasi-periodic or almost-periodic in time, see [KP03, pp. ix, 1–5]. Hence, differently from what we should expect for the Fermi–Pasta–Ulam–Tsingou problem, solutions of the Korteweg–de Vries equation return infinitely often arbitrarily close to their initial states.

Chapter 29

Definitions Come Later On

Given a, b and $c \geqslant 0$, the equation

$$\frac{\partial^2 u}{\partial x^2} - a\frac{\partial^2 u}{\partial t^2} = b\frac{\partial u}{\partial t} + cu \qquad (29.1)$$

is called[1] the telegrapher's equation.

---

[1]The inventor of equation (29.1) was Oliver Heaviside, see Figure 29.1.

Heaviside's uncle was Sir Charles Wheatstone, co-inventor of the first commercially successful telegraph. Following his uncle's advice, Heaviside became a telegraph operator and an electrician, continuing to study and do science while working.

His contributions to mathematics, physics and engineering were significant and remain classical contributions: they include the use of complex numbers to solve circuit analysis differential equations, the calculus of the deformations of an electromagnetic field surrounding a moving charge and the derivation of the magnetic force on a moving charged particle (the first contribution toward the understanding of the Lorentz Force) and the prediction of the existence of a reflective layer of the ionosphere (which allowed radio waves radiated into the sky to return to earth beyond the horizon).

Several works by Heaviside were of great practical use, including the possible exploitation of loading coils in telephone and telegraph lines to increase their self-induction and correct the distortion which they suffered.

Several years later, some American telecommunications companies hired their own scientists to extend Heaviside's work and adapt the use of coils previously introduced by him. Some corporations later offered Heaviside money in exchange for his rights, but he declined to accept any money unless the company were to give him full recognition for his discoveries and inventions (it turns out that Heaviside remained for all his life chronically poor).

Figure 29.1.   Oliver Heaviside (Public Domain image from Wikipedia).

In equation (29.1), we have that $u = u(x,t)$ with $x \in \mathbb{R}$, $t \in [0, +\infty)$. When $a := 0$, $c := 0$ and $b > 0$, equation (29.1) boils down

---

Heaviside was perhaps a bit eccentric too. He was a firm opponent of Einstein's theory of relativity, possibly beyond reasonable scientific arguments, and at the end of his life, he developed a very strong aversion to meeting people and became a recluse. He died at age 74 after falling from a ladder.

See [Nah88] for a very captivating and detailed biography of Oliver Heaviside.

Heaviside said, "Mathematics is an experimental science, and definitions do not come first, but later on. They make themselves, when the nature of the subject has developed itself".

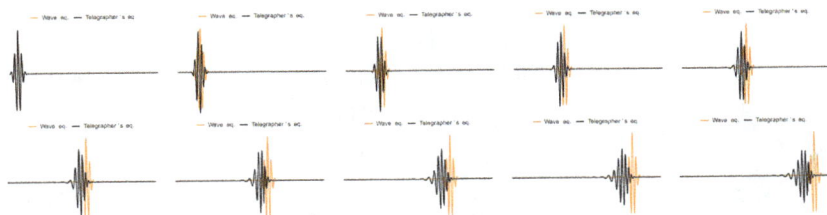

Figure 29.2. Comparison between the evolution in time of the solutions of the wave equation and the telegrapher's equation (author Jacopo Bertolotti, https:// twitter.com/j_bertolotti/status/1172517281374572551, images from Wikipedia, licensed under the Creative Commons CC0 1.0 Universal Public Domain Dedication).

to the heat equation in (1.6). When $b := 0$, $c := 0$ and $a > 0$, it reduces[2] to the wave equation in (7.2).

When a, b and $c > 0$ however the behavior of the solutions of (29.1) are interestingly different from those of the wave equation since solutions of the telegrapher's equation present damping and dispersion effects that are not present in the wave equation. See e.g. Figure 29.2, showing that the evolution of the solution of the telegrapher's equation exhibits a visible dispersion, in which the velocity of travel depends on the frequency, thus enlarging the support of the solution, and loss of intensity, causing[3] the peaks of the traveling front to reduce over time.

The solution u in equation (29.1) describes the electric current intensity (or, in a similar manner, the voltage) along a transmission line. This transmission line can be a telegraph wire, a overhead electrical conductor, a telephone line, etc., see Figure 29.3. See also Figure 29.4 for a peculiar use of transmission lines.

To get to the bottom of the telegrapher's equation (29.1), we consider an electric line made of two parallel electrical wires. There

---

[2]With respect to the terminology in footnote 13 on p. 9, we note that equation (29.1) is hyperbolic when $a > 0$, parabolic when $a := 0 < b$ and elliptic when $a := 0$ and $b := 0$.

[3]These dispersive effects, in which the speed of a signal is not constant but depends on frequency, are typically not good for applications; for instance, when some frequencies travel along long transmission lines at a higher velocity than others, dispersion may cause the signals to become unrecognizable. More about this on p.e 313.

Figure 29.3. (a) Utility pole for a telephone line; (b) electrical wires; (c) overhead lines in Queensland (images from Wikipedia, Public Domain for the first, photo by Novoklimov, licensed under the Creative Commons Attribution-Share Alike 4.0 International, licensed under the Unported license for the second, photo by Pytomelon87 licensed under the Creative Commons Attribution-Share Alike 4.0 International license for the third).

Figure 29.4. White storks nesting on an utility pole (photo by Myrabella; image from Wikipedia, licensed under the Creative Commons Attribution-Share Alike 3.0 Unported license).

is a voltage difference between the wires (e.g. produced by an electric generator "at the end of the wires": since the wires in our idealized model are straight lines, the generator is essentially "at infinity").

Thinking at the two wires simply as parallel straight lines would be however reductive since:

- each straight line is actually made by a conductor which presents some distributed electric resistance (say, modeled by a series resistor) and also some distributed inductance (e.g. due to the magnetic field around the wires and self-inductance, modeled by a series inductor);
- furthermore, the dielectric material separating the two conductors is also not completely neutral from an electric point of view, as it can carry capacitance and conductance (modeled by a shunt capacitor and a resistor located between each "infinitesimal" portion dx of the conductors).

To describe these phenomena, as customary in the physical description of electric phenomena, we reserve the name I for the current, V for the voltage, R for the resistance, L for the inductance and C for the capacity. We also use the letter G to denote the conductance, which is the reciprocal of a resistance. The parameters R, L, C and G will be treated as structural constants, and we will be interested in the description of the functions $I = I(x,t)$ and $V = V(x,t)$ with respect to the position $x \in \mathbb{R}$ on the transmission line and the time $t \geqslant 0$.

More specifically, for concreteness, we focus on the current in the upper conductor, see Figure 29.5; in this setting, the distributed resistance in the infinitesimal elemental length dx of the conductor is denoted by $R\,dx$ and the distributed inductance is denoted by $L\,dx$.

The conductance of the dielectric material separating the two conductors (accounting for bulk conductivity of the dielectric and dielectric loss) is denoted by $G\,dx$ and the capacity by $C\,dx$ (note that, in "real life", there is no wire connecting the top and bottom cables, but Figure 29.5 translates the behavior of the dielectric between the two conductors into the language of electric circuits).

Now, to describe the current flowing through the upper wire, we denote by $I(x,t)$ the current intensity on the left end of the elementary upper conductor in Figure 29.5 and by $I(x + dx, t)$ the one on the right end (say, with the convention that the current is traveling

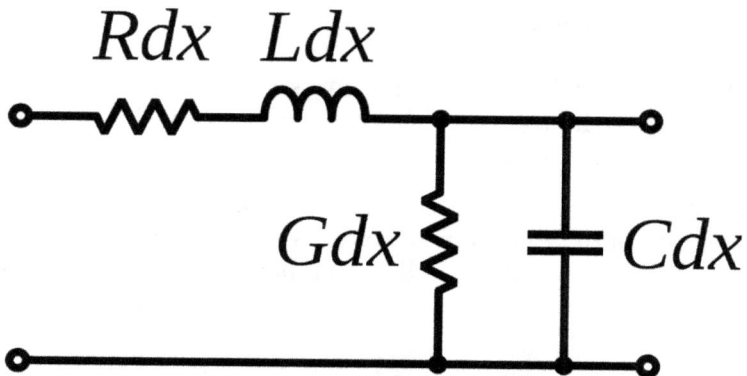

Figure 29.5. Schematic diagram of a transmission line of an elemental length dx (image by Omegatron from Wikipedia, licensed under the Creative Commons Attribution-Share Alike 3.0 Unported license).

from left to right). Similarly, we denote by $V(x, t)$ the voltage on the left end of the elementary upper conductor and by $V(x + dx, t)$ the one on the right end (actually, V would stand for the difference in voltage between the upper and lower conductors; for simplicity[4] one can think that the conductor at the bottom is at zero voltage). Let also ι_1 be the current between the nodes of the upper conductor (oriented left to right), and ι_2 and ι_3 be the currents through the shunt capacitor $C\,dx$ and resistor $G\,dx$, respectively (oriented downward).

---

[4]Actually, telegraphs originally used two wires, utilizing a forward and a return path in a closed circuit to move energy along the transmission line. But it was then observed that earth itself can be used as the return path. In this setting, the bottom wire is simply the ground, which is normalized to be approximately at a constant zero voltage (from practical purposes however, the earth is an adequate, but certainly not optimal, return path). Thus, in the situation in which the transmission line is modeled by one forward cable using the earth as a return path, the poles are also considered dielectric and provide some capacitance and conductance, see Figure 29.6. In this case, the vertical elements related to $G\,dx$ and $C\,dx$ in Figure 29.5 can be thought of as "concrete objects", such as the telegraph poles, located at an "infinitesimal" distance dx at a large scale (in which the transmitter and the receiver are located "at infinity").

See e.g. https://youtu.be/ySuUZEjARPY for a very thorough video about electric transmission.

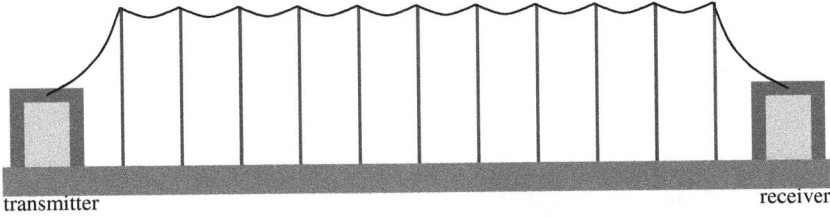

Figure 29.6. Transmission line for a telegraph using the earth as return path.

With this notation, by Kirchhoff's junction rule (or simply by the conservation of charge), we have that

$$I(x,t) = \iota_1 + \iota_3 \quad \text{and} \quad \iota_1 = I(x+dx,t) + \iota_2.$$

Also, by Ohm's and Faraday's laws,

$$V(x,t) = \frac{\iota_3}{G\,dx} \quad \text{and}$$
$$V(x+dx,t) - V(x,t) = -RI(x,t)\,dx - L\partial_t I(x,t)\,dx.$$

Additionally, by the definition of capacity,

$$\partial_t V(x,t) = \frac{\iota_2}{C\,dx}.$$

These considerations lead to

$$\partial_x V(x,t)\,dx \simeq V(x+dx,t) - V(x,t)$$
$$= -RI(x,t)\,dx - L\partial_t I(x,t)\,dx$$

and

$$\partial_x I(x,t)\,dx \simeq I(x+dx,t) - I(x,t) = I(x+dx,t) - \iota_1 - \iota_3$$
$$= -\iota_2 - \iota_3 = -C\partial_t V(x,t)\,dx - GV(x,t)\,dx.$$

Consequently,

$$\begin{cases} \partial_x V = -L\partial_t I - RI, \\ \partial_x I = -C\partial_t V - GV. \end{cases} \tag{29.2}$$

We can take the derivative with respect to x of the first equation and compare it with the derivative with respect to t of the second equation in (29.2), finding that

$$\partial_x^2 V - LC\partial_t^2 V = (RC + GL)\partial_t V + GRV. \qquad (29.3)$$

Similarly, we can take the derivative with respect to t of the first equation and compare it with the derivative with respect to x of the second equation in (29.2), which yields that

$$\partial_x^2 I - LC\partial_t^2 I = (RC + GL)\partial_t I + GRI. \qquad (29.4)$$

We stress that, up to replacing voltage and current, the structures of (29.3) and (29.4) are the same. Also, (29.3) and (29.4) give that both voltage and current are solutions of the telegrapher's equation in (29.1).

It is interesting to note that, in this situation, the structural parameters in the telegrapher's equation in (29.1) correspond to $a := LC$, $b := RC + GL$ and $c := GR$. Thus, a lossless transmission in which the roles of resistances can be neglected corresponds to $R := 0$ and $G := 0$ (this represents a perfect conductor offering no resistance and a perfect electrical insulator as a dielectric allowing for no conductance), which in turn produce $b := 0$ and $c := 0$ in (29.1), thus reducing to the wave equation. With this observation in mind, one can also have a second look at Figure 29.2 to appreciate how the damping and dispersion effects are the outcomes of the imperfect conductors and dielectric in terms of resistance and electrical insulation.

Let us finally return to the dispersion problem (see footnote 3), which annoyingly affects the transmission of signals along extended transmission lines. To this end, given $\kappa > 0$, let us set

$$v_\kappa := \sqrt{\frac{4(GR + \kappa^2)}{(GL + RC)^2 + 4LC\kappa^2}} \quad \text{and} \quad \lambda_\kappa := \frac{(GL + RC)\,v_\kappa}{2},$$

$$(29.5)$$

and observe that the function

$$I(x, t) := e^{-\lambda_\kappa x} \cos\left(\kappa(x - v_\kappa t)\right) \qquad (29.6)$$

is a solution of (29.4).

The function in (29.6) presents an interesting shape dictated by the coefficients involved in this construction. Indeed, on the one hand, the exponential term $e^{-\lambda_\kappa x}$ in (29.6) describes the attenuation of the signal along a transmission line. On the other hand, the cosine term in (29.6) describes a traveling wave, modulated by κ. The speed at which this wave moves is v_κ. The fact that this speed depends on κ is, in principle, a major obstacle to useful signal transmission since different frequencies of a given signal would travel along the line at different rates, producing significant distortion phenomena and quickly making the original signal almost unrecognizable.

This problem was brilliantly solved by Oliver Heaviside, who figured out that a special tuning of the parameters would produce distortion-less transmission lines! Indeed, no dispersion would occur if the line parameters exhibited the following ratio:

$$\frac{R}{L} = \frac{G}{C}. \tag{29.7}$$

As a matter of fact, in this case, $GL = RC$, and we see that (29.5) boils down to

$$v_\kappa = \frac{1}{\sqrt{LC}} \quad \text{and} \quad \lambda_\kappa = \sqrt{GR}.$$

In particular, all frequencies travel with the same speed along the transmission line, and they all present the same attenuation feature. Though condition (29.7) is rather ideal (a bit of dispersion and distortion is unavoidable after all), the mathematical approach adopted by Heaviside and the theoretical solution proposed in (29.7) worked, and long-distance transmission lines finally became feasible.

Chapter 30

Nerd Sniping

"Nerd sniping" is a new sport invented in https://xkcd.com/356/, in which the value of mathematicians is frightfully appreciated, see Figure 30.1. Here, we discuss the solution[1] of the infinite resistor problem presented in the fourth cartoon in Figure 30.1 and its strong relation[2] to elliptic partial differential equations (by which we

---

[1]The infinite resistor problem happened to be studied quite in detail, see in particular [Cse00],

https://www.explainxkcd.com/wiki/index.php/356:_Nerd_Sniping

https://www.mathpages.com/home/kmath668/kmath668.htm

https://physics.stackexchange.com/questions/2072/on-this-infinite-grid-of-resistors-whats-the-equivalent-resistance

and the references therein.

In strict terms, without additional information, the problem is possibly not uniquely posed, but in these pages, we confine ourselves to the case in which the solution is assumed to decay "fast enough" at infinity to validate the formal computations showcased here, though we gloss over any technical aspects related to uniqueness, decay, convergence and infinite cancellations (but let us mention, for instance, that if $\Gamma(k)$ is a solution of (30.3), then so is $\Gamma(k) + k_1$, and the equivalent resistance R in (30.7) would thus be affected by an additional term -4, showing a uniqueness issue unless we add some decay assumption on Γ).

Also, the webcomic xkcd was created by Randall Patrick Munroe, see Figure 30.2.

[2]See also [DS84] for a comprehensive treatment of the deep links between elliptic partial differential equations, random walks and electric networks. In addition, see [PM13, p. 231], [Nah09, Chapter 3] and the references therein for further information about chains of infinitely many resistors.

316   *Elliptic PDEs from an Elementary Viewpoint*

Figure 30.1.   Nerd sniping https://xkcd.com/356/, licensed under the Creative Commons Attribution-NonCommercial 2.5 License.

mean, not the relation between nerds and elliptic partial differential equations, but the one between the infinite resistor problem and a discrete variant of elliptic partial differential equations). Actually, the answer to the problem is

$$\frac{4}{\pi} - \frac{1}{2},\qquad(30.1)$$

but you'll see that it's about the journey, not the destination.

We follow here an approach based on the fundamental solution of a suitable operator reminiscent of the Laplacian. The methods related to fundamental solutions will be refined in Section 2.7 of the companion book [DV23]. For simplicity, we work here in two dimension, but many of the arguments that we present are applicable to higher dimensions and further generality, see [Cse00] and the references therein. Given $\varepsilon > 0$ and a function $u : \mathbb{Z}^2 \to \mathbb{R}$, for all $x \in \mathbb{Z}^2$, we define

$$\mathcal{L}u(x) := \sum_{j=1}^{2} \Big(u(x + e_j) + u(x - e_j) - 2u(x) \Big).\qquad(30.2)$$

Apropos, according to the Google Labs Aptitude Test http://googlesystem.blogspot.com/2005/12/google-labs-aptitude-test.html, Google used the nerd sniping problem as a recruitment question.

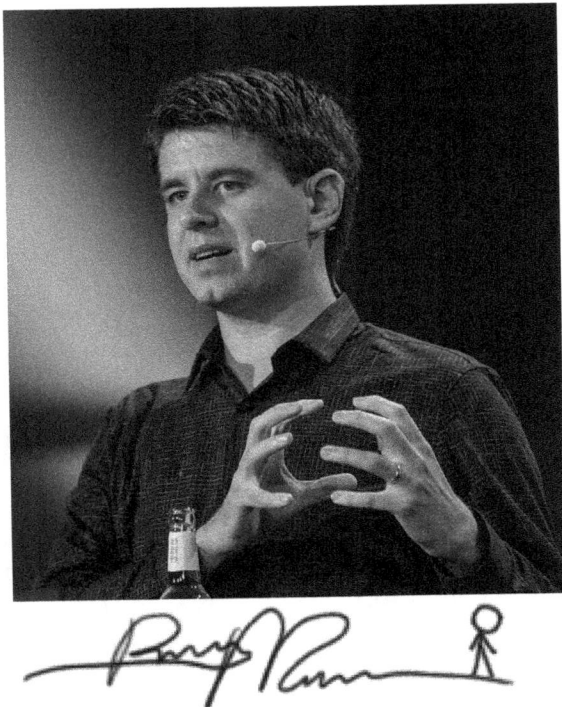

Figure 30.2. Randall Munroe and his signature (images from Wikipedia, the first by re:publica/Jan Zappner, licensed under the Creative Commons Attribution 2.0 Generic license, the second in the Public Domain).

Note that the operator $\mathcal{L}$ can be seen as a discrete version[3] of the Laplacian. Our main goal is to determine the fundamental solution of this operator, namely a function $\Gamma : \mathbb{Z}^2 \to \mathbb{R}$ such that

$$\mathcal{L}\,\Gamma = \delta_0, \tag{30.3}$$

where δ_0 is the Dirac delta function at the origin in the "discrete sense", meaning that

$$\sum_{k \in \mathbb{Z}^2} \varphi(k)\,\delta_0(k) = \varphi(0),$$

for all $\varphi : \mathbb{Z}^2 \to \mathbb{R}$.

---

[3]For more information on the discretization of the Laplace operator (and for the opportunities and dangers entailed by discretization), see e.g. [Str08, Chapter 8].

The convenience of using this fundamental solution is due to the following observation. We put coordinates in $\mathbb{R}^2$ such that the nodes of the infinite resistor problem presented in the fourth cartoon in Figure 30.1 correspond to the lattice $\mathbb{Z}^2$. We can also suppose that the two red points in the lattice in Figure 30.1 correspond to $(0,0)$ and $(2,1)$. Then, to test the resistors, we aim at constructing a distribution of voltage $V : \mathbb{Z}^2 \to \mathbb{R}$ which produces a current with unit intensity from $(0,0)$ to $(2,1)$ and no further net current in the circuit. Hence, we apply Ohm's law to each node k of the lattice, with respect to its first neighbors $k \pm e_j$, for $j \in \{1,2\}$. Namely, by Ohm's law, since all resistors in Figure 30.1 are unitary, the current flowing into the node k from its neighbors $k \pm e_j$ is equal to

$$\sum_{j=1}^{2}(V(k + e_j) - V(k)) + \sum_{j=1}^{2}(V(k - e_j) - V(k)),$$

which in turn equals $\mathcal{L}V(k)$. Hence, the voltage distribution flowing one unit of current from $(0,0)$ to $(2,1)$ is such that

$$\mathcal{L}V(k) = \delta_{(2,1)}(k) - \delta_{(0,0)}(k) = \delta_0(k - (2,1)) - \delta_0(k). \tag{30.4}$$

If we find such a voltage distribution V, we can apply Ohm's law again and determine the requested equivalent resistance R between the two red nodes in Figure 30.1 via the relation

$$R = V(0,0) - V(2,1). \tag{30.5}$$

But to find such a V, it comes in handy to use the fundamental solution Γ; indeed, if we find Γ as in (30.3), it suffices to set

$$V(k) := \Gamma(k - (2,1)) - \Gamma(k) \tag{30.6}$$

and observe that this is a solution of (30.4).

Summarizing, in view of (30.5) and (30.6), once we determine the fundamental solution Γ, we can also find the desired equivalent resistance R using the formal relation

$$R = V(0,0) - V(2,1) = \Big(\Gamma(-2,-1) - \Gamma(0,0)\Big) - \Big(\Gamma(0,0) - \Gamma(2,1)\Big). \tag{30.7}$$

Hence, we focus now on determining the fundamental solution Γ. To this end, we deduce from (30.3) that, for every $\xi \in \mathbb{R}^2$,

$$
\begin{aligned}
1 &= \sum_{k \in \mathbb{Z}^2} \delta_0(k)\, e^{2\pi i k \cdot \xi} \\
&= \sum_{k \in \mathbb{Z}^2} \mathcal{L}\,\Gamma(k)\, e^{2\pi i k \cdot \xi} \\
&= \sum_{\substack{j \in \{1,2\} \\ k \in \mathbb{Z}^2}} \Big(\Gamma(k+e_j) + \Gamma(k-e_j) - 2\Gamma(k) \Big)\, e^{2\pi i k \cdot \xi} \\
&= \sum_{\substack{j \in \{1,2\} \\ m \in \mathbb{Z}^2}} \Gamma(m)\, e^{2\pi i (m - e_j) \cdot \xi} + \sum_{\substack{j \in \{1,2\} \\ m \in \mathbb{Z}^2}} \Gamma(m)\, e^{2\pi i (m + e_j) \cdot \xi} \\
&\quad - 2 \sum_{\substack{j \in \{1,2\} \\ m \in \mathbb{Z}^2}} \Gamma(m)\, e^{2\pi i m \cdot \xi} \\
&= \sum_{\substack{j \in \{1,2\} \\ m \in \mathbb{Z}^2}} \Big(e^{-2\pi i \xi_j} + e^{2\pi i \xi_j} - 2 \Big) \Gamma(m)\, e^{2\pi i m \cdot \xi} \\
&= -\mu(\xi) \sum_{m \in \mathbb{Z}^2} \Gamma(m)\, e^{2\pi i m \cdot \xi},
\end{aligned}
\tag{30.8}
$$

where

$$
\mu(\xi) := 2 \sum_{j=1}^{2} \big(1 - \cos(2\pi \xi_j) \big).
$$

We remark that μ is a periodic function, namely $\mu(\xi + \ell) = \mu(\xi)$, for every $\ell \in \mathbb{Z}^2$ and $\xi \in \mathbb{R}^2$. Thus, $\frac{1}{\mu}$ is also periodic, and then we can write this function as a Fourier Series. For this, to be precise, since μ is nonnegative but may vanish, it is convenient to pick $\delta > 0$ and define $\mu_\delta(\xi) := \max\{\delta, \mu(\xi)\}$, expand the function $\frac{1}{\mu_\delta}$ in Fourier Series and then pass $\delta \searrow 0$ one way or another. Namely, we have that

$$
\frac{1}{\mu_\delta(\xi)} = \sum_{m \in \mathbb{Z}^2} c_{m,\delta}\, e^{2\pi i m \cdot \xi},
$$

where

$$c_{m,\delta} := \int_Q \frac{e^{-2\pi i m \cdot \eta}}{\mu_\delta(\eta)} \, d\eta \quad \text{and} \quad Q := \left(-\frac{1}{2}, \frac{1}{2}\right)^2.$$

Thus, if $\phi : \mathbb{R}^2 \to \mathbb{R}$ is a smooth and periodic function and

$$\widehat{\phi}_m := \int_Q \phi(\xi) \, e^{-2\pi i m \cdot \xi} \, d\xi$$

is the Fourier coefficient of ϕ, we formally deduce from (30.8) that

$$\lim_{\delta \searrow 0} \sum_{m \in \mathbb{Z}^2} c_{m,\delta} \, \widehat{\phi}_{-m} = \lim_{\delta \searrow 0} \int_Q \sum_{m \in \mathbb{Z}^2} c_{m,\delta} \, \phi(\xi) \, e^{2\pi i m \cdot \xi} \, d\xi$$

$$= \lim_{\delta \searrow 0} \int_Q \frac{\phi(\xi)}{\mu_\delta(\xi)} \, d\xi = \int_Q \frac{\phi(\xi)}{\mu(\xi)} \, d\xi \qquad (30.9)$$

$$= -\int_Q \sum_{m \in \mathbb{Z}^2} \Gamma(m) \, e^{2\pi i m \cdot \xi} \, \phi(\xi) \, d\xi = - \sum_{m \in \mathbb{Z}^2} \Gamma(m) \, \widehat{\phi}_{-m}.$$

We can thereby apply (30.9) to the function $\psi(\xi) := \phi(\xi) - \phi(0)$ and find that

$$\lim_{\delta \searrow 0} \sum_{m \in \mathbb{Z}^2} \int_Q \widehat{\psi}_{-m} \frac{e^{-2\pi i m \cdot \eta} \, d\eta}{\mu_\delta(\eta)} = \lim_{\delta \searrow 0} \sum_{m \in \mathbb{Z}^2} c_{m,\delta} \, \widehat{\psi}_{-m}$$

$$= - \sum_{m \in \mathbb{Z}^2} \Gamma(m) \, \widehat{\psi}_{-m}. \qquad (30.10)$$

Also, since

$$\sum_{m \in \mathbb{Z}^2} \widehat{\psi}_{-m} = \sum_{m \in \mathbb{Z}^2} \widehat{\psi}_m = \sum_{m \in \mathbb{Z}^2} \widehat{\psi}_m \, e^{2\pi i m \cdot 0} = \psi(0) = 0,$$

we have that

$$\int_Q \sum_{m \in \mathbb{Z}^2} \widehat{\psi}_{-m} \frac{d\eta}{\mu_\delta(\eta)} = 0.$$

This and (30.10) lead to

$$\lim_{\delta \searrow 0} \sum_{m \in \mathbb{Z}^2} \int_Q \widehat{\psi}_{-m} \frac{(e^{-2\pi i m \cdot \eta} - 1) \, d\eta}{\mu_\delta(\eta)} = - \sum_{m \in \mathbb{Z}^2} \Gamma(m) \, \widehat{\psi}_{-m}. \qquad (30.11)$$

We also point out that μ_δ is an even function and, therefore,

$$\int_Q \frac{\sin(2\pi m \cdot \eta)\, d\eta}{\mu_\delta(\eta)} = 0.$$

From this and (30.11), it follows that

$$\lim_{\delta \searrow 0} \sum_{m \in \mathbb{Z}^2} \int_Q \widehat{\psi}_{-m} \frac{\big(\cos(2\pi m \cdot \eta) - 1\big)\, d\eta}{\mu_\delta(\eta)} = -\sum_{m \in \mathbb{Z}^2} \Gamma(m)\, \widehat{\psi}_{-m}.$$

$$(30.12)$$

The advantage of (30.12) with respect to (30.10) is that the singularity at the denominator in the first integral is compensated by the vanishing of the corresponding numerator, thus allowing us to pass to the limit inside the integral and write that

$$\sum_{m \in \mathbb{Z}^2} \int_Q \widehat{\psi}_{-m} \frac{\big(\cos(2\pi m \cdot \eta) - 1\big)\, d\eta}{\mu(\eta)} = -\sum_{m \in \mathbb{Z}^2} \Gamma(m)\, \widehat{\psi}_{-m}. \quad (30.13)$$

We also observe that

$$\widehat{\psi}_0 = \widehat{\phi}_0 - \phi(0) = \widehat{\phi}_0 - \sum_{m \in \mathbb{Z}^2} \widehat{\phi}_m e^{2\pi i m \cdot 0} = \widehat{\phi}_0 - \sum_{m \in \mathbb{Z}^2} \widehat{\phi}_{-m}$$

and $\widehat{\psi}_m = \widehat{\phi}_m$, for all $m \in \mathbb{Z}^2 \setminus \{0\}$. Plugging this information into (30.13), we find that

$$\sum_{m \in \mathbb{Z}^2} \int_Q \widehat{\phi}_{-m} \frac{\big(\cos(2\pi m \cdot \eta) - 1\big)\, d\eta}{\mu(\eta)} = \sum_{m \in \mathbb{Z}^2} \big(\Gamma(0) - \Gamma(m)\big)\, \widehat{\phi}_{-m}.$$

The arbitrariness of ϕ thus gives that

$$\Gamma(m) - \Gamma(0) = -\int_Q \frac{\big(\cos(2\pi m \cdot \eta) - 1\big)\, d\eta}{\mu(\eta)}$$

$$= \frac{1}{2} \iint_{(-1/2,1/2)\times(-1/2,1/2)} \frac{1 - \cos(2\pi m_1 \eta_1 + 2\pi m_2 \eta_2)}{2 - \cos(2\pi \eta_1) - \cos(2\pi \eta_2)}\, d\eta_1\, d\eta_2.$$

We insert this into the equivalent resistance relation (30.7), and we arrive at

$$R = \iint_{(-1/2,1/2)\times(-1/2,1/2)} \frac{1 - \cos(4\pi \eta_1 + 2\pi \eta_2)}{2 - \cos(2\pi \eta_1) - \cos(2\pi \eta_2)}\, d\eta_1\, d\eta_2.$$

$$(30.14)$$

The good news is that this is an integral involving simple trigonometric functions, so we should feel confident about solving it explicitly thanks to our consolidated calculus skills. For instance, we can proceed as follows. We note that

$$\cos(4\pi\eta_1 + 2\pi\eta_2) = \cos(4\pi\eta_1)\cos(2\pi\eta_2) - \sin(4\pi\eta_1)\sin(2\pi\eta_2)$$

$$= \left(1 - 2\sin^2(2\pi\eta_1)\right)\cos(2\pi\eta_2) - 2\sin(2\pi\eta_1)\cos(2\pi\eta_1)\sin(2\pi\eta_2).$$

$$(30.15)$$

Hence, if $\tau_1 := \tan(\pi\eta_1)$ and $\tau_2 := \tan(\pi\eta_2)$, using in (30.15) the "tangent half-angle formulas"

$$\sin(2\pi\eta_j) = \frac{2\tau_j}{1 + \tau_j^2} \quad \text{and} \quad \cos(2\pi\eta_j) = \frac{1 - \tau_j^2}{1 + \tau_j^2},$$

we find that

$$1 - \cos(4\pi\eta_1 + 2\pi\eta_2) = \frac{2(2\tau_1 + \tau_2 - \tau_1^2\tau_2)^2}{(1 + \tau_1^2)^2 (1 + \tau_2^2)}. \qquad (30.16)$$

Similarly,

$$2 - \cos(2\pi\eta_1) - \cos(2\pi\eta_2) = 2 - \frac{1 - \tau_1^2}{1 + \tau_1^2} - \frac{1 - \tau_2^2}{1 + \tau_2^2}$$

$$= \frac{2(\tau_1^2 + \tau_2^2 + 2\tau_1^2\tau_2^2)}{(1 + \tau_1^2)(1 + \tau_2^2)}. \qquad (30.17)$$

Thus, since

$$d\eta_j = \frac{d\tau_j}{\pi(1 + \tau_j^2)},$$

we infer from (30.16) and (30.17) that

$$\frac{1 - \cos(4\pi\eta_1 + 2\pi\eta_2)}{2 - \cos(2\pi\eta_1) - \cos(2\pi\eta_2)}\, d\eta_1\, d\eta_2$$

$$= \frac{(\tau_1^2\tau_2 - 2\tau_1 - \tau_2)^2}{\pi^2(1 + \tau_1^2)^2 (1 + \tau_2^2) (2\tau_1^2\tau_2^2 + \tau_1^2 + \tau_2^2)}\, d\tau_1\, d\tau_2$$

This and (30.14) lead to

$$R = \iint_{\mathbb{R}\times\mathbb{R}} \frac{(\tau_1^2\tau_2 - 2\tau_1 - \tau_2)^2}{\pi^2(1+\tau_1^2)^2\,(1+\tau_2^2)\,(2\tau_1^2\tau_2^2 + \tau_1^2 + \tau_2^2)}\, d\tau_1\, d\tau_2$$

$$= \iint_{\mathbb{R}\times\mathbb{R}} \frac{(xy^2 - x - 2y)^2}{\pi^2(1+x^2)\,(1+y^2)^2\,(2x^2y^2 + x^2 + y^2)}\, dx\, dy,$$

where we simplified the notation using the variables (x, y) in place of (τ_2, τ_1). Since the latter integral remains invariant under the map $(x, y) \mapsto (-x, -y)$, we can reduce the previous equation to

$$R = \iint_{\mathbb{R}\times(0,+\infty)} \frac{2(xy^2 - x - 2y)^2}{\pi^2(1+x^2)\,(1+y^2)^2\,(2x^2y^2 + x^2 + y^2)}\, dx\, dy. \tag{30.18}$$

Now, one can check that a primitive of the function $\frac{(1+y^2)(xy^2-x-2y)^2}{(1+x^2)(2x^2y^2+x^2+y^2)}$ in the variable x is

$$\Phi(x, y) := 2y(y^2 - 1)\ln\frac{1+x^2}{2x^2y^2 + x^2 + y^2} + (y^4 - 6y^2 + 1)\arctan x$$

$$+ \frac{y(10y^2 - y^4 + 3)\arctan\frac{x\sqrt{2y^2+1}}{y}}{\sqrt{2y^2+1}}.$$

Hence, for every $y \in (0, +\infty)$,

$$\Phi(\pm\infty, y) = 2y(y^2 - 1)\ln\frac{1}{2y^2+1} \pm \frac{\pi}{2}$$

$$\times \left((y^4 - 6y^2 + 1) + \frac{y(10y^2 - y^4 + 3)}{\sqrt{2y^2+1}} \right).$$

As a result, for all $y \in (0, +\infty)$,

$$\int_{\mathbb{R}} \frac{(1+y^2)(xy^2 - x - 2y)^2}{(1+x^2)(2x^2y^2 + x^2 + y^2)}\, dx = \Phi(+\infty, y) - \Phi(-\infty, y)$$

$$= \pi(y^4 - 6y^2 + 1) + \frac{\pi y(10y^2 - y^4 + 3)}{\sqrt{2y^2+1}}$$

and, consequently,

$$\int_{\mathbb{R}} \frac{2(xy^2 - x - 2y)^2}{\pi^2(1 + x^2)\,(1 + y^2)^2\,(2x^2y^2 + x^2 + y^2)}\,dx$$
$$= \frac{2(y^4 - 6y^2 + 1)}{\pi(1 + y^2)^3} + \frac{2y(10y^2 - y^4 + 3)}{\pi(1 + y^2)^3\sqrt{2y^2 + 1}}.$$

Combining this with (30.18), we find that

$$R = \int_0^{+\infty} \frac{2(y^4 - 6y^2 + 1)}{\pi(1 + y^2)^3}\,dy + \int_0^{+\infty} \frac{2y(10y^2 - y^4 + 3)}{\pi(1 + y^2)^3\sqrt{2y^2 + 1}}\,dy$$

$$= \frac{2(y - y^3)}{\pi(1 + y^2)^2}\bigg|_{y=0}^{y=+\infty} - \frac{2}{\pi}\left(\frac{2\sqrt{2y^2 + 1}}{(1 + y^2)^2} + \arctan\sqrt{2y^2 + 1}\right)\bigg|_{y=0}^{y=+\infty}$$

$$= -\frac{2}{\pi}\left(\frac{\pi}{2} - 2 - \arctan 1\right)$$

$$= \frac{\pi}{4} - \frac{1}{2},$$

in accordance with (30.1), as desired.

Chapter 31

The Streetsweeper Problem

Martin Luther King Jr. (see Figure 31.1) once said, "What I'm saying to you this morning, my friends, even if it falls your lot to be a streetsweeper, go on out and sweep streets like Michelangelo painted pictures; sweep streets like Handel and Beethoven composed music; sweep streets like Shakespeare wrote poetry; sweep streets so well that all the host of heaven and earth will have to pause and say: Here lived a great streetsweeper who swept his job well". At the end of the day, whatever we do and whatever the reason, the competence and dedication that we put in what we do makes all the difference in the world, and a large number of tasks and heavy duties are essential for the prosperity of the whole community. For this reason, and many others, the streetsweeper is represented in several monuments worldwide, see e.g. Figure 31.2.

The problem[1] that we present here is also inspired by the noble figure of the streetsweeper. Suppose that the task of a streetsweeper is to clean a finite collection of tiles in the floor (see Figure 31.3 for concrete example of tessellated floors). To derive a mathematical model describing this job, we can consider a tiling of the plane $\mathbb{R}^2$ made of square tiles of the form $\left(k_1 - \frac{1}{2}, k_1 + \frac{1}{2}\right) \times \left(k_2 - \frac{1}{2}, k_2 + \frac{1}{2}\right)$ with $k = (k_1, k_2) \in \mathbb{Z}^2$. In this way, every tile can be indexed by the integer coordinates k of its center. We suppose that initially every

---

[1]This problem was kindly suggested to us by Héctor Chang-Lara, see also [CL23] and the beautiful video: https://www.youtube.com/watch?v=--iUT5IYxZs for further inspiration.

Figure 31.1. Martin Luther King Jr. speaking against the Vietnam War at the University of Minnesota (image of the Minnesota Historical Society from Wikipedia, licensed under the Creative Commons Attribution-Share Alike 2.0 Generic license).

Figure 31.2. Left: Monument of streetsweeper in Saint Petersburg (Public Domain image from Wikipedia). Right: Monument of streetsweeper in Madrid (photo by Alex Pascual Guardia, image from Wikipedia, licensed under the Creative Commons Attribution 2.0 Generic license).

Figure 31.3. Left: natural tessellated pavement at Eaglehawk Neck, Tasmania (photo by J. J. Harrison, image from Wikipedia, licensed under the Creative Commons Attribution-Share Alike 2.5 Generic license). Right: Piazza degli Scacchi, Marostica, Italy (detail of a photo by Rino Porrovecchio, image from Wikipedia, licensed under the Creative Commons Attribution-Share Alike 2.0 Generic license).

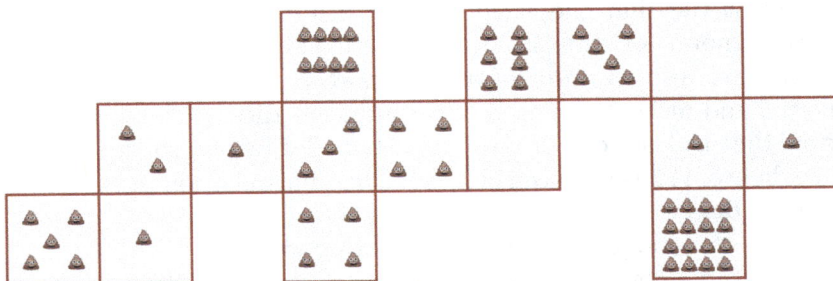

Figure 31.4. A tessellated floor to be cleaned.

tile contains a certain amount of rubbish. For this, we write that the tile centered at $k \in \mathbb{Z}^2$ contains initially a quantity $\mu_0(k)$ of rubbish. In this way, we can think that $\mu_0 : \Omega \to [0, +\infty)$, where $\Omega \subset \mathbb{Z}^2$ is the finite collection of the centers of the tiles, see Figure 31.4. For simplicity, we suppose that

each tile of Ω possesses at least one adjacent tile in Ω. $\qquad$ (31.1)

This is to exclude the somewhat trivial case of having single and isolated tiles.

Then, the cleaning process works as follows: the streetsweeper selects a tile of Ω (say, the one centered at some $k \in \mathbb{Z}^2$) and distributes the rubbish present there into the four adjacent tiles (namely, the ones centered at $k + (0,1)$, $k - (0,1)$, $k + (1,0)$ and $k - (1,0)$), regardless of whether they belong to Ω or not, and then repeats the process by selecting another tile, distributing its rubbish into the adjacent tiles, and so on. The rubbish that ends up outside Ω is left there (in particular, tiles outside Ω but adjacent to tiles in Ω get over time filled in by the rubbish that gets removed from Ω).

The goal of the streetsweeper is thus to clean Ω from all the rubbish. Note that

this cleaning task can only be accomplished "in the limit",

that is only after "infinitely many iterations of the process".

$$(31.2)$$

Indeed suppose, by contradiction, that after a finite number of iterations, the streetsweeper has removed all the rubbish from Ω. Consider the latest tile that the streetsweeper has cleaned, and let $\mu_\star > 0$ be the amount of rubbish located there right before being cleaned. Then, the latest operation of the streetsweeper was to remove $\mu_\star$ from the tile and place $\frac{\mu_\star}{4}$ on each of the adjacent tiles. But, by (31.1), we know that at least one of these adjacent tiles belong to Ω; therefore, the whole of Ω has not been cleaned. This contradiction proves (31.2), as desired.

To better model the cleaning process, we denote by $\mu_j : \mathbb{Z}^2 \rightarrow [0, +\infty)$ the amount of rubbish located at each tile after j iterations of the clean process (in this way, having denoted by μ_0 the initial rubbish distribution, we have that μ_1 stands for the rubbish distribution after one step, and so on). We stress that μ_j does depend on the strategy chosen by the streetsweeper to clean the floor, and this strategy is not "commutative" (that is, cleaning one tile and then another produces, in general, a different result from the same operation performed in the reverse order). As an example for this lack of commutativity, we can consider the simple case in which $\Omega := \{(0,0), (0,1)\}$, $\mu_0(0,0) = 2$ and $\mu_0(0,1) = 1$. If the streetsweeper decided to clean first the tile centered at $(0,0)$ and then the one centered at $(0,1)$, the corresponding result would be $\mu_1(0,0) = 0$, $\mu_1(0,1) = \frac{3}{2}$,

$\mu_2(0,0) = \frac{3}{8}$ and $\mu_2(0,1) = 0$. Instead, if the streetsweeper decided
to clean first the tile centered at $(0,1)$ and then the one centered
at $(0,0)$, the corresponding result would be $\mu_1(0,0) = \frac{9}{4}, \mu_1(0,1) = 0$,
$\mu_2(0,0) = 0$ and $\mu_2(0,1) = \frac{9}{16}$, thus showing that the cleaning process is not commutative, and the cleaning strategy plays a role in the
subsequent rubbish distributions (see Figure 31.5 for a sketch of the
two different cleaning strategies).

However, quite surprisingly,

if the streetsweeper manages to clean Ω in the limit,

then the limit distribution of rubbish outside Ω (31.3)

is independent of the successful cleaning strategy chosen.

This is a rather deep result, and to prove it we will need to introduce
and understand some aspects of elliptic partial differential equations
(at least in the version of discrete mathematics that we have already
encountered in Chapter 30). To this end, it is convenient to denote
by $u_j : \mathbb{Z}^2 \to [0, +\infty)$ the function that keeps a record of "how much
rubbish went out in total from a given tile after j iterations of the

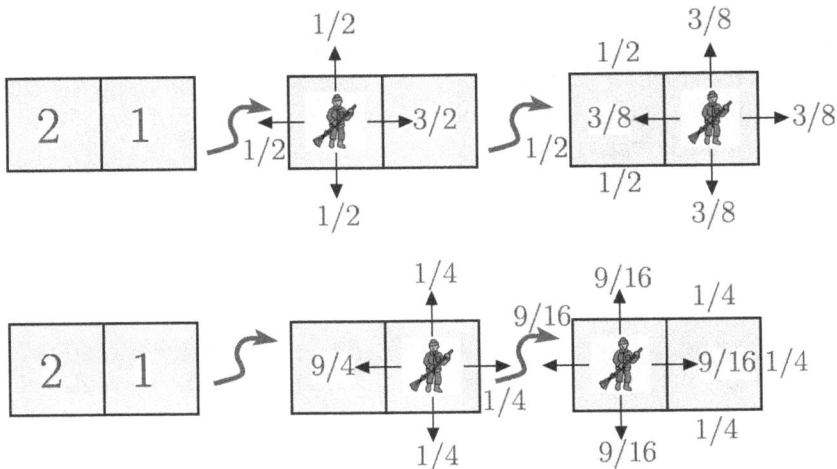

Figure 31.5.   Two different cleaning strategies in a very simple example.

cleaning process" (with $u_0 := 0$). Note that

$$
u_j(k) = \begin{cases} u_{j-1}(k) + \mu_{j-1}(k) & \begin{array}{l} \text{if the tile centered at } k \text{ is the one} \\ \text{selected for the } j\text{th iteration of the} \\ \text{cleaning process}, \end{array} \\ u_{j-1}(k) & \text{otherwise.} \end{cases}
$$

(31.4)

In particular, for a given $k \in \mathbb{Z}^2$, the sequence $u_j(k)$ is monotone; therefore, we can define

$$
u_\infty(k) := \lim_{j \to +\infty} u_j(k). \tag{31.5}
$$

Furthermore, since the cleaning process is only performed at the tiles in Ω, no rubbish is taken outside the tiles centered at $\mathbb{Z}^2 \setminus \Omega$, meaning that $u_j(k) = 0$, whence

$$
u_\infty(k) = 0 \text{ for every } k \in \mathbb{Z}^2 \setminus \Omega. \tag{31.6}
$$

We can also consider $\widetilde{u}_j := u_j - u_{j-1}$ to be the rubbish taken out at the jth step of the cleaning process. Note that $\widetilde{u}_j(k) \geqslant 0$, for all $k \in \mathbb{Z}^2$. We stress that, at the jth iteration, at the tile centered at k, the rubbish can increase or decrease; specifically, it decreases by $\widetilde{u}_j(k)$ (which is strictly positive, coinciding with $\mu_{j-1}(k)$, only if k is the tile selected to be cleaned at the jth step of the cleaning process) and increases by

$$
\frac{1}{4} \sum_{\substack{i \in \{1,2\} \\ \sigma \in \{-1,+1\}}} \widetilde{u}_j(k + \sigma e_i)
$$

(which is the amount of rubbish possibly received by the adjacent tiles). This observation can be formalized through the formula

$$
\mu_j(k) = \mu_{j-1}(k) - \widetilde{u}_j(k) + \frac{1}{4} \sum_{\substack{i \in \{1,2\} \\ \sigma \in \{-1,+1\}}} \widetilde{u}_j(k + \sigma e_i).
$$

This leads to

$$\mu_j(k) + u_j(k) - \frac{1}{4} \sum_{\substack{i \in \{1,2\} \\ \sigma \in \{-1,+1\}}} u_j(k + \sigma e_i)$$

$$= \mu_j(k) + \tilde{u}_j(k) - \frac{1}{4} \sum_{\substack{i \in \{1,2\} \\ \sigma \in \{-1,+1\}}} \tilde{u}_j(k + \sigma e_i) + u_{j-1}(k)$$

$$- \frac{1}{4} \sum_{\substack{i \in \{1,2\} \\ \sigma \in \{-1,+1\}}} u_{j-1}(k + \sigma e_i)$$

$$= \mu_{j-1}(k) + u_{j-1}(k) - \frac{1}{4} \sum_{\substack{i \in \{1,2\} \\ \sigma \in \{-1,+1\}}} u_{j-1}(k + \sigma e_i);$$

therefore, by iteration, for each $\ell \in \{0, \ldots, j\}$,

$$\mu_j(k) + u_j(k) - \frac{1}{4} \sum_{\substack{i \in \{1,2\} \\ \sigma \in \{-1,+1\}}} u_j(k + \sigma e_i)$$

$$= \mu_{j-\ell}(k) + u_{j-\ell}(k) - \frac{1}{4} \sum_{\substack{i \in \{1,2\} \\ \sigma \in \{-1,+1\}}} u_{j-\ell}(k + \sigma e_i).$$

From this, we arrive at

$$\mu_j(k) + u_j(k) - \frac{1}{4} \sum_{\substack{i \in \{1,2\} \\ \sigma \in \{-1,+1\}}} u_j(k + \sigma e_i)$$

$$= \mu_0(k) + u_0(k) - \frac{1}{4} \sum_{\substack{i \in \{1,2\} \\ \sigma \in \{-1,+1\}}} u_0(k + \sigma e_i)$$

$$= \mu_0(k),$$

which can be seen as a "conservation law for the rubbish" (that is, the initial rubbish μ_0 cannot disappear).

Hence, using the notation for the discrete Laplacian introduced in (30.2), we can write that

$$\mu_j(k) - \frac{1}{4}\mathcal{L}u_j(k) = \mu_0(k). \tag{31.7}$$

This and (31.5) also allow us to write that

$$\mu_0(k) + \frac{1}{4}\mathcal{L}u_\infty(k) = \mu_0(k) + \lim_{j \to +\infty} \frac{1}{4}\mathcal{L}u_j(k)$$

$$= \lim_{j \to +\infty} \mu_j(k) =: \mu_\infty(k), \tag{31.8}$$

thus ensuring that the final distribution of rubbish is indeed well-defined.

This observation also entails that

the streetsweeper can always find a successful

strategy to clean any given floor. $\qquad$ (31.9)

Namely, it is always possible to select the sequence of tiles to clean such that $\mu_\infty(k) = 0$, for all $k \in \Omega$. As a matter of fact, it suffices for the streetsweeper to

pick a strategy that cleans each tile infinitely many times.

(31.10)

This means that for every $k \in \Omega$, there exists a sequence $j_m^{(k)}$ such that $j_m^{(k)} \to +\infty$ as $m \to +\infty$ and $\mu_{j_m^{(k)}}(k) = 0$ (note indeed that this condition follows from cleaning the tile centered at k in the $j_m^{(k)}$th step of the process). It thus follows from the existence of the limit in (31.8) that

$$\mu_\infty(k) = \lim_{m \to +\infty} \mu_{j_m^{(k)}}(k) = 0,$$

which establishes (31.9), as desired.

Let us now return to the discrete Laplacian and make the following[2] observation:

> if $w : \mathbb{Z}^2 \to \mathbb{R}$ is such that $\mathcal{L}w(k) \leqslant 0$ for all $k \in \Omega$ and
>
> $w(k) \geqslant 0$ for all $k \in \mathbb{Z}^2 \setminus \Omega$, then $w(k) \geqslant 0$ in Ω. $\quad$ (31.11)

Indeed, suppose not and pick $k_\star \in \Omega$ such that

$$\min_{k \in \Omega} w(k) = w(k_\star) < 0.$$

It thus follows that

$$0 \geqslant \mathcal{L}w(k_\star) = \sum_{i=1}^{2} (w(k_\star + e_i) + w(k_\star - e_i) - 2w(k_\star)). \quad (31.12)$$

Denoting by ℓ the number of tiles that are adjacent to the one centered at $k_\star$ and belonging to Ω, this leads to

$$4w(k_\star) \geqslant \sum_{\substack{i \in \{1,2\} \\ \sigma \in \{-1,+1\} \\ k_\star + \sigma e_i \in \Omega}} w(k_\star + \sigma e_i) + \sum_{\substack{i \in \{1,2\} \\ \sigma \in \{-1,+1\} \\ k_\star + \sigma e_i \in \mathbb{Z}^2 \setminus \Omega}} w(k_\star + \sigma e_i) \geqslant \ell w(k_\star) + 0;$$

therefore,

$$(4 - \ell)w(k_\star) \geqslant 0. \quad (31.13)$$

Note that $\ell \leqslant 4$. If $\ell < 4$, then we would deduce from (31.13) that $w(k_\star) \geqslant 0$, thus obtaining a contradiction.

As a consequence, we conclude that $\ell = 4$, which is the tile centered at $k_\star$ is surrounded by tiles belonging to Ω. As a result, we infer from (31.12) that $w(k_\star + e_i) = w(k_\star - e_i) = w(k_\star)$ for all $i \in \{1, 2\}$. Iterating this argument, we obtain that $k_\star \pm m e_i \in \Omega$ for every $i \in \{1, 2\}$ and every $m \in \mathbb{N}$, thus contradicting the finiteness of Ω, and this contradiction completes the proof of (31.11).

---

[2] Observations such as that in (31.11) will naturally lead to the development of the theory of maximum principles for elliptic partial differential equations. See Section 2.9 of the companion book [DV23] for more details on this topic.

Among the many interesting consequences of (31.11), by creatively exploiting an argument[3] from linear algebra, we deduce an interesting existence and uniqueness result[4] for the discrete Laplacian. Namely, for every $f : \Omega \to \mathbb{R}$ and every $g : \mathbb{Z}^2 \setminus \Omega \to \mathbb{R}$

there exists a unique $U : \mathbb{Z}^2 \to \mathbb{R}$ such that

$$\begin{cases} \mathcal{L}U(k) = f(k) & \text{for all } k \in \Omega, \\ U(k) = g(k) & \text{for all } k \in \mathbb{Z}^2 \setminus \Omega. \end{cases} \qquad (31.14)$$

To check this, we observe that, without loss of generality, we can suppose that

$$g(k) = 0 \text{ for all } k \in \mathbb{Z}^2 \setminus \Omega \qquad (31.15)$$

since solutions of (31.14) correspond to solutions of

$$\begin{cases} \mathcal{L}\widehat{U}(k) = \widehat{f}(k) & \text{for all } k \in \Omega, \\ \widehat{U}(k) = 0 & \text{for all } k \in \mathbb{Z}^2 \setminus \Omega \end{cases} \qquad (31.16)$$

under the transformations defined, for all $k \in \mathbb{Z}^2$, by $\widehat{U}(k) := U(k) - G(k)$ and $\widehat{f}(k) := f(k) - \mathcal{L}G(k)$, where

$$G(k) := \begin{cases} g(k) & \text{if } k \in \mathbb{Z}^2 \setminus \Omega, \\ 0 & \text{otherwise.} \end{cases}$$

Thus, we focus on the establishment of an existence and uniqueness theory of (31.14) using the additional information in (31.15).

---

[3]Other approaches are possible as well, including a finite-dimensional minimization argument. As a matter of fact, to establish (31.3), only the uniqueness result in (31.14) will be exploited, but given the elegance of the argument provided, we decided to state and prove (31.14) in its complete form. Also, we will use the existence result in (31.14) in the forthcoming equation (31.24).

Comparing the terminology with that of Section 2.7 of the companion book [DV23], one can consider (31.14) as a discrete version of the Poisson equation.

[4]For a corresponding existence and uniqueness result about the classical Laplace operator, we will have to introduce more sophisticated mathematical tools and patiently wait for Corollary 3.3.4 of the forthcoming book [DV23].

To achieve this, the main observation is that, after setting $U(k) := g(k) = 0$ for all $k \in \mathbb{Z}^2 \setminus \Omega$, to solve (31.14), one only needs to determine the values of U at the finitely many points $k_1, \ldots, k_N$ of Ω (here, we denote by N the finite cardinality of Ω). Therefore, since the operator $\mathcal{L}$ is linear, we can consider the vector $X := (U(k_1), \ldots, U(k_N)) \in \mathbb{R}^N$ and write the first equation in (31.14) as

$$AX = Y,$$

for a suitable $N \times N$ matrix A and a suitable vector $Y \in \mathbb{R}^N$. We note that Y depends only the values of f, which are given once and for all; more explicitly, in light of (31.15), for all $k \in \Omega$, one can recast the first equation in (31.14) in the form

$$\sum_{i=1}^{2}(U(k+e_i)\chi_\Omega(k+e_i) + U(k-e_i)\chi_\Omega(k-e_i) - 2U(k))$$

$$= \mathcal{L}U(k) = f(k),$$

and we stress that the latter term is given based on f and can be considered as an N-dimensional array Y (with one entry for each $k \in \Omega$), while the first term is a linear map applied to the vector X.

Consequently, to prove that the first equation in (31.14) admits one and only one solution, we need to check that the matrix A is invertible. To this end, by the fundamental theorem of linear algebra (see e.g. [Tay20, Proposition 1.3.6]), it suffices to check that

the linear map associated with A is injective. $\qquad$ (31.17)

For this, suppose that there exists $X_\star = (X_{\star,1}, \ldots, X_{\star,N}) \in \mathbb{R}^N$ such that

$$AX_\star = 0. \qquad (31.18)$$

By letting $U_\star(k_j) := X_{\star,j}$ for every $j \in \{1, \ldots, N\}$, as well as $U_\star(k) := 0$ for every $k \in \mathbb{Z}^2 \setminus \Omega$, we can construct a function $U_\star : \mathbb{Z}^2 \to \mathbb{R}$ satisfying

$$\begin{cases} \mathcal{L}U_\star(k) = 0 & \text{for all } k \in \Omega, \\ U_\star(k) = 0 & \text{for all } k \in \mathbb{Z}^2 \setminus \Omega. \end{cases}$$

By applying (31.11) to both $U_\star$ and $-U_\star$, we thus conclude that $U_\star(k) = 0$ for each $k \in \mathbb{Z}^2$, and hence $X_\star = 0$, which completes the proof of (31.17).

Now, with this preliminary work, we can complete the proof of the surprising result in (31.3) by arguing as follows. We consider two different successful cleaning strategies (the existence of a cleaning strategy being warranted by (31.9)), and we denote by μ_j and μ'_j the corresponding rubbish densities after j steps of each strategy. Note that $\mu'_0 = \mu_0$ since the initial rubbish distribution is given. Since the strategies are successful, we know that $\mu_\infty(k) = 0 = \mu'_\infty(k)$ for every $k \in \Omega$ (recall that these limit densities are well-defined owing to (31.8)).

Thus, to prove (31.3), we need to show that

$$\mu_\infty(k) = \mu'_\infty(k) \text{ for every } k \in \mathbb{Z}^2. \tag{31.19}$$

For this, we let u_j and u'_j be the functions accounting for the total rubbish taken out after j steps of the cleaning process, corresponding to the two cleaning strategies (as formalized in (31.4)). Thanks to (31.5), we can also consider their limit configurations: u_∞ and u'_∞. We observe that, as a consequence of (31.7), for all $k \in \Omega$,

$$-\frac{1}{4}\mathcal{L}u_\infty(k) = \mu_\infty(k) - \frac{1}{4}\mathcal{L}u_\infty(k)$$

$$= \lim_{j \to +\infty}\left(\mu_j(k) - \frac{1}{4}\mathcal{L}u_j(k)\right) = \mu_0(k) \tag{31.20}$$

and, similarly,

$$-\frac{1}{4}\mathcal{L}u'_\infty(k) = \mu_0(k).$$

These observations and (31.6) yield that both u_∞ and u'_∞ are solutions of

$$\begin{cases} \mathcal{L}U(k) = -4\mu_0(k) & \text{for all } k \in \Omega, \\ U(k) = 0 & \text{for all } k \in \mathbb{Z}^2 \setminus \Omega. \end{cases}$$

Hence, by the uniqueness result in (31.14), it follows that $u_\infty = u'_\infty$. From this and (31.8), we deduce that

$$\mu_\infty(k) = \mu_0(k) + \frac{1}{4}\mathcal{L}u_\infty(k) = \mu_0(k) + \frac{1}{4}\mathcal{L}u'_\infty(k) = \mu'_\infty(k).$$

This completes the proof of (31.19) and, therefore, the surprising claim in (31.3).

It is now very tempting to switch on our personal computer and write down a short code to perform some simulations of such a sweeping process, nicely checking that different successful cleaning strategies do produce the same final distribution, so as to have a "numerical confirmation" for the rigorous result obtained in (31.3). It is however also useful to work out at least a simple example using only pencil and paper; after all, according to Albert Einstein, "computers are incredibly fast, accurate, and stupid. Human beings are incredibly slow, inaccurate, and brilliant. Together they are powerful beyond imagination".

A simple and explicit example presents the additional boon to clearly relate the final distribution to the geometry of the domain. For instance, leaving no room for doubt, one can consider the case in which $\Omega := \{(0,0),\ (0,1),\ (1,1)\}$ and

$$\mu_0(k) := \chi_{\{(0,1)\}}(k) = \begin{cases} 1 & \text{if } k = (0,1), \\ 0 & \text{otherwise.} \end{cases}$$

In this case, using only pencil and paper, we see that

$$\mu_\infty(k) = \begin{cases} 2/7 & \text{if } k \in \{(-1,1),\ (0,2)\}, \\ 1/7 & \text{if } k = (1,0), \\ 1/14 & \text{if } k \in \{(-1,0),\ (0,-1),\ (2,1),\ (1,2)\}, \end{cases} \tag{31.21}$$

see Figure 31.6. To check (31.21), one can follow the cleaning strategy depicted in Figure 31.7 (for peace of mind, one can recall that this is

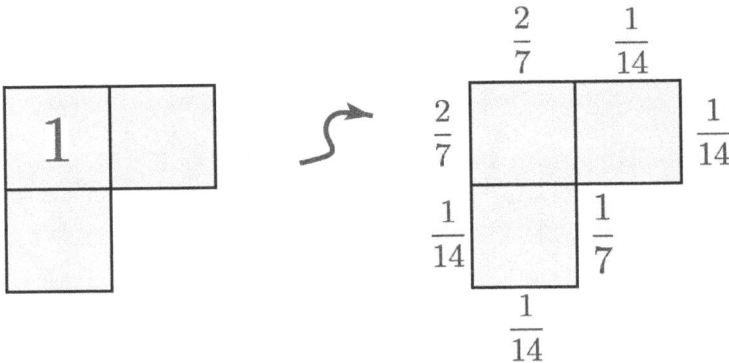

Figure 31.6.   A simple example workable by pencil and paper.

indeed a successful cleaning strategy, thanks to (31.10), and any other successful cleaning strategy would produce, in view of (31.3), the same final distribution). Combining together the three steps of the strategy in Figure 31.7, one obtains the iterative situation sketched in Figure 31.8: namely, the corner square in Ω is filled by an amount a_n of rubbish, with $a_{n+1} = \frac{a_n}{8}$ and $a_0 = 1$; the two squares outside Ω adjacent to it are filled by an amount b_n of rubbish, with $b_{n+1} = b_n + \frac{a_n}{4}$ and $b_0 = 0$; then we see four squares outside Ω filled by an amount c_n of rubbish, with $c_{n+1} = c_n + \frac{a_n}{16}$ and $c_0 = 0$; and finally, one square filled by an amount d_n of rubbish, with $d_{n+1} = d_n + \frac{a_n}{8}$ and $d_0 = 0$.

As a consequence, we find that $a_n = \frac{1}{8^n}$ (confirming that $a_n \to 0$ as $n \to +\infty$, hence providing a successful cleaning strategy),

$$b_n = b_0 + \frac{1}{4}\sum_{i=0}^{n-1} a_i = \frac{1}{4}\sum_{i=0}^{n-1}\frac{1}{8^i} \longrightarrow \frac{1}{4}\sum_{i=0}^{+\infty}\frac{1}{8^i} = \frac{2}{7},$$

$$c_n = c_0 + \frac{1}{16}\sum_{i=0}^{n-1} a_i = \frac{1}{16}\sum_{i=0}^{n-1}\frac{1}{8^i} \longrightarrow \frac{1}{16}\sum_{i=0}^{+\infty}\frac{1}{8^i} = \frac{1}{14}$$

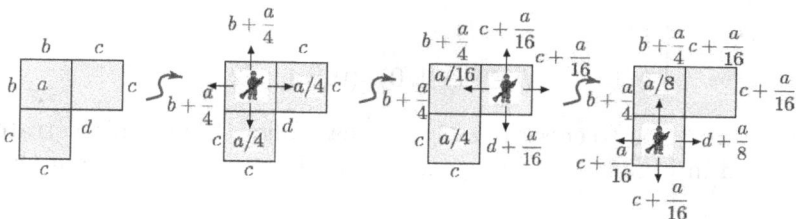

Figure 31.7.   A cleaning strategy in a simple example.

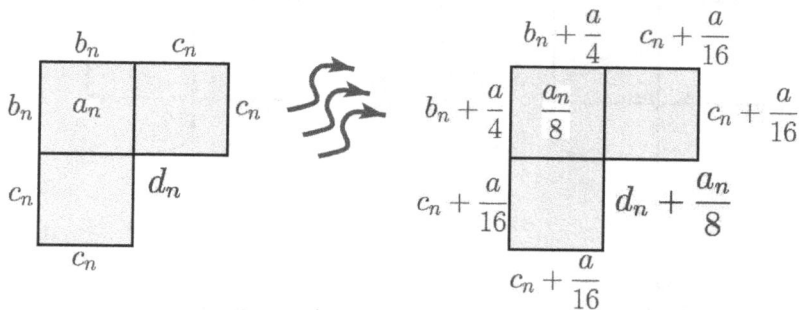

Figure 31.8.   Summary of the cleaning strategy shown in Figure 31.7.

and

$$d_n = d_0 + \frac{1}{8}\sum_{i=0}^{n-1} a_i = \frac{1}{8}\sum_{i=0}^{n-1}\frac{1}{8^i} \longrightarrow \frac{1}{8}\sum_{i=0}^{+\infty}\frac{1}{8^i} = \frac{1}{7}.$$

These observations give (31.21), as desired.

This sweeping problem provides additional intuition for many other methodologies typical of elliptic partial differential equations. We give indeed another example of a remarkable construction that can be performed in this (apparently innocent) discrete setting. Given $\ell \in \Omega$, we let $G_\ell : \mathbb{Z}^2 \to \mathbb{R}$ be the function u_∞ obtained in (31.5) when the initial rubbish distribution has the form $\mu_0(k) := \frac{1}{4}\chi_{\{\ell\}}(k)$. Then, we claim that the function

$$u(k) := \sum_{\ell \in \Omega} f(\ell)\, G_\ell(k) \tag{31.22}$$

is the unique solution[5] of

$$\begin{cases} -\mathcal{L}u(k) = f(k) & \text{for all } k \in \Omega, \\ u(k) = 0 & \text{for all } k \in \mathbb{Z}^2 \setminus \Omega. \end{cases} \tag{31.23}$$

To check this, it suffices to prove that u in (31.22) is a solution of (31.23) since the uniqueness claim follows from (31.14).

Also, $G_\ell(k) = 0$ for all $k \in \mathbb{Z}^2 \setminus \Omega$, owing to (31.6); consequently, $u(k) = 0$ for all $k \in \mathbb{Z}^2 \setminus \Omega$.

Furthermore, by (31.20), if $k \in \Omega$, then

$$-\mathcal{L}G_\ell(k) = \chi_{\{\ell\}}(k)$$

and, therefore, by (31.22),

$$\mathcal{L}u(k) = \sum_{\ell \in \Omega} f(\ell)\,\mathcal{L}G_\ell(k) = -\sum_{\ell \in \Omega} f(\ell)\,\chi_{\{\ell\}}(k) = -f(k).$$

These considerations establish the validity of (31.23), as desired.

---

[5]The function $G_\ell(k)$ can be seen as a discrete counterpart of an object which is called the "Green function" in the continuous setting and which will be addressed in further detail in Section 2.10 of the companion book [DV23].

Along these lines, exploiting the existence and uniqueness theory of (31.14), given $\ell \in \mathbb{Z}^2 \setminus \Omega$, one can consider the unique function $P_\ell : \mathbb{Z}^2 \to \mathbb{R}$ such that

$$\begin{cases} \mathcal{L}P_\ell(k) = 0 & \text{for all } k \in \Omega, \\ P_\ell(k) = \chi_{\{\ell\}}(k) & \text{for all } k \in \mathbb{Z}^2 \setminus \Omega. \end{cases} \tag{31.24}$$

This function also provides a useful representation formula, somewhat related to (31.22) and (31.23); more specifically, we have that the function

$$v(k) := \sum_{\ell \in \mathbb{Z}^2 \setminus \Omega} g(\ell) \, P_\ell(k) \tag{31.25}$$

is the unique solution[6] of

$$\begin{cases} \mathcal{L}v(k) = 0 & \text{for all } k \in \Omega, \\ v(k) = g(k) & \text{for all } k \in \mathbb{Z}^2 \setminus \Omega. \end{cases} \tag{31.26}$$

For this, once again, in light of the uniqueness result in (31.14), it is sufficient to check that v is a solution of (31.26). But this follows by linearity since, by combining (31.24) and (31.25), we see that when $k \in \Omega$,

$$\mathcal{L}v(k) = \sum_{\ell \in \mathbb{Z}^2 \setminus \Omega} g(\ell) \, \mathcal{L}P_\ell(k) = 0,$$

and when $k \in \mathbb{Z}^2 \setminus \Omega$,

$$v(k) = \sum_{\ell \in \mathbb{Z}^2 \setminus \Omega} g(\ell) \, \chi_{\{\ell\}}(k) = g(k).$$

---

[6]Comparing with the framework in Section 2.11 of the companion book [DV23], one can consider $P_\ell(k)$ as a discrete version of the Poisson kernel.

Chapter 32

Image Processing

A common operation nowadays is to process a given (say, for simplicity, grayscale) image in order to remove noise or enhance some visual effect.

To describe this procedure, one can denote by u the brightness of the image at a certain pixel of the screen, and we may consider that $u = u(x, t)$, where x lies in some domain of $\mathbb{R}^2$ (or even $\mathbb{R}^n$, for the sake of generality), which represents the position on the screen (which we consider, in the limit, as a continuum of pixels) and t is time. With this notation, the initial image $u(x, 0)$ undergoes some evolution, with the aim of improving its aspects, according to some standards that we can choose.

The simplest possibility would be to consider a heat equation for the brightness function u. This would have the advantage of possibly removing the imperfection from the image manifested, e.g. in abrupt variations of its brightness caused by impurities. Also, due to the smoothing effect of the heat equation, such a process does not introduce additional spurious details.

These benefits of using the heat equation, however, present an inconvenient drawback caused by the coarsening of the image resolution. While this downside is perhaps unavoidable after all, since some kind of averaging effect is necessary precisely to get rid of noises and impurities, the most significant hiccup is that this blurring does not necessarily respect the natural boundaries of the original image, see Figure 32.1 for a clear sketch[1] of this pitfall.

---

[1]The choice of using a portrait of Gauß here to check the different image processing procedures is completely arbitrary. For many years, research articles used for this scope a 512×512px standard test version of model Lena Forsén (stage name "Lenna", previously "Lena Soderberg", born Sjööblom). This standard image was a closeup detail of Lenna wearing a hat (the complete original image being a full length portrait of Lenna wearing a hat, published as the centerfold of the November 1972 issue of the *Playboy* magazine).

Allegedly, the Lenna standard image has become one of the most widely used images in computer history. After an initial concern about copyright infringement, it seems that over time Playboy has implicitly decided to overlook the wide and free use of the Lenna image for scientific purposes – however, at the beginning, Playboy tried to restrain the use of its copyrighted material e.g. by writing a letter to the journal *Optical Engineering*, see https://doi.org/10.1117/12.60707, whose content was: "It has come to our attention that you have used a portion of the centerfold photograph of our November 1972 PLAYBOY PLAYMATE OF THE MONTH Lenna Sjööblom, in your July 1991 issue of Optical Engineering magazine... Playboy Enterprises, Inc., the publisher of PLAYBOY magazine, owns the copyright in and to this photograph. As fellow publishers, we're sure you understand the need for us to protect our proprietary rights..."; later on, anyway, Playboy seemed to be pleased about this phenomenon, see e.g. http://www.lenna.org/playboy_backups/index.html and http://www.lenna.org/playboy_backups/lena.html, since the story was included in "The World History of Playboy" and the magazine seemed proud of the perspective that "the image of this Playboy Playmate can remain the standard reference image for comparing compression technologies into the 21st century".

Due to the popularity of this image, Lena Forsén was invited as a guest at the 50th Annual Conference of the Society for Imaging Science and Technology in 1997 and as a guest of honor at the banquet of IEEE ICIP in 2015; on the latter occasion, she also delivered a speech and chaired the best paper award ceremony.

The use of the Lenna picture has however become controversial, often considered as being degrading to women and of detrimental to aspiring female students in computer science. Several scientific journals nowadays explicitly and strongly discourage the use of the Lenna image, while others do not consider new submissions containing the image.

In January 2019, in an interview to the monthly American magazine *Wired*, Lena Forsén seemed to have declared that she was "really proud of that picture" ("the only note of regret she expressed was that she wasn't better compensated", according to https://www.wired.com/story/finding-lena-the-patron-saint-of-jpegs/), but when asked "if she had heard anything about the recent controversy around her image, she seemed alarmed at the thought that she could have a part in hurting or discouraging young women". In November of the same year, Lena

Figure 32.1. Portrait of Gauß by artist Siegfried Detlev Bendixen published in *Astronomische Nachrichten* in 1828 (Public Domain images from Wikipedia) and a Gaußian filtered image of it (obtained via Mathematica).

It would be instead more desirable to accomplish a reduction of image noise without removing significant content from the image, such as edges and details which are relevant for the interpretation of the image.

Though several methods have been proposed for this purpose, we do not intend here to give an exhaustive presentation of them but rather recall a classical method proposed by Pietro Perona and

---

Forsén also took part in a short documentary titled "Losing Lena", by Australian cinematographer Anna Howard, which aimed at galvanizing efforts to end the use of her image in technology research. On that occasion, Lena Forsén explicitly stated, *"I retired from modeling a long time ago. It's time I retired from tech, too [...] Let's commit to losing me"*.

For the sake of completeness, we also mention that the Lenna picture was shot by photographer Dwight Hooker and the "unknown researcher" who first scanned the Lenna image is sometimes (see e.g. https://www.ee.cityu.edu.hk/~lmpo/lenna/Lenna97.html) reported to be William K. Pratt.

Also, though it could simply be a coincidence, allegedly the Lenna issue (i.e. November 1972) is said to be *Playboy*'s best selling issue ever, with more than seven million copies sold.

Figure 32.2. Pietro Perona and Jitendra Malik in action (Public Domain images from Wikipedia).

Jitendra Malik [PM90] (see Figure 32.2). The gist of their technique consists of replacing the classical heat equation with a "shape-adapted" smoothing process in which, in rough terms, the diffusion coefficient, instead of being a constant, is a function induced by the image itself. This allows one to remove noise from digital images without blurring edges since one can somewhat enhance the smoothing effects within regions in which the original image is already sufficiently smooth, while suppressing these diffusion effects across strong edges, along which the brightness function changes abruptly.

To make this strategy clear, we remark that in a perfectly black-and-white image (without any gray tones), we can model u as a step function, taking values in $\{0, 1\}$. In this ideal situation, the gradient of u vanishes both in the white regions (corresponding, say, to $u = 1$) and in the black regions (where $u = 0$). The gradient is instead "concentrated" along the sharp separations between the black and white areas, along which it formally attains an infinite value.

For grayscale images, we can therefore interpret edges and details as thin lines along which the gradient of the brightness function reaches its maximal norm. The idea of the Perona–Malik method is thus to force the diffusion coefficient to be small for large values of $|\nabla u|$. If possible, it could be also convenient to have an evolution equation with a divergence structure (e.g. to interpret it as a gradient descent). The Perona–Malik equation is therefore a nonlinear diffusion equation of the form

$$\partial_t u(x, t) = \operatorname{div}(g(|\nabla u(x, t)|)\, \nabla u(x, t)). \tag{32.1}$$

While the choice $g := 1$ gives back the standard heat equation, to preserve the region boundaries of the images by suppressing diffusion at the sharp edges, the Perona–Malik method consists of choosing g to be a nonnegative monotonically decreasing function approaching zero at infinity. As models for this type of functions, it is proposed in [PM90] to take, for instance,

$$g(r) := \frac{1}{1 + (r/K)^2},$$ (32.2)

for some $K > 0$.

The Perona–Malik method manages to maintain some of the advantage of the classical heat equation (for example, a suitable maximum principle still avoids the creation of new features in the image when passing from fine to coarse scale, see [PM90, Appendix A]); however, some instabilities can arise for large gradients (though we skate around these difficulties at this level).

To convince ourselves of the fact that the Perona–Malik method tends to preserve, and possibly sharpen, the brightness edges, we can assume for simplicity that one of these high gradient edges passes through the origin, with the gradient vector in the direction of e_n. At a very small scale, we can suppose that the image presents an edge orthogonal to e_n, whence we approximate the brightness function u near the origin with a one-dimensional function depending only on x_n and with the Taylor expansion

$$u(x) = c + Ax_n + \frac{Mx_n^2}{2} + \frac{Lx_n^3}{6} + o(x_n^3),$$

for some $c, M, L \in \mathbb{R}$ and a "very large" slope $A > 0$.

We observe that the condition that $|\nabla u|$ is maximal (hence critical) at the origin entails that

$$0 = \partial_n \left. \frac{|\nabla u(x)|^2}{2} \right|_{x=0} = \nabla u(0) \cdot \nabla \partial_n u(0) = AM;$$

therefore, $M = 0$ and accordingly, near the origin,

$$u(x) = c + Ax_n + \frac{Lx_n^3}{6} + o(x_n^3).$$

Furthermore, since $|\nabla u|$ is actually maximal at the origin, we have that

$$0 \geqslant \partial_n^2 \frac{|\nabla u(x)|^2}{2}\bigg|_{x=0} = |\nabla \partial_n u(0)|^2 + \nabla u(0) \cdot \nabla \partial_n^2 u(0) = AL;$$

hence, $L \leqslant 0$ and supposed to be in a "nondegenerate" situation. Therefore,

$$L < 0. \tag{32.3}$$

Thus, setting $\phi(r) := g(r)r$, we deduce from (32.1) that

$$\partial_t u = \operatorname{div}\left(\phi(|\nabla u|)\frac{\nabla u}{|\nabla u|}\right) = \partial_n(\phi(\partial_n u)).$$

As a result, setting $U := \partial_n u = A + \frac{Lx_n^2}{2} + o(x_n^2)$, we see that

$$\partial_t U = \partial_n^2(\phi(U)) = \phi''(U)(\partial_n U)^2 + \phi'(U)\partial_n^2 U = \phi'(U)L + o(1).$$

In the model case (32.2), we have that $\phi(r) = \frac{r}{1+(r/K)^2}$; consequently, in this situation,

$$\partial_t U = \frac{K^2 L(K^2 - U^2)}{(K^2 + U^2)^2} + o(1) > 0$$

when $U(0) = A > K$ due to (32.3). This shows that if the slope of the edge is large enough, then the slope itself will increase with time and the edge becomes sharper (for small slope, instead, the smoothing effect of diffusion may prevail).

The remarkable effect of sharpening edges while denoising smooth regions is depicted in Figure 32.3 (to be compared with Figure 32.1).

We remark that the Perona–Malik equation (32.1) is also related to (but structurally different from) the level-set description of the mean curvature flow

$$\partial_t u(x,t) = |\nabla u| \operatorname{div}\left(\frac{\nabla u(x,t)}{|\nabla u(x,t)|}\right), \tag{32.4}$$

see e.g. [ES91].

To show, at least formally, that the level sets of a smooth and nondegenerate solution u of (32.4) evolve with normal velocity equal

Figure 32.3. Portrait of Gauß by artist Siegfried Detlev Bendixen published in Astronomische Nachrichten in 1828 (Public Domain images from Wikipedia) and a Perona-Malik filtered image of it (obtained via Mathematica).

to their mean curvature, one can proceed as follows. We suppose that, say, the level set $\{u(\cdot,t)=0\}$ corresponds to a smooth and bounded hypersurface M_t given by an embedding x_t. In this framework, we aim at showing that the normal velocity (pointing inward near convex portions) of M_t is equal to its mean curvature H_t, namely that

$$\nu_t \cdot \partial_t x_t = -H_t, \tag{32.5}$$

where ν_t is the exterior normal to $M_t = \partial\{u(\cdot,t) > 0\}$ at the point identified by x_t.

In order to check this, we pick a time t and a point, say the origin, on M_t. and we set normal coordinates around it. In this way, $-\frac{\nabla u(0,t)}{|\nabla u(0,t)|} = \nu_t(0) = e_n$, and for every $j \in \{1, \ldots, n-1\}$, we have that $\partial_j x_t \cdot e_n = 0$.

As a result, denoting by $\eta = (\eta_1, \ldots, \eta_{n-1})$ the coordinates of the embedding x_t (hence $\partial_j x_t = \frac{\partial x_t}{\partial \eta_j}$, and we assume that $x_t(0) = 0$), we obtain that if ϖ is any smooth vector field such that $|\varpi(x)| = 1$ for all $x \in \mathbb{R}^n$ and $\varpi(0) = \nu_t(0)$, then

$$0 = \partial_n \frac{|\varpi(x)|^2}{2}\bigg|_{x=0} = \varpi(0) \cdot \partial_n \varpi(0) = \partial_n \varpi_n(0)$$

and thus

$$\nabla\left(\frac{\nabla u(x,t)}{|\nabla u(x,t)|}\cdot\varpi(x)\right)\cdot\varpi(x)\bigg|_{x=0} = \sum_{k=1}^{n}\partial_n\left(\frac{\partial_k u(x,t)\,\varpi_k(x)}{|\nabla u(x,t)|}\right)\bigg|_{x=0}$$

$$= \frac{\partial_{nn}u(0,t)}{|\nabla u(0,t)|} - \frac{(\partial_n u(0,t))^2\partial_{nn}u(0,t)}{|\nabla u(0,t)|^3} = 0 \qquad (32.6)$$

and, as a byproduct,[2] we arrive at

$$H_t(0) = -\operatorname{div}\left(\frac{\nabla u(x,t)}{|\nabla u(x,t)|}\right)\bigg|_{x=0}.$$

Figure 32.4. Portrait of Gauß by artist Siegfried Detlev Bendixen published in *Astronomische Nachrichten* in 1828 (Public Domain images from Wikipedia) and a mean curvature flow filtered image of it (obtained via Mathematica).

---

[2]Note indeed that the quantity in (32.6) corresponds to the normal component of the divergence of the normal vector field.

A more precise description of the mean curvature will be given in Section 2.3 of the companion book [DV23]. Here, we are essentially taking for granted that the mean curvature is the tangential divergence of the normal field, a concept that will be better clarified in Section 2.3 of the companion book [DV23].

Hence, since

$$0 = \partial_t(u(x_t(\eta), t)) = \nabla u(x_t(\eta), t) \cdot \partial_t x_t(\eta) + \partial_t u(x_t(\eta), t),$$

we have that

$$\nu_t(0) \cdot \partial_t x_t(0) = -\frac{\nabla u(0, t)}{|\nabla u(0, t)|} \cdot \partial_t x_t(0) = \frac{\partial_t u(0, t)}{|\nabla u(0, t)|}$$

$$= \text{div}\left(\frac{\nabla u(x, t)}{|\nabla u(x, t)|}\right)\bigg|_{x=0} = -H_t(0).$$

This gives (32.5), as desired.

Other geometric evolution equations will appear in Chapter 35.

See Figure 32.4 for an application of the (mean) curvature flow filter. Once again, by comparing Figures 32.1, 32.3 and 32.4, we have a practical confirmation of how sharply the Perona–Malik method outperforms both classical diffusion and curvature flows in maintaining definite meaningful edges while smoothing intermediate brightness regions.

Chapter 33

Artificial Intelligence and Machine Learning

We have already encountered in Chapters 30 and 31 some fascinating problems related to discrete versions of the Laplace operator. Here, we provide a concrete situation in which these discrete Laplacians naturally surface.

The setting is that of image classification (not to be confused with the image processing presented in Chapter 32): the idea is to train a computer to identify specific objects in photos. For this, the simplest model to keep in mind is to have a bunch of (say, grayscale, for simplicity) images which represent some classes of targets to be identified (again, for simplicity, we can think of two targets: "cats" and "dogs", see Figure 33.1). The goal is thus to train the machine to *distinguish cats from dogs* by looking at the photos.

Without aiming to be exhaustive, we mention some hints on how to implement such a type of learning. First of all, let us transform the photos of cats and dogs into well-defined mathematical concepts. Suppose that all the images have the same size, with, say each image consisting of d pixels. Then, each image can be considered as an element of $X := [0, 1]^d$, in which, for every $m \in \{1, \ldots, d\}$, the mth coordinate of an element $x \in X$ would correspond to the brightness level of the mth pixel (scaled to be a number between 0 and 1). Also, each image has a label attached to it, which specifies its content; we can think such a label as an element of the set $Y := [-1, +1]$ (for example, the value -1 corresponding to "dog" and $+1$ to "cat", as an example of a binary class).

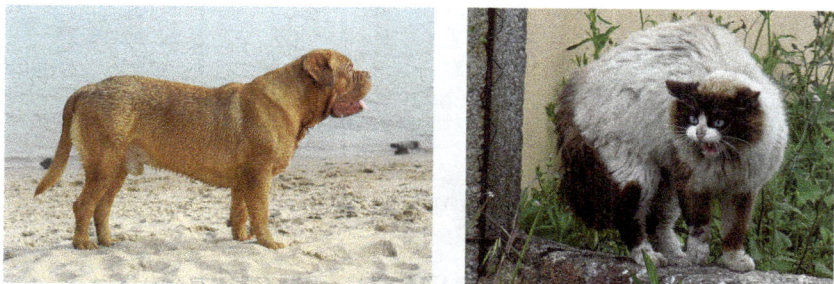

Figure 33.1. Distinguishing a dog from a cat (images from Wikipedia; left image by DJT, licensed under the Creative Commons Attribution-Share Alike 2.5 Generic license, right image by Luis Miguel Bugallo Sánchez, licensed under the Creative Commons Attribution 2.0 Generic license).

The gist is now to show to the machine some data sample S of these labeled images, say

$$(x_1, y_1), \ldots, (x_M, y_M) \in X \times Y, \tag{33.1}$$

which can be used as useful training for the machine to learn how to independently label images.

For instance, the machine could be provided access to a certain class $\mathcal{F}$ of functions $f : X \to Y$ that could be used to label new photos. In this way, upon seeing a new image $x \in X$, the machine would produce a value $f(x)$ which is the label "guessed" as appropriate; in our example, $f(x) = -1$ (or $f(x)$ very close to -1) would say that photo x is identified to be a dog, $f(x) = +1$ (or $f(x)$ very close to $+1$) that photo x is identified to be a cat, and values such as $f(x) = 0$ (or $f(x)$ quite far from both -1 and $+1$) would denote that the machine is uncertain about the attribution.

In order to make the machine select correctly the function needed for this identification process, we can think of using the sample S to build some empirical loss function which penalizes wrong attributions. Consider, for example, the case of a quadratic loss function,

$$L_f(x_1, \ldots, x_M, y_1, \ldots, y_M) := \sum_{j=1}^{M} (y_j - f(x_j))^2,$$

which penalizes any possible choices of $f \in \mathcal{F}$ unless it produces the correct identification on the sample S (which is $f(x_j) = y_j$).

To let the previous example look more independently at the number of elements of the sample, one can also consider

$$L_f(x_1, \ldots, x_M, y_1, \ldots, y_M) := \frac{1}{M} \sum_{j=1}^{M} (y_j - f(x_j))^2.$$

Another example to keep in mind can be that of a "cross-entropy loss function" of the type

$$L_f(x_1, \ldots, x_M, y_1, \ldots, y_M)$$

$$:= \sum_{j=1}^{M} \left((1 + y_j) \ln \frac{2}{1 + f(x_j)} + (1 - y_j) \ln \frac{2}{1 - f(x_j)} \right).$$

To develop an intuition of this loss function, one can observe that if, say, the jth image is a dog, then $y_j = -1$ and the corresponding addend boils down to $\ln \frac{2}{1 - f(x_j)}$, which is minimized indeed by the correct identification $f(x_j) = -1$, producing a positive (possibly infinite) term whenever $f(x_j) \in (-1, +1]$; differently from the previous cases, however, this loss function would produce an infinite value when $f(x_j) = +1$, thus penalizing heavily the predictions that are confident but wrong.

Now, in light of these discussions, one could argue that to efficiently train a machine, it suffices to find the minimizer f of some empirical loss function L_f (e.g. one of those mentioned above, or a similar one). But there's a catch. Working in this way, our minimizer could simply be any function $f_\star$ such that

$$f_\star(x_j) = y_j, \text{ for all } j \in \{1, \ldots, M\}. \tag{33.2}$$

In this way, our machine has just "memorized" all the given pictures, but it has discovered no patterns at all which can be used for future predictions. Actually, it is quite likely that in front of a new image $\tilde{x} \notin S$, the value $f_\star(\tilde{x})$ may have little to do with the object represented in the image $\tilde{x}$; indeed, without imposing any additional restrictions, there are simply "too many functions" satisfying a finite number of pointwise constraints, as in (33.2).

This situation, in which the machine fits excellently against its training data without[1] being able to generalize to new sets of data, is called "overfitting".

To overcome this difficulty, it is therefore convenient to appropriately select our functional space $\mathcal{F}$ (called in jargon as the "hypothesis space") and to somehow invite the machine to select, among the possible optimizing functions, the "simplest" one (taking the *ansatz* that nature, after all, complies with Ockham's razor, and among possible explanations, the simpler one is to be preferred, see Figure 33.3 to meet William of Ockham). Though dealing with the fine problems posed by the suitable selection of hypothesis spaces goes well beyond the scope of these notes (and relies on clever uses of functional analysis), we can still try to get our hands in some aspects of this theory.

---

[1] As an evident example of overfitting, one can also consider a sample set given by the points in the plane

$$\{(0,0),\ (0,1),\ (0,2),\ (0,3),\ (0,4),\ (0,5),\ (0,6),\ (0,7),\ (0,8),\ (0,9)\}.$$

Then, both the graph of the function f_0 constantly equal to zero and that of the function

$$\overline{f}(t) := t^{10} - 45t^9 + 870t^8 - 9450t^7 + 63273t^6 - 269325t^5 + 723680t^4$$
$$- 1172700t^3 + 1026576t^2 - 362880t$$

would fit the data (see Figure 33.2).

However, the function f_0 constantly equaling zero would probably "generalize": indeed, in our everyday experience, we may expect the next element to be taken into account to be the point $(0,10)$, which would actually lie on the graph of f_0. Instead, the prediction of the unnecessarily and oscillatory function $\overline{f}$ would produce $\overline{f}(10) = 3628800$, which would be quite far away from the correct value.

This is an indication that simpler functions should be preferred to avoid overfitting, also because oscillatory functions tend to produce values outside the interpolation points which are almost completely unrelated to the prescribed interpolation.

First and foremost, machine learning is an excellent playground to understand human learning too: for machines, as well as for humans, any sort of at least partially supervised learning may stem from some training on a finite number of examples; however, after that, to provide solid outcomes, it needs to generalize to a broader class from which the examples are drawn. Otherwise it is not real learning; it is mere case-by-case memorization.

In a sense, for machines as well as for humans, the ultimate goal of learning is not to learn what to learn but rather to learn how to learn.

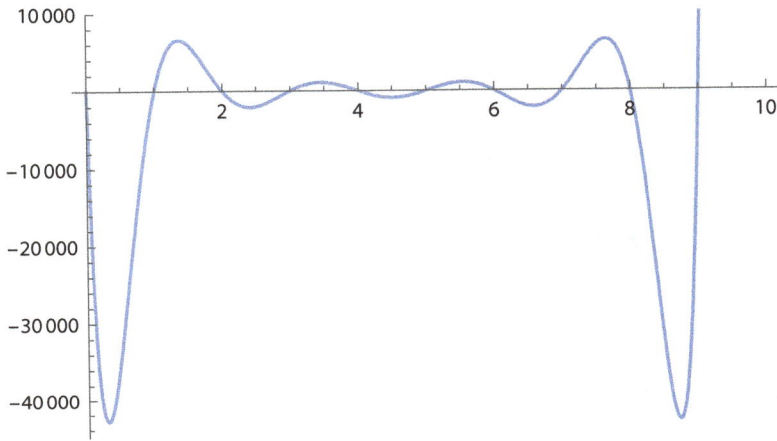

Figure 33.2. Plot of the function $t \mapsto t^{10} - 45t^9 + 870t^8 - 9450t^7 + 63273t^6 - 269325t^5 + 723680t^4 - 1172700t^3 + 1026576t^2 - 362880t$.

Figure 33.3. A stained glass window at an English church allegedly representing William of Ockham (image by Moscarlop from Wikipedia, licensed under the Creative Commons Attribution-Share Alike 3.0 Unported license).

To make things as simple as possible, though the method is quite general, we can focus only on a concrete case in which we try to approximate a given location by a Gaußian, e.g. a point (p, q) in the plane by the function $t \mapsto q e^{-\pi(t-p)^2}$. In analogy with this idea, one can define[2] a kernel, given $\xi_1, \xi_2 \in X$, as

$$K(\xi_1, \xi_2) := e^{-\pi(\xi_1 - \xi_2)^2}$$

and pick a large family of elements of X, say $\{\xi_1, \ldots, \xi_D\}$ and define, for all $i \in \{1, \ldots, D\}$,

$$\phi_i(x) := K(x, \xi_i). \tag{33.3}$$

Then, a natural space of functions capable of extracting information from this family is that of the functions f that can be written as a linear combination of $\phi_1, \ldots, \phi_D$, say the functions of the form

$$X \ni x \longmapsto f(x; \theta) := \sum_{i=1}^{D} \theta_i \, \phi_i(x), \tag{33.4}$$

for some $\theta = (\theta_1, \ldots, \theta_D) \in \mathbb{R}^D$, see e.g. Figure 33.4 for a simplified visual sketch of such a "linear expansion[3] of a nonlinear function".

---

[2]See e.g. [Min10] for more information about Gaußian kernels. The choice of π in the exponent of the kernel is not relevant; we chose this normalization here simply because it would produce a total integral equal to 1.

[3]After all, expressions like those in (33.4) are not too different from a truncation of a Fourier series: at heart, an approximating Fourier sum is nothing but a linear superposition of nonlinear functions, with the aspiration of reconstructing "essentially" every interesting function.

Instead of sines and cosines, maybe peaked functions as in (33.3) have better chances of reconstructing functions out of pointwise constraints, as in (33.2) and hence can better serve to the scope of machine learning.

To build a serious theory out of the nonsense presented here, one has to leverage the analysis of reproducing kernel Hilbert spaces, see e.g. [PR16].

Also, here, we surf over the important problem of which type of functions can be reconstructed by the superposition of given functions (i.e. whether the sums of linear combinations of suitable translations are dense in interesting functional spaces). This type of question often relies on functional analysis methods (such as the Stone–Weierstraß theorem, the Hahn–Banach theorem, the Riesz representation theorem, the Wiener Tauberian theorem, and the Kolmogorov–Arnol'd superposition theorem). See e.g. [Cyb89, Fun89, HSW89, Hor91, Cyb92, Hor93, LLPS93, Pin99, Ste02, Min10, BDVRV21] for more information about density and approximation results (for instance, see Theorems 3.1 and 5.1 in [Pin99] for specific approximation and interpolation results).

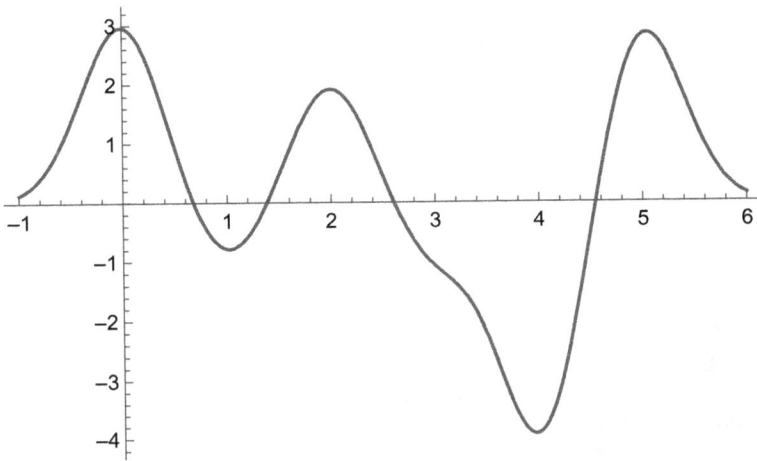

Figure 33.4. Plot of the function $x \mapsto 3e^{-\pi x^2} - e^{-\pi(x-1)^2} + 2e^{-\pi(x-2)^2} - e^{-\pi(x-3)^2} - 4e^{-\pi(x-4)^2} + 3e^{-\pi(x-5)^2}$.

With this, a natural choice of the hypothesis space $\mathcal{F}$ is that of functions f as in (33.4), and the problem of minimizing a loss functional becomes a finite-dimensional[4] minimization problem over the set of parameters $\theta \in \mathbb{R}^D$.

With this setting of the hypothesis space $\mathcal{F}$, we can also return to the overfitting problem, trying now to avoid it by adding to the empirical loss function L_f a term which penalizes unnecessary complicated oscillations. For example, one can consider a penalized loss

---

[4]In general, the dimension D here is larger than the dimension $d + 1$ of the space in which the sample S is taken (recall (33.1)) in order to allow the machine to compare with the available data in the training sample and have sufficient parameters to optimize.

The convenience of embedding the data into a larger-dimensional space is indeed quite common in machine learning, and it is sometimes explained via a simple geometric analogy. For example, let us suppose that we want to train our machine to recognize the best kiwis in our garden, and for simplicity, suppose that the choice is made only in virtue of the ripening of the fruit. Of course, kiwis that are either too unripe or gone off should be avoided, and the ones that are to be selected have a level of ripening lying in some appropriate interval. A structured way for a machine to select such an interval therefore consists of embedding these ripening data into a higher-dimensional space and then separating the convenient ones by using a simple linear function, see Figure 33.5.

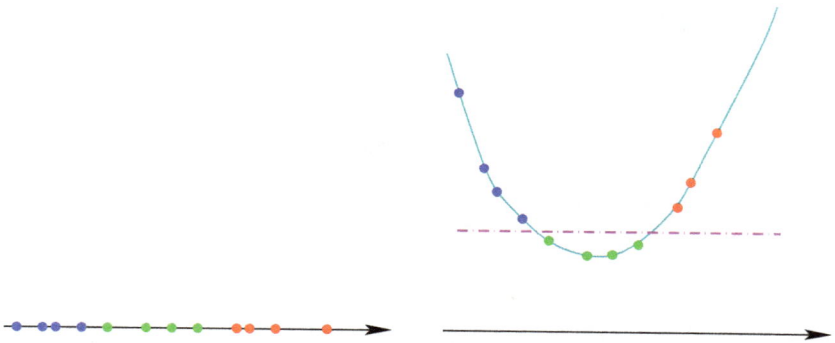

Figure 33.5. A machine trying to learn how to pick good kiwis by embedding methods and separation by linear functions: the kiwis with the appropriate level of ripening are depicted here in green color, the ones in blue are still unripe and the ones in red are gone off.

function given by

$$L_f(x_1, \ldots, x_M, y_1, \ldots, y_M) + \lambda|\theta|^2, \tag{33.5}$$

for some $\lambda > 0$, where, e.g. $|\theta|$ is simply the Euclidean norm in $\mathbb{R}^D$; in this way, functions with "more economical" expansions would be preferred, and the larger the value of λ, the larger the penalty on the "complexity" of f (other choices of penalization are certainly possible, and methods of this sort are sometimes referred to as "regularizations").

Minimizing methods of empirical functionals as in (33.5) are indeed the core of the so-called *supervised machine learning*, in which only labeled data sets are utilized for the training. Despite the fact that this procedure is often quite satisfactory, the method presents marginal improvements. As a matter of fact, in many natural processes, learning occurs in a rather *semi-supervised* regime; for example, a child is exposed to many new phenomena, but only few of them are directly connected to a specific label, yet a relatively small amount of feedback is sufficient to allow the child to make consistent learning progress.

This suggests that unlabeled data can also be usefully processed and used to extract valuable information. To build a semi-supervised learning algorithm, one can modify the setting in (33.4) by including in the minimization problem some unlabeled examples as well: $x_{M+1}, \ldots, x_{M+P} \in X$ (note that these examples are not labeled; namely, differently from the sample in (33.1), they are not attached to any element of Y). To use these unlabeled examples, one must somewhat quantify how "close any element is to any other". To this end, given $i, j \in \{1, \ldots, M + P\}$, it is customary to endow the pair $(x_i, x_j) \in X \times X$ with a weight $w_{ij} \geq 0$ representing their "similarity" (the case of $w_{ij} = 0$ corresponding to dissimilar objects and[5] higher values of w_{ij} indicating a higher degree of kinship). The similarity relation is supposed to be symmetric, hence $w_{ji} = w_{ij}$.

In this setting, one can impose on the empirical loss function an additional cost if the assessment of x_i is far from that of x_j anytime the objects x_i and x_j are strongly related (e.g. they have a high value of the similarity weight w_{ij}); for example, one can consider an additional cost of the form

$$\frac{\gamma}{(M+P)^2} \sum_{i,j=1}^{M+P} w_{ij}(f(x_i) - f(x_j))^2, \tag{33.6}$$

for some $\gamma > 0$.

From this and (33.5), we may conclude that an interesting loss function, to be minimized within a suitable set of parameters that describes a convenient hypothesis space (recall (33.4)), in the framework of semi-supervised machine learning takes the form

$$L_f(x_1, \ldots, x_M, y_1, \ldots, y_M) + \lambda |\theta|^2$$

$$+ \frac{\gamma}{(M+P)^2} \sum_{i,j=1}^{M+P} w_{ij}(f(x_i) - f(x_j))^2.$$

From the perspective of partial differential equations, the term in (33.6) is of particular interest since the minimization of loss functions containing this term produces discrete equations involving the

---

[5]Of course, extracting sufficient features to assign automatically a precise degree of similarity to unstructured data like images is a rather sophisticated task in itself, ultimately linked to pattern recognition, see e.g. [VL00]. For the purposes of this simplified explanation, we are assuming that our data set is endowed with these similarity weights (regardless of whether these weights were assigned by humans or by machines); what matters at this level is that this unlabeled data set can now be used for the semi-supervised learning algorithm that we are presenting.

operator

$$\sum_{j=1}^{M+P} w_{ij}(f(x_i) - f(x_j)).$$

In particular, when

$$w_{ij} = \begin{cases} 1 & \text{if } |i-j| = 1, \\ 0 & \text{otherwise,} \end{cases}$$

this operator reduces to the discrete Laplacian studied in Chapters 30 and 31 (compare with (30.2)).

Once we understand the training process of a machine as the minimization over a set of parameters θ of a suitable loss function, it is interesting to mention a fashionable direction of research related to the use of parameters and nonlinear functions, which is somewhat modeled based on the way natural organisms "learn", hence having interesting connections with neurology and cognitive sciences, and which paved the way[6] to the theory of "deep learning", where the adjective "deep" refers to a performance improvement due to the presence of additional layers in the artificial neural architecture.

Namely, rather than restricting our attention only to the kernel methods described on p. 356, a convenient way to obtain a simple, but possibly effective, hypothesis space is to produce a "rich family" of functions by composing repeatedly an affine function and a nonlinear one. After all, this is the way in which natural organisms develop their perceptions; for instance, if a neuron receives a sufficiently intense stimulus (corresponding to a significant voltage change over a short interval), it generates an electrochemical pulse, which can reach another neuron; these traveling signals may be excitatory or

---

[6]On the one hand, it is truly fascinating to compare machine and human learning, and this can certainly enrich cross-disciplinary research involving mathematics, engineering, computer sciences, biology and neuroscience. On the other hand, one should also be aware that, as pointed out in [GBC16, p. 164], "modern neural network research, however, is guided by many mathematical and engineering disciplines, and the goal of neural networks is not to perfectly model the brain. It is best to think of feedforward networks as function approximation machines that are designed to achieve statistical generalization, occasionally drawing some insights from what we know about the brain, rather than as models of brain function".

inhibitory, increasing or reducing the voltage. That is, very roughly, when a neuron receiving the information can either suppress it or transmit it (more or less in an "all-or-nothing" fashion, according to, say, the intensity of the stimulus), and in the reception and transmission process to another neuron, the signal can be "adjusted" (that is increased, decreased or modulated).

Thanks to mathematics, one can reproduce this method for a silicon-based intelligence. In this situation, a "neuron" is replaced by a function, sometimes called a nonlinear activation function (after all, a neuron is simply a function which gives different values to model different reactions, or no reaction at all, to a given stimulus).

The literature related to machine learning presents plenty of nonlinear activation functions related to neural networks. In some sense, the ones closest to the intuition coming from biological neurons are activators which "switch from 0 to 1", such as functions $\sigma : \mathbb{R} \to \mathbb{R}$ such that $\sigma(t) \to 0$ as $t \to -\infty$ and $\sigma(t) \to 1$ as $t \to +\infty$. This type of functions are sometimes called "sigmoids".

However, for computational purposes, it is often convenient to pick simpler activation functions; in this respect, a very popular choice is to select as activation function the "positive part" of a real number, namely the function

$$\mathbb{R} \ni t \longmapsto t_+ := \max\{t, 0\}.$$

This function is sometimes dubbed with fancy names, such as "rectified linear unit", or "ReLU" for short.

Roughly, a sigmoid mimics a neural response of the type "all-or-nothing", up to a possible small transition between the regime in which "nothing happens" to that in which "something happens", while the ReLU reproduces a "ramp-type" reaction in which "nothing happens below a threshold, after which the reaction is linear".

The technical advantage of considering a ReLU, besides its simplicity, is that it is a convex and piecewise linear function. Additionally, a sigmoid can be easily constructed as a linear superposition of two ReLUs since

$$t_+ - (t-1)_+ = \begin{cases} 1 & \text{if } t \geqslant 1, \\ t & \text{if } t \in (0,1), \\ 0 & \text{if } t \leqslant 0; \end{cases}$$

hence, using the ReLUs may end up being technically simpler but roughly "as good as" using the (perhaps conceptually more intuitive) sigmoids, thus obtaining a performance boost[7] of the method.

The specific choice of nonlinear activation functions is in some way related to the choice of the loss function, and these two design elements are connected since the configuration of the neural layer frames the prediction problem of the machine, while the choice of the loss function calculates the predictive error for a given frame of the problem.

One of the strengths of deep learning consists of building multiple processing layers by placing a number of "hidden" computational levels between the initial input and the final predictive output. In this way, perhaps an artificial network can "imitate" the way humans acquire sophisticated types of knowledge since these hidden layers in the artificial neural structure may approximately reproduce multiple levels[8] of "abstraction" in which the system represents and analyzes the data.

Broadly, the construction of the additional layers added to enhance the possibility for a machine to learn from the data is sketched in Figure 33.6, depicting a simple[9] neural network, with an input consisting of a vector (x_1, x_2), represented by the yellow dots, three "hidden" layers, portrayed by the green, cyan and red dots, and an output, given by a scalar and illustrated by the magenta dot.

---

[7]For higher accuracy, the ReLU is also sometimes replaced by the so-called "leaky ReLU"

$$\begin{cases} t & \text{if } t \geqslant 0, \\ -ct & \text{if } t < 0, \end{cases}$$

for $c \in (0, +\infty)$.

[8]In rough terms, one can imagine that the hidden layers in a neural network correspond to different levels of perception; for instance, if, given a collection of pictures, the task of the machine is to distinguish a kangaroo from a shark, one can imagine that the first hidden layer focuses on the identification of the edges of the picture, the second on the detection of the corners (recognizable as a collection of edges), the third on the perception of object parts (detectable as a collection of corners and contours), the fourth on the mutual interaction between object parts, and so on. But of course, this approach to describing the layers of a neural network may well be too simplistic and too much "human oriented". See, however, the forthcoming description of convolutional neural networks on p. 366.

[9]We strongly recommend to try to understand computer science, particularly deep learning, first with your computer switched off and using only pencil and paper instead. In this spirit, for instance, a simple learning algorithm is presented in detail in Section 6.1 of [GBC16].

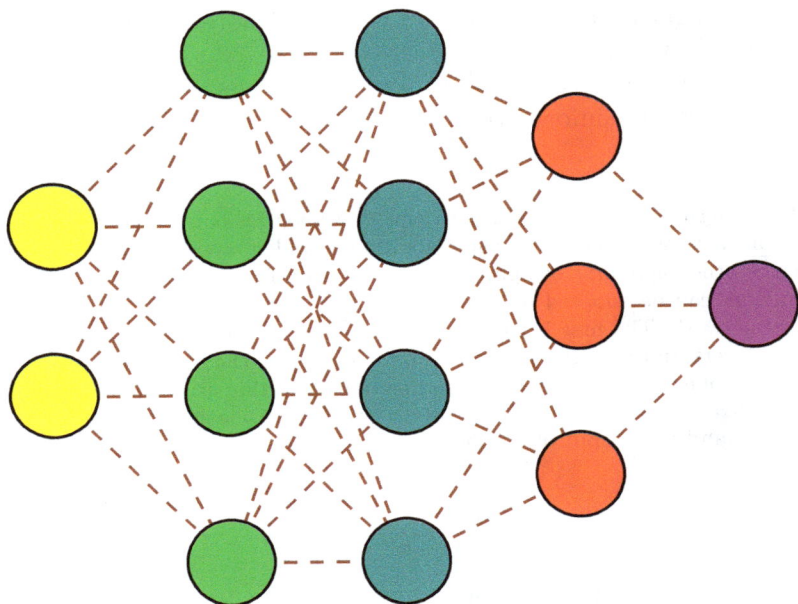

Figure 33.6.   Scheme of a neural network.

The brown dotted lines in Figure 33.6 represent the actions of matrices of weights, and the composition with a nonlinear activator function. For instance, suppose that the brown dotted lines entering the top green dot correspond to the weights $\frac{1}{2}$, coming from the upper yellow dot, and $\sqrt{\pi}$, coming from the bottom yellow dot, then the value assigned to the top green dot corresponds to $g\left(\frac{i_1}{2} + \sqrt{\pi}i_2\right)$. In general, if $k \in \{1, \ldots, 4\}$, the kth green dot is assigned[10] to some value $h_{1,k} := g(\theta_{1,1,k}x_1 + \theta_{1,2,k}x_2)$, for some weights $\theta_{1,1,k}$ and $\theta_{1,2,k}$.

And so on, looking at the subsequent hidden layer of cyan dots, the kth cyan dot is assigned to some value

$$h_{2,k} := g(\theta_{2,1,k}h_{1,1} + \theta_{2,2,k}h_{1,2} + \theta_{2,3,k}h_{1,3} + \theta_{2,4,k}h_{1,4}),$$

for some weights $\theta_{2,1,k}$, $\theta_{2,2,k}$, $\theta_{2,3,k}$ and $\theta_{2,4,k}$, or, in a short vectorial notation,

$$h_{2,k}(x) = g(\theta_{2,k} \cdot h_1(x)) = g\left(\theta_{2,k} \cdot g(\theta_1 \cdot x)\right),$$

---

[10]More generally, one can consider affine transformations such as $\theta_{1,1,k}x_1 + \theta_{1,2,k}x_2 + \theta_{1,3,k}$ instead of the linear one $\theta_{1,1,k}x_1 + \theta_{1,2,k}x_2$. But this boils down mostly to a notational convention since it would suffice to replace (x_1, x_2) with $(x_1, x_2, x_3) := (x_1, x_2, 1)$ to reduce the affine setting to the linear one.

with the slight notational abuse in which g denotes both a scalar and a vectorial function.

This network structure[11] is sometimes referred to as "feedforward" since the information flows[12] from the input through the

---

[11]Revisiting the approximation problems described in footnote 3 on p. 356, we mention that the possibility of using additional hidden layers in a neural network enhances the capability of approximating a given function. That is, two basic techniques can be adopted to approximate a given function by the outputs of a neural network. The first is to build a network with many neurons but without hidden layers. In this case, the approximation relies on the linear superposition of a large number of nonlinear activation functions, and this situation, often denoted the "arbitrary-width case", corresponds to having only one nonlinear step between the input and the output of the form

$$\sum_{i=1}^{N} C_i \sigma \left(\sum_{1 \leqslant j \leqslant n} \theta_{ij} x_j + b_i \right),$$

see e.g. [Cyb89, Cyb92, Pin99, Fun89, HSW89, Hor91, Hor93]. In this case, the "width" is given, for instance, by the number N of superpositions of neurons considered here above.

Note that the introduction of hidden layers amplifies, in principle, the possibility of approximating a given function, without spoiling the possibility of using a network with no hidden layers because one can simply approximate the identity function with additional later layers, for which it suffices to pick a point t_0 for which $\sigma'(t_0) \neq 0$ and note that

$$\frac{1}{\varepsilon} \left[\sigma \left(\frac{\varepsilon t}{\sigma'(t_0)} + t_0 \right) - \sigma(t_0) \right] = t + O(\varepsilon).$$

In this spirit, the other basic approximation method, called "arbitrary-depth case", consists of approximating a given function by using a neural network with an arbitrary number of hidden layers, each with at most some prescribed number of neurons, see e.g. [LPW+17, Han19, KL20, DXZ20]. In this situation, the "depth" is given by the number of hidden layers.

All these "universal approximation" results put perhaps some pressure on the scientists utilizing machine learning: as stated in [HSW89, Section 3], "standard multilayer feedforward networks are capable of approximating any measurable function to any desired degree of accuracy, in a very specific and satisfying sense [...] This implies that any lack of success in applications must arise from inadequate learning, insufficient numbers of hidden units or the lack of a deterministic relationship between input and target".

[12]Enhanced algorithms allowing the information to go back from the loss functional to the network itself are instead called "backpropagation".

intermediate layers and finally to the output; interestingly, the training procedure typically occurs only at the level of the output, which needs to be as close as possible to the real data, as prescribed by the loss function (in this sense, the learning algorithm uses the hidden layers to produce an optimal output, but the training data do not specify directly what individual layer should do).

We also remark that, while the machine learning process typically relies on optimized sequences of linear and nonlinear operations, and the precise choice of nonlinearity may play a role in the efficiency of the algorithm (recall the discussion about sigmoids and ReLUs on p. 361), the choice of the linear operation can also contribute to the achievement of performant learning outcomes. Without going into the details, for example, we recall that the standard linear superposition gets sometimes replaced, or complemented, by a convolution operation (see [GBC16, Chapter 9]). Roughly, the idea for such convolutional neural networks can be understood by considering for simplicity the pixels of a black-and-white image as our input (corresponding, according to the color of the pixel, to either the value 0 or 1). A feature detector is applied to this input by applying a discrete convolution: say, that the pixels of the input image are labeled $x_{ij} \in \{0, 1\}$ with $i \in \{1, \ldots, N_1\}$ and $j \in \{1, \ldots, N_2\}$, the action of a "feature detector" produces a value $\widetilde{x}_{\ell m} \in \mathbb{R}$ with $\ell \in \{1, \ldots, \widetilde{N}_1\}$ and $m \in \{1, \ldots, \widetilde{N}_2\}$ via the relation

$$\widetilde{x}_{\ell m} = \sum_{\substack{1 \leqslant i \leqslant N_1 \\ 1 \leqslant j \leqslant N_2}} K_\theta(\ell - i, m - j)\, x_{ij}.$$

Namely, kernels K_θ are used to perform linear convolutional operations, the result of which is called a "feature map" (note that in our example, the input image is $(N_1 \times N_2)$-dimensional, while the feature map is $(\widetilde{N}_1 \times \widetilde{N}_2)$-dimensional); therefore, a convolution operation may be convenient not only to relate the information encoded into a pixel with that of the neighboring pixels but also to reduce the size of the input image whenever convenient, thus accelerating the algorithm (yet, possibly at the cost of losing some information, but sometimes to recognize a pattern one uses some broadly seen features, without needing to dig into their minutiae). The linear operation of convolution is then combined with some nonlinear activation function, possibly within an architecture of hidden neural layers dealing

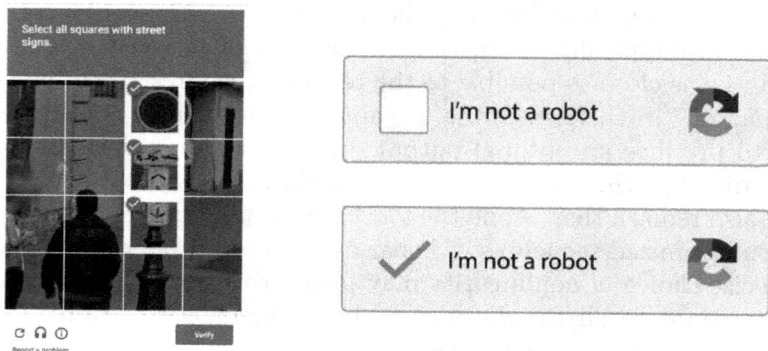

Figure 33.7. CAPTCHA (Completely Automated Public Turing test to tell Computers and Humans Apart): I'm not robot (am I?).

with collections of feature maps, and as above, the parameters θ describing the kernels are optimized by the minimization of a suitable loss functional, with the goal of training the machine against a given sample.

In terms of actions performed on the image, different convolution kernels can correspond to blurring, sharpening contours and detection of edges; thus, in some sense, the training process of the kernel parameters allows the machine to detect and interpret features such as lines, corners and contours by selecting the optimal feature detector for the appropriate task at the convenient layer of the network, with the nonlinear activation function maximizing the effect of the feature determinant for the required task. In this respect, convolutional networks are intimately related to their neuroscientific counterpart[13] in the primary visual cortex of the brain (see [GBC16, Section 9.10] for further details on the relation between neuroscience and convolutional networks).

Returning to the topic of overfitting in machine learning, an interesting phenomenon has been recently discovered, named "grokking",

---

[13] And nowadays convolutional networks can outperform humans in several visual recognition tasks. For instance, machine learning outperforms humans by a factor of two on a traffic sign recognition benchmark, see [CMS12, Section 3.5].

So, when on the internet some websites waste our time asking us to prove that we are not a robot by clicking on the images of traffic signs, we'd better ask the help of a robot, see Figure 33.7.

see [PBE+22, LKN+22]. Quite surprisingly, there are situations in which, long after severely overfitting, the machine suddenly becomes capable of an almost perfect generalization, i.e. of correctly adapting to new, previously unseen data. In a sense, the machine suddenly flips from brutally memorizing and reproducing training data to correctly generalizing on unseen inputs.

From a cognitive point of view, it is interesting that this capacity of "understanding the pattern" in the data can happen well past the point of overfitting; it is possible that the machine first memorizes the training dataset and then only much later[14] "groks" the task, undergoing a "phase transitions".

A structurally different but somewhat related phenomenon is that of "benign overfitting" due to "overparametrization", see [MRSY19]. In this situation, the number of parameters available to the machine largely exceeds the number of training points, i.e. the neural network possesses much more model parameters than necessary.

Possibly, though a complete explanation has remained elusive, these phenomena are due to a further "regularization" provided by the additional, much longer, training period, or by the overabundant class of functions utilized, which allows the machine to match the training data using simpler, more regular, functions, which are more prone to generalization.

Let us also mention that in the numerical simulations of partial differential equations, neural networks have been widely exploited in view of their efficiency in terms of computational time. Typically, in

---

[14]By the way, the origin of the term "grokking" is intriguing in its own right. It was coined by the American author Robert Anson Heinlein in his science-fiction masterpiece, *Stranger in a Strange Land*. In this context, "grok" was a Martian word with no perfect counterpart in Earthling terms that could be loosely associated with various literal meanings such as "water", "to drink", "life" or "to live". Indeed, given the scarcity of water on Mars, the act of drinking water held a central significance for Martians: they had a unique practice of merging their bodies with water to form a new entity that transcended the sum of its parts, the water becoming an integral part of the drinker, and the drinker becoming intertwined with the water itself. Therefore, to "grok" can be simplified in Earthian terms as fully comprehending, assimilating or even "digesting" a topic or concept, making it an intrinsic part of oneself. See Figure 33.8 for a portrait of Heinlein.

Figure 33.8. Portrait of Robert A. Heinlein, as depicted in the science fiction magazine "Amazing Stories" in 1953 (Public Domain image from Wikipedia).

these simulations, neural networks do not replace the classical methods of numerical analysis (such as finite elements or finite difference algorithms); indeed, not only do these classical methods continue to produce more accurate results than those obtained through machine learning, but they are also used precisely to produce the numerical solutions used to train the machine via the neural network. In this, the machine learning approach to simulations of partial differential equations sometimes aims not at producing more precise results but rather to obtain quick answers, even at the risk of being slightly suboptimal.

Very interestingly, neural networks can also be used to suggest new perspectives, expanding our mathematical knowledge of partial differential equations and helping mathematicians draw a road map toward the proof of very difficult theorems, see e.g. https://www.quantamagazine.org/deep-learning-poised-to-blow-up-famed-fluid-equations-20220412/ and the references therein.

For additional and more comprehensive information[15] about machine learning, seen from various perspectives, refer for instance Adam Oberman's lecture https://www.youtube.com/watch?v=s-rMfJsf35Y, Dejan Slepcev's lecture https://www.youtube.com/watch?v=bLlgdzg094g, the textbooks [GBC16, MRT18], the articles [BNS06, vEH20], the set of notes by Leonardo Ferreira Guilhoto https://math.uchicago.edu/~may/REU2018/REUPapers/Guilhoto.pdf and the references therein.

---

[15]We do not address in these pages all the latest developments of artificial intelligence and machine learning. It suffices to mention that some typical issues of the classical machine learning methods are related to vanishing gradient problems (affecting the gradient descents used, e.g. for minimizing the loss function), slow learning algorithms (due to the many weights to be optimized) and high learning costs (due to the power consumption employed by a machine while training large datasets). Also, an additional challenge for classical networks is to deal with data which vary with time.

A fashionable answer to these problems is given by reservoir computing. Roughly, a reservoir is a high-dimensional nonlinear dynamical system, playing the role of a recurrent network of interconnected neurons. The reservoir typically receives an input, e.g. through a weighted linear combination of time-series data, which are processed by the nonlinear recursive features of the reservoir, thus producing an output, e.g. again via a weighted linear combination.

In this architecture, typically only the weights related to the linear combinations entering and exiting the reservoir are trained against the data sample, while the reservoir itself is not subject to any specific training (though some of the parameters of the reservoir typically require a bespoke choice for high performances).

The learning algorithms can be sped up in this way since the size of the training set is often downsized, and the necessary computing resources can be reduced, as the number of adjustable weights to be optimized is lower than in the classical neural architectures.

Interestingly, a physical system can be used as a reservoir, thus replacing a step usually performed by silicon chips by possibly cheaper physical systems and devices (also, a link with quantum computing may arise since the nonlinear system utilized can be borrowed from either classical or quantum mechanics).

From the perspective of cognitive neurosciences, one can also observe similarities (as well as differences) between the use of reservoirs in artificial intelligence and some traits of the prefrontal cortex in the human brain, see e.g. Figure 2 in [CAC+22]. See also [GBGB21] and the references therein for further information about reservoir computing.

Chapter 34

Cutting Networks

Other interesting applications of the discrete Laplacians presented in Chapters 30, 31 and 33 surface when studying networks. In rough terms, a network (also known as graph) is a (finite) collection of objects in some mutual relation. To better visualize this structure, these objects are depicted as vertices (also called nodes) and vertices that are related to each other are connected by an edge (also called link). Formally, a network thus consists of a pair (V, E), where V is a collection of vertices and E is a collection of edges.

We note that an edge in the network is identified by its endpoints, which are vertices of the network. Namely, given v, $w \in V$, an edge connecting v and w is denoted by (v, w); in our setting, the link between two vertices has no preferred direction; therefore, the edge (v, w) coincides with (w, v).

If the vertices v and w are connected[1] by an edge, i.e. if $(v, w) \in E$, we will write $v \sim w$.

We tacitly assume that the network under consideration is connected, i.e. for each pair of vertices v and w, there is a sequence of edges $(v_0, v_1), (v_1, v_2), \ldots, (v_{k-1}, v_k) \in E$ such that $v_0 = v$, $v_k = w$, $v_i \neq v_j$ unless $i = j \in \{0, \ldots, k\}$, and $(v_i, v_{i+1}) \neq (v_j, v_{j+1})$ unless $i = j \in \{0, \ldots, k-1\}$ (that is, we can connect any two vertices in our connected network).

---

[1]For simplicity, here we do not allow edges from a vertex to itself (loops), so $v \sim v$ never holds.

An edge $(v, w) \in E$ is said to be incident to a vertex $z \in V$ if either $v = z$ or $w = z$.

The degree of a vertex of a network is the number of edges that are incident to the vertex (for instance, in Figure 34.1, the vertices v and w are connected by an edge, and so are the vertices w and z; the vertex v has degree 2, the vertex w has degree 4, and the vertex z has degree 6). If $v \in V$ is a vertex, the degree of v is often denoted by $\deg(v)$.

The volume of a collection $S \subseteq V$ of vertices is the sum of the degrees of vertices in S, namely

$$|S| := \sum_{v \in S} \deg(v).$$

In a sense, the volume of S describes "how big" the collection S of vertices is, by taking into account not simply the number of vertices in S but rather weighing this number by the "importance" of these vertices, as encoded by their degrees.

A cut of a network is a partition of its vertices into two subsets, say S and $T := V \setminus S$. The boundary (also called a "cut-set") of such a cut is the set

$$\partial S := \big\{ (v, w) \in E \text{ s.t. } v \in S \text{ and } w \in T \big\},$$

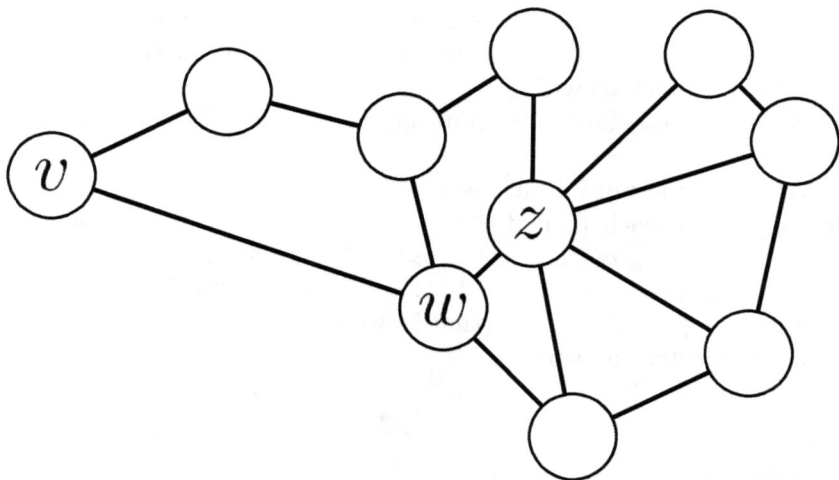

Figure 34.1.   A network.

which is the collection of edges that have one endpoint in S and the other endpoint in T. We observe that, by construction, $\partial S = \partial T$.

The size of such a cut is the number of edges crossing the cut, which is the cardinality of ∂S, denoted by $|\partial S|$. For instance, if we take $S := \{v, w, z\}$ in Figure 34.1, we have that $|S| = 12$ and $|\partial S| = 8$.

In a social network, every vertex represents a person and an edge represents some link between people (e.g. the edge (v, w) can represent the fact that two people know each other). In this sense, one may be interested in using the notion of cut for a social network to split the community V into two subgroups S and T such that people in the same subgroup are "likely to know each other" and people in different subgroups are "less likely to know each other".

Another example of a network consists of a collection of handwritten digits. In this situation, for instance, we can consider vertices corresponding to handwritten "threes" or "eights", and two vertices are linked by an edge if they "look sufficiently similar to each other". Here, we may want to implement an efficient algorithm to tell handwritten threes and eights apart, i.e. to cut the network consisting of these digits into two subsets, one collecting the digits which are "likely to be three" and the other collecting the digits which[2] are "likely to be eight".

Of course, in choosing a cut in these examples, one would like to reduce the size of the cut as much as possible, but a greedy choice in this can be highly counterproductive, in view of possible outliers. To get into the swing of this phenomenon, one can look at the case depicted in Figure 34.2. In this situation, one would like to distinguish "blue" and "yellow" colors, and vertices are linked if the colors are sufficiently similar. Note the presence of a "green" vertex (which looks sufficiently similar to both a couple of pale blue vertices and a couple of intense yellow vertices) and of a "magenta" vertex (which looks sufficiently similar a dark blue vertex but different from all the yellow vertices).

---

[2]To make things more accurate, one could even consider networks with weights, assigning a higher importance to a link between people who know each other well, or to handwritten digits which are particularly similar to each other. For simplicity, we do not consider this case here. A procedure such as this was already mentioned in the context of semi-supervised machine learning, see p. 315.

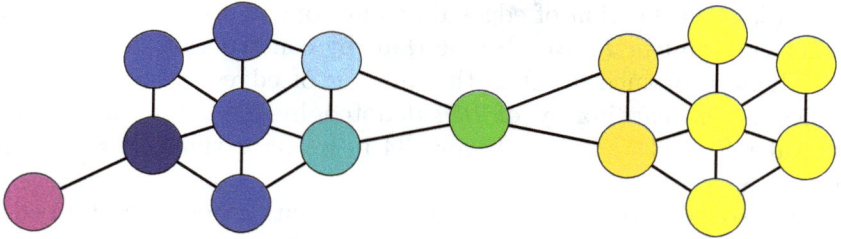

Figure 34.2.   A colorful network.

In this case, the "greedy" choice of simply minimizing the size of the cut would lead to consider the magenta vertex as separate from all the others since this cut has size 1, but this operation would fail badly in telling blues and yellows apart. It would be instead more appropriate to select a cut which groups together, for instance, all the blues (maintaining the magenta vertex in this group as a negligible error) and separately all the yellows (keeping perhaps the green vertex in this group as a negligible error); this cut has indeed a slightly larger size, equal to 2, but has the advantage of splitting the network into two groups of equal numerousness.

To implement this idea, it is convenient to introduce the Cheeger ratio of a cut, given by

$$h(S) := \frac{|\partial S|}{\min\{|S|, |T|\}}, \tag{34.1}$$

and we observe that $h(S) = h(T)$ since $T = V \setminus S$.

The Cheeger constant of the network (also known as the conductance of the network) is then defined by

$$h := \min h(S), \tag{34.2}$$

where the minimum above is taken among all possible choices of subsets S of V.

We observe that finding a cut that attains the Cheeger constant may actually be much more successful in the task of separating blues and yellows, especially when one knows to start with the network that collects two classes of individuals of approximately equal sizes, since the minimization method in (34.2) finds more "balanced" separations. For instance, in Figure 34.2, the cut corresponding to the isolation of the magenta vertex produces a Cheeger ratio of 1, while the cut dividing the blues (magenta included) and the yellows (green included) provides a Cheeger ratio strictly smaller than 1.

For practical purposes, the task of finding a cut attaining the Cheeger constant may be very hard to accomplish since, in principle, for a network consisting of n vertices, one has 2^n possible cuts to take into account (or about 2^{n-1} if one exploits the symmetry between the cut identified by S and the one identified by $T = V \setminus S$), making the procedure unfeasible for large (and interesting) networks, such as the ones of prohibitively large size arising in social sciences, biology or internet data. It could be therefore convenient to reduce the problem to a smaller and more manageable one, possibly producing an explicit algorithm, in such a way that this new problem maintains similar features to that proposed by the Cheeger constant. In particular, it would be ideal to replace the Cheeger constant by another more tractable quantity which is close to zero if and only if the Cheeger constant is close to zero, so as to somewhat maintain the efficiency of the cut in the spirit of the Cheeger constant.

An interesting solution for this question is provided by the spectral gap of a network, given by[3]

$$\lambda := \inf R(g), \quad \text{with} \quad R(g) := \frac{\sum_{v \sim w} (g(v) - g(w))^2}{\sum_{v \in V} \deg(v)\, g^2(v)}, \qquad (34.3)$$

where the above infimum is taken over the functions $g : V \to \mathbb{R}$ (not identically equal to zero) satisfying

$$\sum_{v \in V} \deg(v)\, g(v) = 0. \qquad (34.4)$$

For our purposes here, the interest of the spectral gap λ is that it detects the Cheeger constant within a quadratic factor of the

---

[3]The quantity $R(g)$ is sometimes called the Rayleigh quotient. From a computational viewpoint, a good approximation for the optimizers of the Rayleigh quotient can be obtained through power iteration algorithms, see e.g. [Dem97].

As a notational remark, in the sum in the numerator of (34.3), we have that $v \sim w$ if and only if $w \sim v$, but the vertices v and w are counted only once in the sum.

optimum since we have that

$$2h \geqslant \lambda \geqslant \frac{h^2}{2},$$
(34.5)

which is called the (discrete) Cheeger inequality[4] for networks.

Also, the understanding of this inequality entails the detection of efficient methods in the search of "good cuts", according to the following algorithm. One first seeks an optimizer $g_\star$ for λ in (34.3). Then, if the network consists of n vertices, one reorders them in the form $v_1, \ldots, v_n$ such that

$$g_\star(v_1) \geqslant g_\star(v_2) \geqslant \ldots \geqslant g_\star(v_n).$$
(34.6)

Figure 34.3.   Jeff Cheeger in 2007 (closeup of an image from Wikipedia, licensed under the Creative Commons Attribution-Share Alike 2.0 Germany license).

---

[4]See the following for a refinement of this inequality and for its proof. This type of estimate first appeared in [Che70] in the context of geometric analysis. Interestingly, mathematical topics happen to be intensively intertwined: in [Che70], Cheeger "wishes to thank J. Simons for helpful conversations, etc." See Figure 23.1 for a picture of Simons and Figure 34.3 for a picture of Cheeger.

For every $j \in \{1, \ldots, n\}$, one considers the cut induced by

$$S_j := \{v_1, \ldots, v_j\} \tag{34.7}$$

and, in the notation of (34.1), looks at the optimizer

$$\alpha := \min_{j \in \{1, \ldots, n\}} h(S_j). \tag{34.8}$$

Now, it turns out that the quantity α, as detected by this explicit algorithm, is already a "sufficiently good" estimate for an optimal cut in terms of the Cheeger constant since we have that

$$2h \geqslant \lambda \geqslant \frac{\alpha^2}{2} \geqslant \frac{h^2}{2}, \tag{34.9}$$

which can be considered as a refined version of (34.5).

We also remark that the link between the spectral gap and the discrete Laplacian discussed in Chapters 30, 31 and 33 becomes clear by assuming, up to multiplying $g_\star$ by a scalar factor, that

$$\sum_{v \in V} \deg(v)\, g_\star^2(v) = 1$$

and by noticing that, by the minimality of $g_\star$, for every $\phi : V \to \mathbb{R}$ satisfying (34.4) and every small ε, we have that

$$0 \geqslant R(g_\star) - R(g_\star + \varepsilon\phi)$$

$$= R(g_\star) - \frac{\sum_{v \sim w} \left((g_\star(v) - g_\star(w)) + \varepsilon(\phi(v) - \phi(w)) \right)^2}{\sum_{v \in V} \deg(v)\, (g_\star(v) + \varepsilon\phi(v))^2}$$

$$= R(g_\star) - \frac{\sum_{v \sim w}(g_\star(v) - g_\star(w))^2 \atop +2\varepsilon \sum_{v \sim w}(g_\star(v) - g_\star(w))(\phi(v) - \phi(w)) + o(\varepsilon)}{\sum_{v \in V} \deg(v)\, g_\star^2(v) \atop +2\varepsilon \sum_{v \in V} \deg(v)\, g_\star(v)\phi(v) + o(\varepsilon)}$$

$$= R(g_\star) - \frac{R(g_\star) + 2\varepsilon \sum_{v \sim w}(g_\star(v) - g_\star(w))(\phi(v) - \phi(w)) + o(\varepsilon)}{1 + 2\varepsilon \sum_{v \in V} \deg(v)\, g_\star(v)\phi(v) + o(\varepsilon)}$$

$$= -2\varepsilon \left(\sum_{v \sim w}(g_\star(v) - g_\star(w))(\phi(v) - \phi(w)) \right.$$

$$\left. - \lambda \sum_{v \in V} \deg(v)\, g_\star(v)\phi(v) \right) + o(\varepsilon).$$

and, therefore,

$$\sum_{v \sim w} (g_\star(v) - g_\star(w))(\phi(v) - \phi(w)) = \lambda \sum_{v \in V} \deg(v)\, g_\star(v)\phi(v). \quad (34.10)$$

In particular, for a given $\bar{v} \in V$, choosing

$$\phi(v) := \begin{cases} \delta(\bar{v}) & \text{if } v = \bar{v}, \\ -\deg(\bar{v}) & \text{if } v \neq \bar{v}, \end{cases}$$

with

$$\delta(\bar{v}) := \sum_{v \neq \bar{v}} \deg(v),$$

we see that ϕ fulfills (34.4), and thus (34.10) reduces to

$$\big(\delta(\bar{v}) + \deg(\bar{v})\big) \sum_{w \sim \bar{v}} (g_\star(\bar{v}) - g_\star(w))$$

$$= \lambda \left(\deg(\bar{v})\, g_\star(\bar{v})\, \delta(\bar{v}) - \deg(\bar{v}) \sum_{v \in V \setminus \{\bar{v}\}} \deg(v)\, g_\star(v) \right)$$

$$= \lambda \deg(\bar{v})\big(\delta(\bar{v}) + \deg(\bar{v})\big) g_\star(\bar{v}),$$

that is,

$$\mathcal{L}g_\star(\bar{v}) = -\lambda g_\star(\bar{v}),$$

where[5]

$$\mathcal{L}g(v) := \frac{1}{\deg(v)} \sum_{\substack{w \in V \\ w \sim v}} (g(w) - g(v)) = \frac{1}{\deg(v)} \sum_{w \sim v} g(w) - g(v), \quad (34.11)$$

to be compared with (30.2).

---

[5]Up to a possible sign convention, the operator

$$\sum_{\substack{w \in V \\ w \sim v}} g(w) - \deg(v)g(v)$$

is sometimes called the Laplacian of a network, see e.g. [Nic18, p. 64]. In this sense, the operator in (34.11) plays the role of a normalized Laplacian on a network, see e.g. equation (1) in [BJ08]. We also observe that if g has a minimum at v, then $\mathcal{L}g(v) \geqslant 0$.

It is now time to prove (34.9) (which in turn entails (34.5)). To this end, we first consider a partition of vertices $S_\star$ that achieves the Cheeger constant in (34.2). Up to replacing $S_\star$ with $V \setminus S_\star$, we can assume that $|S_\star| \leqslant |V \setminus S_\star|$; therefore,

$$|S_\star| \leqslant \frac{|V|}{2} \tag{34.12}$$

and

$$h = h(S_\star) = \frac{|\partial S_\star|}{|S_\star|}.$$

Suppose also

$$\tilde{g} := \chi_{S_\star} - \frac{|S_\star|}{|V|}.$$

We remark that

$$\sum_{v \in V} \deg(v)\, \tilde{g}(v) = \sum_{v \in S_\star} \deg(v) - \frac{|S_\star|}{|V|} \sum_{v \in V} \deg(v) = |S_\star| - \frac{|S_\star|}{|V|} |V| = 0;$$

therefore, $\tilde{g}$ fulfills (34.4).

As a result, by (34.3),

$$\lambda \leqslant \frac{\sum_{v \sim w} (\tilde{g}(v) - \tilde{g}(w))^2}{\sum_{v \in V} \deg(v)\, \tilde{g}^2(v)}. \tag{34.13}$$

Moreover,

$$\sum_{v \sim w} (\tilde{g}(v) - \tilde{g}(w))^2 = \sum_{v \sim w} (\chi_{S_\star}(v) - \chi_{S_\star}(w))^2$$

$$= \sum_{S_\star \ni v \sim w \in V \setminus S_\star} 1 = |\partial S_\star|$$

and, by (34.12),

$$\sum_{v \in V} \deg(v)\, \tilde{g}^2(v) = \sum_{v \in V} \deg(v) \left[\left(1 - \frac{2|S_\star|}{|V|} \right) \chi_{S_\star} + \frac{|S_\star|^2}{|V|^2} \right]$$

$$= \left(1 - \frac{2|S_\star|}{|V|} \right) \sum_{v \in S_\star} \deg(v) + \frac{|S_\star|^2}{|V|^2} \sum_{v \in V} \deg(v)$$

$$= \left(1 - \frac{2|S_\star|}{|V|} \right) |S_\star| + \frac{|S_\star|^2}{|V|^2} |V|$$

$$= |S_\star| - \frac{|S_\star|^2}{|V|} \geqslant \frac{|S_\star|}{2}.$$

By plugging this information into (34.13), we arrive at

$$\lambda \leqslant \frac{2|\partial S_\star|}{|S_\star|} = 2h. \qquad (34.14)$$

Now, we recall the notations in (34.7) and (34.8), and we pick the largest $r \in \{1, \dots, n\}$ for which $|S_r| \leqslant \frac{|V|}{2}$ (this is possible since $S_n = \{v_1, \dots, v_n\} = V$). This gives that

$$|S_j| \leqslant |V \setminus S_j| \quad \text{for all } j \in \{1, \dots, r\}$$
$$\text{and} \quad |S_j| > |V \setminus S_j| \quad \text{for all } j \in \{r+1, \dots, n\}. \qquad (34.15)$$

We also observe that, for every $c \in \mathbb{R}$ and any function $g : V \to \mathbb{R}$ satisfying (34.4),

$$\sum_{v \in V} \deg(v)(g(v) - c)^2 = \sum_{v \in V} \deg(v)g^2(v) + c^2|V| \geqslant \sum_{v \in V} \deg(v)g^2(v);$$

therefore, we see that

$$\lambda = \frac{\sum_{v \sim w}(g_\star(v) - g_\star(w))^2}{\sum_{v \in V} \deg(v)\, g_\star^2(v)} \geqslant \frac{\sum_{v \sim w}(g_\star(v) - g_\star(w))^2}{\sum_{v \in V} \deg(v)\,(g_\star(v) - g_\star(v_r))^2}.$$

$$(34.16)$$

Now, we set

$$g_1(v) := \begin{cases} g_\star(v) - g_\star(v_r) & \text{if } g_\star(v) \geqslant g_\star(v_r), \\ 0 & \text{otherwise,} \end{cases}$$

$$\text{and} \quad g_2(v) := \begin{cases} g_\star(v_r) - g_\star(v) & \text{if } g_\star(v) \leqslant g_\star(v_r), \\ 0 & \text{otherwise.} \end{cases} \qquad (34.17)$$

We point out that $g_\star(v) - g_\star(v_r) = g_1(v) - g_2(v)$ and $g_1(v)g_2(v) = 0$, hence

$$(g_\star(v) - g_\star(v_r))^2 = (g_1(v) - g_2(v))^2 = g_1^2(v) + g_2^2(v). \qquad (34.18)$$

Furthermore,

$$\begin{aligned}(g_\star(v) - g_\star(w))^2 &= (g_\star(v) - g_\star(v_r) + g_\star(v_r) - g_\star(w))^2 \\ &= (g_1(v) - g_2(v) - g_1(w) + g_2(w))^2 \\ &= (g_1(v) - g_1(w))^2 + (g_2(v) - g_2(w))^2 \\ &\quad -2(g_1(v) - g_1(w))(g_2(v) - g_2(w)).\end{aligned}$$

We also observe that $g_1(v) \geqslant 0$ and $g_2(v) \geqslant 0$; therefore,

$$(g_1(v) - g_1(w))(g_2(v) - g_2(w)) = -g_1(v)g_2(w) - g_1(w)g_2(v) \leqslant 0.$$

As a consequence,

$$(g_\star(v) - g_\star(w))^2 \geqslant (g_1(v) - g_1(w))^2 + (g_2(v) - g_2(w))^2.$$

From this observation, (34.16) and (34.18), we conclude that

$$\lambda \geqslant \frac{\sum_{v\sim w}\left((g_1(v) - g_1(w))^2 + (g_2(v) - g_2(w))^2\right)}{\sum_{v\in V}\deg(v)(g_1^2(v) + g_2^2(v))}. \qquad (34.19)$$

We now recall the mediant inequality[6] for all $a, b, c, d \geqslant 0$

$$\frac{a+b}{c+d} \geqslant \min\left\{\frac{a}{c}, \frac{b}{d}\right\}, \qquad (34.20)$$

and we thereby deduce from (34.19) that

$$\lambda \geqslant \min\left\{\frac{\sum_{v\sim w}(g_1(v) - g_1(w))^2}{\sum_{v\in V}\deg(v)g_1^2(v)}, \frac{\sum_{v\sim w}(g_2(v) - g_2(w))^2}{\sum_{v\in V}\deg(v)g_2^2(v)}\right\}$$
$$= \min\{R(g_1), R(g_2)\}. \qquad (34.21)$$

---

[6]To prove (34.20), one can suppose that $\frac{a}{c} \leqslant \frac{b}{d}$ and note that
$$\frac{a+b}{c+d} - \frac{a}{c} = \frac{bc - ad}{c(c+d)} = \frac{d}{c+d}\left(\frac{b}{d} - \frac{a}{c}\right) \geqslant 0.$$

Now, we assume that $R(g_1) \leqslant R(g_2)$, with the other case being similar, and rewrite (34.21) in the form

$$\lambda \geqslant R(g_1). \tag{34.22}$$

Now, we point out that, by the Cauchy–Schwarz inequality, for every $g : V \to \mathbb{R}$,

$$\sum_{v \sim w} |g^2(v) - g^2(w)| = \sum_{v \sim w} \Big(|g(v) + g(w)|\,|g(v) - g(w)| \Big)$$

$$\leqslant \sqrt{\sum_{v \sim w} (g(v) + g(w))^2} \sqrt{\sum_{v \sim w} (g(v) - g(w))^2}$$

and, therefore,

$$\begin{aligned}
R(g) &= \frac{\sum_{v \sim w} (g(v) - g(w))^2}{\sum_{v \in V} \deg(v)\, g^2(v)} \\[2mm]
&= \frac{\sum_{v \sim w} (g(v) + g(w))^2 \sum_{v \sim w} (g(v) - g(w))^2}{\sum_{v \sim w} (g(v) + g(w))^2 \sum_{v \in V} \deg(v)\, g^2(v)} \\[2mm]
&\geqslant \frac{\left(\sum_{v \sim w} |g^2(v) - g^2(w)| \right)^2}{\sum_{v \sim w} (g(v) + g(w))^2 \sum_{v \in V} \deg(v)\, g^2(v)}.
\end{aligned} \tag{34.23}$$

Moreover, if $i < j \in \{1, \dots, n\}$, using a telescoping sum, we have that

$$g^2(v_i) - g^2(v_j) = \sum_{k=i}^{j-1} (g^2(v_k) - g^2(v_{k+1}))$$

and thus, exchanging the order of the summations,

$$\begin{aligned}
\sum_{v \sim w} |g^2(v) - g^2(w)| &= \sum_{\substack{v \sim w \\ g(v) \geqslant g(w)}} (g^2(v) - g^2(w)) \\[2mm]
&= \sum_{\substack{1 \leqslant i < j \leqslant n \\ g(v_i) \geqslant g(v_j)}} \chi_E\big((v_i, v_j)\big)(g^2(v_i) - g^2(v_j))
\end{aligned}$$

$$= \sum_{\substack{1 \leqslant i < j \leqslant n \\ g(v_i) \geqslant g(v_j)}} \sum_{i \leqslant k \leqslant j-1} \chi_E\big((v_i, v_j)\big)\big(g^2(v_k) - g^2(v_{k+1})\big)$$

$$= \sum_{1 \leqslant k \leqslant n-1} \sum_{\substack{1 \leqslant i \leqslant k \\ k+1 \leqslant j \leqslant n \\ g(v_i) \geqslant g(v_j)}} \chi_E\big((v_i, v_j)\big)\big(g^2(v_k) - g^2(v_{k+1})\big). \qquad (34.24)$$

Besides, it is interesting to observe that, by (34.6) and (34.17),

$$g_1(v_j) = \begin{cases} g_*(v_j) - g_*(v_r) & \text{if } j \in \{1, \ldots, r-1\}, \\ 0 & \text{otherwise,} \end{cases}$$

and therefore $g_1(v_i) \geqslant g_1(v_j)$ if and only if $i \leqslant j$.

Hence, we can specialize (34.24) in the case of $g := g_1$, obtaining the simpler expression

$$\sum_{v \sim w} |g_1^2(v) - g_1^2(w)|$$

$$= \sum_{1 \leqslant k \leqslant n-1} \sum_{\substack{1 \leqslant i \leqslant k \\ k+1 \leqslant j \leqslant n}} \chi_E\big((v_i, v_j)\big)\big(g_1^2(v_k) - g_1^2(v_{k+1})\big). \qquad (34.25)$$

Note also that the expression $\sum_{\substack{1 \leqslant i \leqslant k \\ k+1 \leqslant j \leqslant n}} \chi_E\big((v_i, v_j)\big)$ counts all the edges joining vertices in $\{v_1, \ldots, v_k\}$ and in $\{v_{k+1}, \ldots, v_n\}$, whence, recalling the notation in (34.7),

$$\sum_{\substack{1 \leqslant i \leqslant k \\ k+1 \leqslant j \leqslant n}} \chi_E\big((v_i, v_j)\big) = |\partial S_k|.$$

This and (34.25) yield that

$$\sum_{v \sim w} |g_1^2(v) - g_1^2(w)| = \sum_{1 \leqslant k \leqslant n-1} |\partial S_k|\big(g_1^2(v_k) - g_1^2(v_{k+1})\big). \qquad (34.26)$$

In addition, recalling (34.8),

$$\alpha \leqslant h(S_k) = \frac{|\partial S_k|}{\mu_k}, \quad \text{where} \quad \mu_k := \min\big\{|S_k|, |V \setminus S_k|\big\}.$$

As a result,

$$\sum_{1\leqslant k\leqslant n-1}|\partial S_k|(g_1^2(v_k)-g_1^2(v_{k+1})) \geqslant \alpha \sum_{1\leqslant k\leqslant n-1}\mu_k(g_1^2(v_k)-g_1^2(v_{k+1}))$$

$$= \alpha\left[\sum_{1\leqslant k\leqslant n-1}\mu_k g_1^2(v_k) - \sum_{2\leqslant k\leqslant n}\mu_{k-1}g_1^2(v_k)\right]$$

$$= \alpha\left[\sum_{2\leqslant k\leqslant n-1}(\mu_k-\mu_{k-1})g_1^2(v_k) + \mu_1 g_1^2(v_1) - \mu_{n-1}g_1^2(v_n)\right].$$

We also observe that $V\setminus S_n = \varnothing$, whence $\mu_n = 0$. Therefore, setting $S_0 := \varnothing$, we also have that $\mu_0 = 0$, and thus we obtain the compact expression

$$\sum_{1\leqslant k\leqslant n-1}|\partial S_k|(g_1^2(v_k)-g_1^2(v_{k+1})) \geqslant \alpha \sum_{1\leqslant k\leqslant n}(\mu_k-\mu_{k-1})g_1^2(v_k)$$

$$= \alpha \sum_{1\leqslant k\leqslant r-1}(\mu_k-\mu_{k-1})g_1^2(v_k). \tag{34.27}$$

It is now useful to observe that, by (34.15), $\mu_j = |S_j|$ for all $j \in \{1,\ldots,r\}$. Therefore, if $k \leqslant r-1$, then

$$\mu_k-\mu_{k-1} = |S_k|-|S_{k-1}| = \sum_{j=1}^{k}\deg(v_j) - \sum_{j=1}^{k-1}\deg(v_j) = \deg(v_k).$$

Thus, we deduce from (34.27) that

$$\sum_{1\leqslant k\leqslant n-1}|\partial S_k|(g_1^2(v_k)-g_1^2(v_{k+1}))$$

$$\geqslant \alpha \sum_{1\leqslant k\leqslant r-1}\deg(v_k)g_1^2(v_k) = \alpha\sum_{v\in V}\deg(v)g_1^2(v).$$

Combining this inequality with (34.26), we have that

$$\sum_{v\sim w}|g_1^2(v)-g_1^2(w)| \geqslant \alpha\sum_{v\in V}\deg(v)g_1^2(v).$$

As a consequence, recalling (34.23),

$$R(g_1) \geqslant \frac{\alpha^2 \sum_{v \sim w} \deg(v) g_1^2(v)}{\sum_{v \sim w} (g_1(v) + g_1(w))^2}. \qquad (34.28)$$

Also, for every $h : V \times V \to \mathbb{R}$ such that $h(v, w) = h(w, v)$, we have that

$$\sum_{v \sim w} h(v, w) = \frac{1}{2} \sum_{1 \leqslant i \leqslant n} \sum_{\substack{1 \leqslant j \leqslant n \\ v_i \sim v_j}} h(v_i, v_j)$$

where the factor $\frac{1}{2}$ is used to avoid the double counting of the edge $(v_i, v_j) = (v_j, v_i)$; therefore, for all $g : V \to \mathbb{R}$,

$$\sum_{v \sim w} (g(v) + g(w))^2 \leqslant 2 \sum_{v \sim w} (g^2(v) + g^2(w))$$

$$= \sum_{1 \leqslant i \leqslant n} \sum_{\substack{1 \leqslant j \leqslant n \\ v_i \sim v_j}} (g^2(v_i) + g^2(v_j))$$

$$= 2 \sum_{1 \leqslant i \leqslant n} \sum_{\substack{1 \leqslant j \leqslant n \\ v_i \sim v_j}} g^2(v_i) = 2 \sum_{1 \leqslant i \leqslant n} \deg(v_i) g^2(v_i) = 2 \sum_{v \in V} \deg(v) g^2(v).$$

This and (34.28) give

$$R(g_1) \geqslant \frac{\alpha^2}{2}.$$

Hence, in light of (34.22),

$$\lambda \geqslant \frac{\alpha^2}{2}. \qquad (34.29)$$

Also, comparing (34.2) and (34.8), we see that $\alpha \geqslant \lambda$. Combining this information with (34.14), and (34.29), we obtain (34.9), as desired.

See [Chu10] for further information about the Cheeger inequality, its connection to network partitions, random walks and algorithm used to rank web pages by search engines (such as the PageRank algorithm employed by Google Search). For more information about the problem of detecting communities in a network, see e.g. [New06]. See also [Chu97, Nic18] and the references therein for further readings on related topics.

Chapter 35

The Wear of Rolling Stones

Let us now deal with rolling stones. No, not The Rolling Stones (see Figure 35.1); our objective here is to understand the wearing process of stones on beaches and riverbeds, which are pounded by waves beautifully shaping their forms (see Figure 35.2).

This model was introduced by William J. Firey in [Fir74] and describes the boundary of the stone at time t as a two-dimensional, smooth, bounded, strictly convex surface M_t, corresponding to an embedding x_t. In this framework, Firey's equation for rolling stones reads

$$\partial_t x_t = -K_t \nu_t, \tag{35.1}$$

where ν_t is the exterior normal to M_t at the point identified by x_t and K_t is the Gauß curvature.

Equation (35.1) is known in jargon as the "Gauß curvature"[1] flow.

---

[1]Note that (35.1) prescribes that the normal velocity of the evolving surface coincides with its Gauß curvature (i.e. the product of its principal curvatures). For planar curves, both the Gauß curvature and the mean curvature flows coincide with the curve-shortening flow. See [ACGL20] and the references therein for thorough presentations of geometric flows.

It is also interesting to point out that mean curvature flows, Gauß curvature flows, and more general types of geometric flows also play a pivotal role in the description of bushfire spread, see [WWMS15].

A remarkable feature of stones is that, as observed in [Fir74], "often stones on beaches pounded by waves wear into quite smooth, regular shapes, sometimes apparently ellipsoidal and even spherical". It was indeed established in [And99]

Figure 35.1.   A 1965 trade ad for a tour of a popular English rock band (Public Domain image from Wikipedia).

To understand the rationale behind (35.1), one can imagine that the wear of a stone comes from its tumbling over an abrasive plane. Suppose that each direction has equal likelihood of specifying a contact position between the stone and this ideal abrasive plane. Given a point p on the surface of the stone, the wear is then proportional to

the ratio between the (infinitesimal) measure of the set of directions

for which the abrasive plane touches the stone in an (infinitesimal)

neighborhood of p and the (infinitesimal) measure of the stone's

surface of this (infinitesimal) neighborhood of p.   (35.2)

---

that surfaces moving according to Gauß curvature become spherical as they contract to a point.

Figure 35.2. Stones in a riverbed (photo by Fir0002, image from Wikipedia, licensed under the Creative Commons Attribution-Share Alike 3.0 Unported license).

For the sake of simplicity, in this process, we can disregard any possible dynamic effect of the global shape of the stone on the tumbling process.

Specifically, up to rigid motion, we can suppose that p is the origin of our reference frame and that the tangent plane to the stone in 0 is horizontal. In this way, the (n-dimensional) stone is parameterized in a small neighborhood of 0 by the sublevel sets of some function $\varphi :\mathbb{R}^{n-1} \to \mathbb{R}$, with $\varphi(0) = 0$ and $\nabla\varphi(0) = 0$.

We observe that, when $x \in \mathbb{R}^{n-1}$ lies near 0,

$$\varphi(x) = \frac{D^2\varphi(0)x \cdot x}{2} + o(|x|^2).$$

Also, the exterior normal to the stone in the vicinity of 0 takes the form

$$\nu(x) = \frac{(-\nabla\varphi(x), 1)}{\sqrt{1 + |\nabla\varphi(x)|^2}} = \frac{(-D^2\varphi(0)x + o(|x|), 1)}{\sqrt{1 + |D^2\varphi(0)x + o(|x|)|^2}}$$

$$= \frac{(-D^2\varphi(0)x, 1)}{\sqrt{1 + |D^2\varphi(0)x|^2}} + o(|x|) = -D^2\varphi(0)x + o(|x|).$$

(35.3)

Accordingly, the measure of the set of directions for which the abrasive plane touches the stone near the origin (i.e. the measure of the possible tangent planes in the vicinity of the origin) is the $(n-1)$-dimensional measure of $\nu(\mathcal{B}_\varepsilon)$ in ∂B_1, where

$$\mathcal{B}_\varepsilon := \{x \in \mathbb{R}^{n-1} \text{ s.t. } |x| < \varepsilon\},$$

see Figure 35.3.

It is thus convenient to write the surface $\nu(\mathcal{B}_\varepsilon)$ in a graphical form: namely, since $|\nu| = 1$, the points $(y, y_n) \in \mathbb{R}^{n-1} \times \mathbb{R}$ on the surface $\nu(\mathcal{B}_\varepsilon)$ can be written in the form

$$y_n = \sqrt{1 - |y|^2} =: \psi(y),$$

with, owing to (35.3),

$$y = -D^2\varphi(0)x + o(|x|),$$

for $x \in \mathcal{B}_\varepsilon$.

Thus, denoting by $\widetilde{\mathcal{B}_\varepsilon}$ the above set of y's, and assuming for definiteness that the Hessian matrix $D^2\varphi(0)$ is invertible, it follows that

$$|\nu(\mathcal{B}_\varepsilon)| = \int_{\widetilde{\mathcal{B}_\varepsilon}} \sqrt{1 + |\nabla\psi(y)|^2}\, dy = \int_{\widetilde{\mathcal{B}_\varepsilon}} \sqrt{1 + \left|\frac{y}{\sqrt{1-|y|^2}}\right|^2}\, dy$$

$$= \int_{\widetilde{\mathcal{B}_\varepsilon}} \frac{1}{\sqrt{1-|y|^2}}\, dy.$$

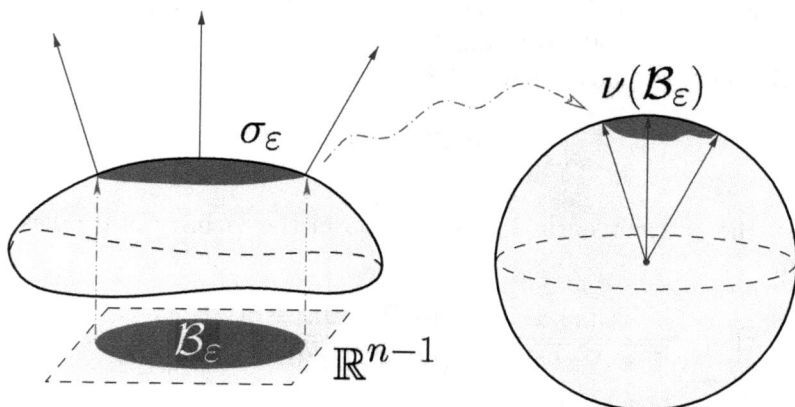

Figure 35.3. The normal map of a surface (sometimes called a Gauß map).

$$= \int_{\mathcal{B}_\varepsilon} \frac{1}{\sqrt{1 - |-D^2\varphi(0)x + o(|x|)|^2}} \left| \det \left(-D^2\varphi(0) + o(1) \right) \right| dx$$

$$= (1 + o(1)) \left| \det D^2\varphi(0) \right| |\mathcal{B}_\varepsilon|.$$

If we assume that the stone is a convex body, then $D^2\varphi(0) \geqslant 0$ and, accordingly,

$$|\det D^2\varphi(0)| = \det D^2\varphi(0) = K(0),$$

where K is the Gauß curvature. From this, we arrive at

$$|\nu(\mathcal{B}_\varepsilon)| = (1 + o(1)) K(0) |\mathcal{B}_\varepsilon|$$

$$= K(0) \int_{\mathcal{B}_\varepsilon} \sqrt{1 + |\nabla\varphi(x)|^2}\, dx + o(\varepsilon^{n-1})$$

$$= K(0) |\sigma_\varepsilon| + o(\varepsilon^{n-1})$$

where σ_ε is the $(n-1)$-dimensional surface describing the stone in the vicinity of the origin (that is, the $(n-1)$-dimensional surface described by $\{(x, \varphi(x)), \text{ with } x \in \mathcal{B}_\varepsilon\}$).

Thus, by (35.2), the wear inducing the normal velocity is proportional to

$$\lim_{\varepsilon \searrow 0} \frac{|\nu(\mathcal{B}_\varepsilon)|}{|\sigma_\varepsilon|} = K(0).$$

This justifies equation (35.1).

Of course, equation (35.1) presents a natural generalization of the case in which an additional normal velocity takes part, for instance due to a function g, leading to

$$\partial_t x_t = (g - K_t)\, \nu_t.$$

Stationary solutions correspond in this case to solutions of

$$K = g, \tag{35.4}$$

which is a form of prescribed Gauß curvature equation.

In particular, if we are interested in a graphical framework in which the n-dimensional surface under consideration is (locally) the graph of some function $u : \mathbb{R}^n \to \mathbb{R}$ (and thus, up to a slight change in notation, also $g : \mathbb{R}^n \to \mathbb{R}$), we have that

$$K = \frac{\det D^2 u}{(1 + |\nabla u|^2)^{\frac{n+2}{2}}},$$

see e.g. [Kaz85, equation (4.15)], and thus (35.4) can be recast in the form

$$\det D^2 u(x) = g(x)(1 + |\nabla u(x)|^2)^{\frac{n+2}{2}}.$$

This type of prescribed Gauß curvature equation is in fact a particular case of the so-called Monge–Ampère equation

$$\det D^2 u = f,$$

which arises in several areas of science, including antenna design, image processing, and optimal mass transport.

We refer the reader to [Gut01, TW08, DPF14, Fig17] for further information about the prescribed Gauß curvature equation and the Monge–Ampère equation.

It is interesting to observe the similarities and differences between the Monge–Ampère equation and the Poisson equation. First of all, while the Laplacian operator appearing in the Poisson equation can be interpreted as the *trace* of the Hessian matrix of the solution, the leading operator in the Monge–Ampère equation is the *determinant* of the Hessian matrix.

With respect to this algebraic structure, the Monge–Ampère equation and the Poisson equation are special cases of the so-called k-Hessian equations, focusing on the kth elementary symmetric polynomials[2] of the Hessian matrix, see [Wan09].

---

[2]Specifically, the kth elementary symmetric polynomial of n variables $\lambda = (\lambda_1, \ldots, \lambda_n)$ is defined by

$$S_k(\lambda) := \sum_{1 \leqslant i_1 < \cdots < i_k \leqslant n} \lambda_{i_1} \ldots \lambda_{i_k}.$$

The k-Hessian equation thus deals with $S_k(\lambda)$ when λ is the collection of the eigenvalues (with multiplicity) of the Hessian matrix $D^2 u$.

In particular, $S_1(\lambda) = \lambda_1 + \cdots + \lambda_n$, which produces the Laplace operator, while $S_n(\lambda) = \lambda_1 \ldots \lambda_n$, which produces the determinant operator and leads to the Monge–Ampère equation.

Differently from the Poisson equation, the Monge–Ampère equation shows a severe nonlinear dependence on the Hessian of the solution, and it is indeed an interesting case of a fully nonlinear equation. It is however instructive to look at its linearization (see [CG97]): namely, if A is an invertible matrix, B is a matrix and ε a small parameter,

$$\det(A + \varepsilon B) = \det A \left(1 + \varepsilon \operatorname{tr}(A^{-1} B)\right) + o(\varepsilon);$$

therefore, if one considers a function $u = u_0 + \varepsilon v$ and assumes that $D^2 u_0$ is invertible, then

$$\begin{aligned}
\det D^2 u &= \det(D^2 u_0 + \varepsilon D^2 v) \\
&= \det(D^2 u_0)(1 + \varepsilon \operatorname{tr}((D^2 u_0)^{-1} D^2 v)) + o(\varepsilon) \\
&= \det(D^2 u_0) + \varepsilon \det(D^2 u_0) \sum_{i,j=1}^{n} a_{ij} \partial_{ij} v + o(\varepsilon), \quad (35.5)
\end{aligned}$$

where

$$a_{ij} \text{ is the } (i,j)\text{th entry of the inverse matrix of } D^2 u_0. \quad (35.6)$$

Considering the first order in ε in (35.5) and dividing by $\det(D^2 u_0)$ (which is nonzero since we assumed $D^2 u_0$ to be invertible), the linearized Monge–Ampère equation is therefore written as

$$\sum_{i,j=1}^{n} a_{ij} \partial_{ij} v = h, \quad (35.7)$$

for a suitable h and with the notation in (35.6).

A particularly interesting structure in equation (35.7) arises under a suitable "uniform convexity" assumption on u_0: namely, if we assume that $D^2 u_0$ is bounded and strictly positive definite, then its inverse is also bounded and strictly positive definite. This observation and (35.6) yield that, in this situation, equation (35.7) becomes elliptic (according to the classification in footnote 13 on p. 9).

We will get a glimpse of linear elliptic equations, like (35.7), in Chapters 4 and 5 of the companion book [DV23].

Chapter 36

Bushrangers and Outlaws

Criminal activities are a common concern for states and citizens (though criminals, bushrangers and heroic outlaws are sometimes glamorized with a romantic aura, see Figure 36.1, and supervillains are popular stock characters in many comic books, see Figure 36.2). Mathematics can certainly play a role in understanding criminal activities; partial differential equations lend themselves well to a realistic description of several situations related to transgression of law, and several research articles have been recently devoted to this topic (see e.g. [SDP$^+$08, BN10]).

These models are actually closely intertwined with the mathematical descriptions of epidemics (see Chapter 40) since they aim at describing "collective[1] behaviors" and "social phenomena", which

---

[1]The mathematical analysis of collective behaviors has an interesting predecessor in science fiction literature. Namely, the eminent writer (and biochemist) Isaac Asimov in some of his novels discussed about *psychohistory*, i.e. a fictional science combining history, sociology, statistics and mathematics, in order to make predictions, on average, about the behavior of large groups of people.

Like all disciplines based on mathematics, psychohistory possessed its own axioms (such as that the population whose behavior is being modeled should be sufficiently large and that the population should remain ignorant of the results of the application of psychohistorical analyses, as awareness would cause the group to changes its behavior). And like all disciplines based on mathematics, psychohistory possessed its own limitations (while disciplines possessing no limitations are a bunch of crooked claptrap!). In Asimov's novels, the limitations of psychohistory are showcased by its incapability of predicting the appearance of the Mule (mutant warlord endowed with immensely powerful psychic skills,

Figure 36.1. (a) John Dillinger, (b) Ned Kelly, (c) Bonnie and Clyde (Public Domain images from Wikipedia).

Figure 36.2. Batman villains: the Penguin, the Riddler, the Catwoman and the Joker in the 1966 film Batman (Public Domain image from Wikipedia).

suffering from megalomania and psychopathic paranoia, and aiming to conquer the galaxy).

Perhaps the catch in psychohistory, which was supposed to foresee even thousands of years into the future, happens to be the role of single individuals (as a statistical science, it deals with large groups of people, while the Mule is a "statistical outlier") and its inability to quickly evolve to include situations in which its own axioms are overturned (e.g. not taking into account the probability of the Mule's mutation). All in all, real, as well as fictitious, predictive sciences are of great importance, but they must be taken with a pinch of salt.

present several traits that are reminiscent of the spread of an infection. Remarkably, similar models have also been introduced to describe, for instance, the diffusion of new ideas and scientific knowledge [BCAKCC06, CHJ16, HS16], the diffusion of a product on the market [Bas69], the emergence and establishment of a political party [JHRW16], the propagation of riots [BNR22], etc.

Here, we recall a model for criminal activity put forth in [BN10]. Such a model leverages two basic *ansätze* from criminology: namely, the *repeated victimization theory*, postulating that places at which certain crimes have been committed incur a higher risk of this crime being repeated, leading to a situation in which crime in an area produces more crime, and the *routine activity theory*, according to which motivating agents, potential benefits and opportunities are key factors in the occurrence of a crime.

To translate these ingredients into a mathematical framework, one can consider the evolution of two functions, namely the function $v = v(x, t)$, which describes the "propensity for people to commit a crime" at position $x \in \mathbb{R}^n$ at time $t > 0$, and the function $u = u(x, t)$, accounting for the number of crimes really occurring at position $x \in \mathbb{R}^n$ at time $t > 0$. Note that while u can be concretely observed and measured in some explicit, empiric or objective way (e.g. by the number of incidents reported by the police), the function v is mostly a lumped quantity arising from many complex social interactions. Nonetheless, the susceptibility function v plays a decisive role in activating and modulating the growth of u, and conversely the activity level u induces a feedback from v.

More specifically, to model the number of crimes, one can assume that the positive values of v correspond to a tendency to induce criminal activities and postulate that the number of crimes u, as time flows, would approach zero exponentially fast (say, like $\exp(-\lambda_u t)$ for some $\lambda_u > 0$) in the absence of any propensity to commit a crime, but would increase in the presence of positive values of v (say, by a number $\Lambda(v)$, where Λ is an increasing function in $(0, +\infty)$ such that $\Lambda = 0$ in $(-\infty, 0]$). One can also suppose that the occurrence of a crime in a given location may trigger unlawful enterprises in neighboring areas, and thus it is suggestive to model this diffusion process via a heat equation, say with a diffusion coefficient $d_u \geqslant 0$. These considerations lead to an evolution equation for the criminal activities of the form

$$\partial_t u(x,t) = d_u \, \Delta u(x,t) + \Lambda\big(v(x,t)\big) - \lambda_u \, u(x,t). \qquad (36.1)$$

As for the tendency v of committing crimes, one can assume that there is an "innate propension" of the population for felonies and quantify it by some function σ. This σ may indeed depend on space to allow for spatial inhomogeneities, such as bad and dangerous neighborhoods, as well as on time, since for instance improved life conditions, access to healthcare and education systems and promotion of cultural activities may contribute to reducing the value of σ. Note that while positive values of σ correspond to a natural inclination toward criminal activities, negative values indicate a natural anti-crime tendency. Thus, in the absence of feedback from u, one could suppose that, as time progresses, v would approach σ exponentially fast (say, like $\exp(-\lambda_v \, t)$ for some $\lambda_v > 0$). One can also suppose that the crime propension in a given location may trigger criminal tendencies in neighboring areas as well, and again one could model this diffusion process via a heat equation, say with a diffusion coefficient $d_v \geqslant 0$.

Also, in light of the repeated victimization and routine activity theories, the feedback of criminal activities on criminal inclination can be modeled by supposing a growth in criminal intention that is directly proportional to the number of critical events, say by a proportionality coefficient ϱ, possibly varying in time and space (concretely, this ϱ could measure the benefit of committing the crime, with positive values corresponding to an advantageous payoff of the criminal activity while negative values accounting for the negative consequences of unlawful actions). These observations lead to an evolution equation for the criminal tendency v of the form

$$\partial_t v(x,t) = d_v \, \Delta v(x,t) - \lambda_v \, \big(v(x,t) - \sigma(x,t)\big) + \varrho(x,t) \, u(x,t). \qquad (36.2)$$

The system of equations (36.1) and (36.2) can be seen as an activity/susceptibility model linking criminal activities with social tendencies and dispositions, and it could also be complemented with an evolution equation for the benefit function ϱ.

Chapter 37

Fighting Cancer Using
Differential Equations

While the body's cells in a normal condition grow and multiply in a controlled way, some cells can become abnormal and keep growing, possibly ending up forming a lump called tumor. The name "cancer" thus comprises this group of diseases involving abnormal cell growth with the potential to spread to other parts of the body.

There are more than 100 types of cancers which affect humans, and cancer is a leading cause of death worldwide, usually reported to be the second most common cause of death, after heart diseases. Also, as the world population is growing and aging, the global number of cancer deaths appears to be increasing.

In view of its tragic impact on our lives and the obvious importance of improving early detection methods and treatments, the support for research into curing cancer is usually prioritized. For instance, in 2021, the expenditure specifically targeting cancer research by the Australian National Health and Medical Research Council amounted to 153.7 million Australian dollars, i.e. more than 27% of the total investment,[1] see Figure 37.1.

---

[1]These figures can be compared, for instance, with those coming from the Australian Research Council Discovery Projects: e.g. the total funds allocated for this type of scheme to approved (less than 20% of the total) applications in all disciplines of mathematics, physics, chemistry and earth sciences, amounted in 2021 to about 53.7 million Australian dollars, see Figure 37.2.

Elliptic PDEs from an Elementary Viewpoint

NHMRC expenditure by Former National Health Priority Areas 2013 to 2021

Former National Health Priority Areas	2013 $ million	2014 $ million	2015 $ million	2016 $ million	2017 $ million	2018 $ million	2019 $ million	2020 $ million	2021 $ million
Arthritis and Osteoporosis	23.7	22.7	24.7	19.3	18.9	17.5	18.3	16.1	14.8
Asthma	21.5	23.6	22.7	15.3	13.3	15.7	13.8	13.3	14.1
Cancer	179.2	188.3	191.4	170.6	175.8	178.9	181.6	170.2	153.7
Cardiovascular Disease	117.1	129.4	130	114.9	111.4	105.3	112.6	107.6	102.5
Dementia[1]	24.9	31.5	33.4	45.6	50.2	60.9	71.2	64.1	55.3
Diabetes	65.2	70.2	70.3	65.0	57.7	50.7	46.5	45.6	42.6
Injury	45.4	58.4	61.5	45.8	44.2	49.9	51.1	49.8	46.6
Mental Health[2]	85.1	95.9	100	91.1	93.4	104.9	110.2	103.9	102.3
Obesity	41.7	40.7	39.0	28.1	27.6	23.0	23.5	24.3	23.1

1 Funding for Dementia excludes research activities that are outside the Medical Research Endowment Account (MREA) such as the Clem Jones Centre for Ageing Dementia Research and the Dementia Collaborative Research Centres.

2 Includes research into addiction.

Figure 37.1. National Health and Medical Research Council funding for major diseases, table from https://www.nhmrc.gov.au/funding/data-research/research-funding-statistics-and-data.

NHMRC expenditure by Former National Health Priority Areas 2013 to 2021

Former National Health Priority Areas	2013 $ million	2014 $ million	2015 $ million	2016 $ million	2017 $ million	2018 $ million	2019 $ million	2020 $ million	2021 $ million
Arthritis and Osteoporosis	23.7	22.7	24.7	19.3	18.9	17.5	18.3	16.1	14.8
Asthma	21.5	23.6	22.7	15.3	13.3	15.7	13.8	13.3	14.1
Cancer	179.2	188.3	191.4	170.6	175.8	178.9	181.6	170.2	153.7
Cardiovascular Disease	117.1	129.4	130	114.9	111.4	105.3	112.6	107.6	102.5
Dementia[1]	24.9	31.5	33.4	45.6	50.2	60.9	71.2	64.1	55.3
Diabetes	65.2	70.2	70.3	65.0	57.7	50.7	46.5	45.6	42.6
Injury	45.4	58.4	61.5	45.8	44.2	49.9	51.1	49.8	46.6
Mental Health[2]	85.1	95.9	100	91.1	93.4	104.9	110.2	103.9	102.3
Obesity	41.7	40.7	39.0	28.1	27.6	23.0	23.5	24.3	23.1

1 Funding for Dementia excludes research activities that are outside the Medical Research Endowment Account (MREA) such as the Clem Jones Centre for Ageing Dementia Research and the Dementia Collaborative Research Centres.

2 Includes research into addiction.

Figure 37.2. Australian Research Council funding for Discovery Projects, table from https://www.arc.gov.au / grants / grant-outcomes / selection-outcome-reports/selection-report-discovery-projects-2021.

Though it is not realistic that mathematics by itself could provide the keystone to defeat cancer, it (particularly partial differential equations) has provided several tools for attaining a deeper and quantitative understanding of many issues related to cancer. Among them, we present here a research direction related to cell adhesion as a form of fighting the malignant progression of cancers.

More specifically, in most cases, the fatal outcome of cancer is caused not by the primary tumor itself, but due to the invasion by malignant cells into a different body part from where the tumor started (this phenomenon is named metastasis, the newly pathological body locations being called metastasis sites, or metastases). Hence, to expand our knowledge about cancer, it could be of great importance to understand how malignant cells can migrate and how fast, and ideally it would be exceptionally useful to detect conditions which can reduce or suppress tumor invasiveness.

Among these conditions, a pivotal role could be played by cell adhesion. This process allows cells to interact and each cell to attach itself to neighboring ones through specialized molecules of the cell surface. In normal conditions, cell adhesion maintains cells together and strengthens contact between them. However, cancer metastasis can leverage the dysfunction of cell adhesion, which allows pathological cells to escape their site of origin and spread across the organism, see e.g. [OPLH04].

Therefore, one of the directions undertaken to fight cancer is to investigate the role of cell adhesion as metastasis suppressors by following the idea that

cell adhesion is involved in limiting tumor cell migration,   (37.1)

see e.g. [MS09].

In relation to this, here we describe a mathematical model, first introduced in [APS06] for the one-dimensional case, which analyzes cell spread as induced by standard diffusion and moderated by cell adhesion, based on the equation

$$\partial_t u(x,t) = \kappa \Delta u(x,t) - \text{div}\left(u(x,t)\,\mathcal{K}_u(x,t)\right),$$

$$\text{with} \quad \mathcal{K}_u(x,t) := \int_{B_R} u(x+y,t)\,\frac{w(|y|)y}{|y|}\,dy.$$

(37.2)

Here, $u = u(x,t)$ denotes the density of cells at some point $x \in \mathbb{R}^n$ (or, more generally, at some point in a domain of $\mathbb{R}^n$) at time $t > 0$. Also, $\kappa \in (0, +\infty)$, and w is a nonnegative function.

The idea behind establishing the model in (37.2) is to revisit the derivation of the heat equation put forth in Chapter 1 by including in that setting the effect of cell adhesion too. To this end, we assume that there is no cell birth or death in our system. The conservation of cell mass thus allows us to follow the strategy described in (1.1): namely, we suppose that the variation in density in a given region of space Ω is the flux of some heat flux vector $B(x, t)$, that is,

$$\partial_t \int_\Omega u(x, t)\, dt = - \int_{\partial\Omega} B(x, t) \cdot \nu(x)\, d\mathcal{H}_x^{n-1}. \qquad (37.3)$$

To keep things as simple as possible, one can assume that this flux vector arises as the superposition of two independent effects: classical diffusion (which can be modeled via the Fick's law in (1.3)) and an additional adhesive flux vector $A(x, t)$, that is,

$$B(x, t) = -\kappa \nabla u(x, t) + A(x, t), \qquad (37.4)$$

for some (say, constant) $\kappa > 0$.

To describe the adhesive flux vector A, we can assume that it arises due to molecule interactions within a given distance from a reference point x, say, occurring in a ball $B_R(x)$. Thus, we suppose that A is proportional to the adhesive force acting on the cells. This force comes from the interaction of the biological material located at the point x with the one located somewhere else in the space (and that, say, it is proportional to such biological materials and directed along the vector joining their positions). Hence, if we assume, for simplicity, that this interaction takes place within a finite distance, this leads to

$$A(x, t) = \int_{B_R(x)} u(x, t)\, u(y, t)\, w(|x - y|)\, \frac{(x - y)\, dy}{|x - y|}$$

$$= \iint_{(0,R)\times(\partial B_1)} u(x, t)\, u(x + re, t)\, r^{n-1} w(r)\, e\, dr\, d\mathcal{H}_e^{n-1},$$

$$(37.5)$$

where w is a nonnegative function weighting the interaction force with respect to the distance. For instance, one could consider w to be a decreasing function of the distance, or, alternatively, to keep the model as simple as possible, one could also take w to be equal to $\frac{1}{|B_R|}$, in which case the adhesion force contribution would be simply the product of the density at the point x and the average density in the ball $B_R(x)$.

From (37.4) and (37.5), we arrive at

$$B(x,t) = -\kappa \nabla u(x,t)$$
$$+ \iint_{(0,R)\times(\partial B_1)} u(x,t)\, u(x+re,t)\, r^{n-1} w(r)\, e\, dr\, d\mathcal{H}_e^{n-1}.$$

Hence, in light of (37.3),

$$\int_\Omega \partial_t u(x,t)\, dt = \kappa \int_{\partial\Omega} \nabla u(x,t) \cdot \nu(x)\, d\mathcal{H}_x^{n-1}$$
$$- \iiint_{(\partial\Omega)\times(0,R)\times(\partial B_1)} u(x,t)\, u(x+re,t)\, r^{n-1} w(r)\, e$$
$$\cdot \nu(x)\, d\mathcal{H}_x^{n-1}\, dr\, d\mathcal{H}_e^{n-1}.$$

$$(37.6)$$

As a side comment, we note that the sign difference between the two terms on the right-hand side of (37.6) has an important practical outcome. To appreciate this difference, one can focus, for instance, on the case of a region Ω with a high density of cells. Indeed, recalling that ν is the external unit normal (by the notation introduced on p. 11), the first term on the right-hand side of (37.6) is the one produced by Fick's law and describes the tendency, induced by the gradient, for *high-density regions to push cells out* because high-density regions correspond to gradients pointing inward, yielding a negative contribution to the first term on the right-hand side of (37.6), which in turn tries to reduce the density (with the time derivative of the density present on the left-hand side of (37.6)).

Instead, for the second term on the right-hand side of (37.6), coming from cell adhesion, we can make these considerations. When Ω corresponds to a high-density region, given $x \in \partial\Omega$, we can imagine $u(x+re,t)$ to be large when e points inward and small when e points outward. Hence, the integrand in the second term on the right-hand side of (37.6) is large when $e \cdot \nu(x) \leqslant 0$ and small when $e \cdot \nu(x) \geqslant 0$. That is, in the case of the high-density region Ω, the last integral in (37.6) provides a negative contribution. But given the minus sign in front of it, this says that the cell adhesion is trying to *increase the density of a high-density region and pull cells in*, which is consistent with the idea that adhesion forces try to make cells to cluster together.

This comment clarifies that Fick's law and adhesion forces end up having opposite tendencies and can be considered as a first indication that the model presented is consistent with the fact highlighted in (37.1) (see p. 406 for additional confirming evidence).

Now, to obtain (37.2), one can write (37.6) in the form

$$\int_\Omega \partial_t u(x,t)\, dt = \kappa \int_{\partial\Omega} \nabla u(x,t) \cdot \nu(x)\, d\mathcal{H}^{n-1}_x$$

$$- \iint_{(\partial\Omega)\times B_R} u(x,t)\, u(x+y,t)\, w(|y|) \frac{y}{|y|} \cdot \nu(x)\, d\mathcal{H}^{n-1}_x\, dy$$

and then combine this with the divergence theorem, finding that

$$\int_\Omega \partial_t u(x,t)\, dt = \kappa \int_\Omega \Delta u(x,t)\, dx$$

$$- \iint_{\Omega\times B_R} \mathrm{div}_x \left(u(x,t)\, u(x+y,t)\, \frac{w(|y|)\, y}{|y|} \right) dx\, dy.$$

Since Ω is arbitrary, we thereby find that

$$\partial_t u(x,t) = \kappa \Delta u(x,t) - \int_{B_R} \mathrm{div}_x \left(u(x,t)\, u(x+y,t)\, \frac{w(|y|)\, y}{|y|} \right) dy,$$

which corresponds to (37.2).

Some further comments on (37.2) are in order. First of all, the one-dimensional case of (37.2) reduces to

$$\partial_t u = \kappa \partial_{xx} u - \partial_x \left(u\, \mathcal{K}_u \right),$$

$$\text{with} \quad \mathcal{K}_u(x,t) := \int_{-R}^{R} u(x+y,t)\, \varpi(y)\, dy, \tag{37.7}$$

for an odd symmetric function $\varpi : \mathbb{R} \to \mathbb{R}$, with $\varpi \geqslant 0$ in $(0,+\infty)$.

Furthermore, equation (37.2) (as well as (37.7)) has a nonlocal character due to the last term. Indeed, differently from the other terms involved, to compute $\mathcal{K}_u$ (and therefore its derivatives) at a given (x,t), it does not suffice to know u in an arbitrarily small neighborhood of (x,t) since a full knowledge of the values of u in the whole of $B_R(x)$ (or in a neighborhood of it, when we want to compute derivatives) is needed.

There are however convenient approximations of (37.2) that attempt to reduce the model to a more standard, and local, partial differential equation. For example, assuming small oscillation densities, one could formally look at the expansion

$$u(x + y, t) \simeq u(x, t) + \nabla u(x, t) \cdot y + \frac{1}{2} D^2 u(x, t) y \cdot y + \cdots ,$$

where higher orders are dismissed.

This approximation, via the definition of $\mathcal{K}_u$ in (37.2), translates into

$$\mathcal{K}_u(x, t) \cdot e_j \simeq \int_{B_R} \left(u(x, t) + \nabla u(x, t) \cdot y + \frac{1}{2} D^2 u(x, t) y \cdot y \right)$$

$$\times \frac{w(|y|)y_j}{|y|} \, dy + \cdots$$

$$= \int_{B_R} (\nabla u(x, t) \cdot y) \frac{w(|y|)y_j}{|y|} \, dy + \cdots$$

$$= \xi_j \cdot \nabla u(x, t) + \cdots ,$$

where

$$\xi_j := \int_{B_R} y \frac{w(|y|)y_j}{|y|} \, dy \in \mathbb{R}^n .$$

Correspondingly, we have that

$$\operatorname{div} \left(u \, \mathcal{K}_u \right) = \sum_{j=1}^{n} \partial_j \left(u \, \mathcal{K}_u \cdot e_j \right) + \cdots$$

$$= \sum_{j=1}^{n} \xi_j \cdot \partial_j \left(u \nabla u \right) + \cdots$$

$$= \sum_{j,k=1}^{n} \xi_{j,k} \, \partial_j \left(u \, \partial_k u \right) + \cdots ,$$

where

$$\xi_{j,k} := \xi_j \cdot e_k = \int_{B_R} y_k \frac{w(|y|)y_j}{|y|} \, dy = \delta_{jk} \int_{B_R} \frac{w(|y|)y_j^2}{|y|} \, dy .$$

Therefore,

$$\operatorname{div} \left(u \, \mathcal{K}_u \right) = \sum_{j=1}^{n} c_j \partial_j \left(u \, \partial_j u \right) + \cdots ,$$

with

$$c_j := \int_{B_R} \frac{w(|y|)y_j^2}{|y|} \, dy \geq 0. \tag{37.8}$$

We observe that, by symmetry, $c_1 = \cdots = c_n$. Thus, to ease notation, we write $c_j = 2$.

As a result, if one is willing to take such an approximation, the nonlocal equation (37.2) gets replaced by the local partial differential equation

$$\partial_t u = \kappa \Delta u - 2 \sum_{j=1}^n \partial_j \left(u \, \partial_j u \right) = \Delta u - \sum_{j=1}^n \partial_j \left(\partial_j u^2 \right) = \Delta u - \Delta u^2.$$

$$\tag{37.9}$$

It is also interesting to observe that this approximate equation also confirms the agreement between the model presented here and the fact showcased in (37.1). To see this, let us consider a point x_0 and a time t_0 when the density achieves a local maximum. We are going to check that the effect of the first term on the right-hand side of (37.9) is to cause the density at x_0 to decrease, thus sending out cells, but conversely the tendency of the second term on the right-hand side of (37.9) is to cause this density to increase, hence trying to keep the cells in their original high-density location, and thus providing another confirmation of (37.1).

For this, we suppose that $x_0 = 0$, and thus, up to higher orders, for x in the vicinity of 0, using our local maximality assumption, we write

$$u(x, t_0) = u(0, t_0) - \frac{Mx \cdot x}{2} + \cdots ,$$

for a symmetric and nonnegative definite matrix M; therefore, (37.9) gives that

$$\partial_t u(0, t_0) = -\kappa \operatorname{Tr} M + 2 \operatorname{Tr} M. \tag{37.10}$$

By our assumptions on M and (37.8) (and the fact that $u(0, t_0) \geq 0$, being a density), it follows that the first term on the right-hand side of (37.10) is indeed less than or equal to zero, while the second term is greater than or equal to zero, as claimed.

Chapter 38

When You Are a Mathematician

Don Pedro Calderón was a pragmatic person. Urologist by profession, he had five children: three daughters, Margarita Isabel (known as Nenacha), Matilde Jr. and María Teresa, and two sons, Alberto Pedro and Calixto Pedro.

Don Pedro had a natural affinity for mathematics that he tried to transmit (together with its own name!) to his sons at an early age by challenging them with rapid mental calculations at the dinner table. Nonetheless, he was a pragmatic person; as such, he firmly believed that a decent person could not make a living as a mathematician and that, to have a good career, one needed to study practical subjects.

Don Pedro happened to be quite a convincing person too, since, despite his true love for mathematics, Alberto ended up graduating in civil engineering and, soon after his graduation, working as an engineer for the geophysical division of the Yacimentos Petrolíferos Fiscales, or YPF for short, the Argentinian state oil company (see Figure 38.1 for the high-rise building located in the Puerto Madero barrio of Buenos Aires, hosting nowadays the headquarters of this company, and Figure 38.2 for a group of YPF workers in the Patagonian Chubut Province).

Unfortunately, or fortunately for Alberto's real passion and for the progress of mathematics, Alberto's work at YPF, according to his recollection, was very interesting, but he was not well treated there; therefore, he changed his mind and decided to become a professional mathematician, despite his father's resolute advice.

Figure 38.1. YPF Tower (photo by Allan Aguilar, image from Wikipedia, licensed under the Creative Commons Attribution 3.0 Unported license).

Don Pedro's prediction turned out to be inaccurate: Alberto soon became a prominent mathematician, with a good salary paid by the University of Chicago.

Don Pedro's expectations from his sons remained unfulfilled since Calixto, who firstly enrolled in an engineering degree, also rapidly switched to a mathematics degree (thanks to the strong moral and financial support by his elder brother as well, who at that point had already become a professor). Thus, Calixto became a professional mathematician (and happened to become the thesis advisor of Luis Caffarelli).

The moral of the story is perhaps that parents don't always know what's best for their sons, as well as for their daughters.

Figure 38.2. Group of YPF workers at an oil well (Public Domain images from Wikipedia).

Or perhaps mathematics can very often offer rewarding careers and a pleasant lifestyle.

Or perhaps when you are a mathematician (and you can well be a mathematician even with a degree in civil engineering, or with no degree at all, since being a mathematician has to do with one's mind and heart and little to do with paperwork), you end up creating beautiful mathematics even while you are working for a national oil company that treats you badly. This is precisely what happened to Alberto Calderón during his employment period with the Yacimentos Petrolíferos Fiscales – let's see why (and see also Section 5.2 in the companion book [DV23] for more about Alberto Calderón).

The electrical conductivity of an inhomogeneous body varies significantly due to changes in temperature, density and ionic concentration. Therefore, variations in the electrical conductivity of the earth may reveal the presence of oil or gas, helping with the identification of productive zones to drill. No wonder this was an interesting topic for YPF!

The issue is therefore to measure the electrical conductivity of some regions in the earth in a way that is as simple, affordable and efficient as possible. The idea of drilling the ground to examine samples, for instance, does not really fulfill these requirements, as it is rather expensive, risky and invasive, as well as environmentally problematic. Instead, how about performing a geophysical prospection for oil and gas exploration[1] only through simple measurements on the surface of a region? What if one could use some cheap electrical methods for this goal, such as voltage and current measurements on the surface of the earth?

To provide a mathematical model for the problem under consideration, we recall that the electrical conductivity measures how easily electric current can flow through a given material. More specifically,

---

[1]Not only can this method be useful for the detection of oil and gas, it can also provide valuable information about the water contaminated with hydrocarbons, which is produced during the extraction. This is called in jargon "produced water" and needs to be managed properly to avoid serious environmental consequences. Therefore, it is important, before the extraction, to have an estimate on the proportion of water that will be present in the so-called "multiphase flow" (i.e. the mixed combination of different substances, such as oil, gas and water) which will be taken out of the ground.

Actually, besides its environmental importance, the methods presented here have also an economic impact since water is typically separated from the oil and gas by a company specializing in this process, not by the extractor company itself; therefore, a careful estimate of these additional costs plays a role in the extractor company's financial plan.

Another important application of the determination of the electrical conductivity is its use, along with other measurements, to enhance drilling safety by preventing blowout. In this framework, a high variation in the values of the conductivity may indicate that the properties of the rock at a certain depth have changed, and actions have to be taken to prevent a retaining wall from collapsing.

Once again, it would be ideal to detect relative proportions of oil, gas and water without going through complicated, expensive or intrusive procedures, such as sampling methods or gamma rays produced by radioactive sources. The method that Calderón studied and appropriate modifications of it have the potential to provide helpful strategies.

To appreciate how electrical conductivity can be efficiently utilized to tell different substances apart, let us compare the different orders of magnitude (measured in siemens per meter) of several media, such as gold 4.11×10^7, sea water 4.8, drinking water 5×10^{-4}–5×10^{-2} and air 10^{-15}–10^{-9}. In general, gas on its own has very little ability to conduct electricity, but its conductivity increases in the presence of ions; also, electric charge carriers are absent in oil, hence it has a very small conductivity too. The electric conductivity of the soil changes considerably according to its salinity, but to get a rough idea, sand has a relatively low conductivity $(1 \times 10^{-3}$–$10m \times 10^{-3})$, silt has a medium conductivity $(8 \times 10^{-3}$–$8 \times 10^{-2})$ and clay has a high conductivity $(2 \times 10^{-2}$–$8 \times 10^{-1})$.

the resistivity ρ at a particular point of a given material is defined[2] as the ratio of the electric field E to the density of the current J that E creates at that point. The conductivity σ is thus defined as the reciprocal of the resistivity ρ.

In this way,

$$\rho = \frac{E}{J} \quad \text{and} \quad J = \frac{E}{\rho} = \sigma E.$$

Accordingly, for a given electric field E, a higher-conductivity material produces more current flow J than a low-conductivity material (hence the name of "conductivity").

Since the electrostatic force is conservative, we can also write $E = \nabla u$ for some potential function u. Also, by Ampère's law, we know that the curl of the magnetic field H is equal to the electric current density J and, consequently,

$$\operatorname{curl} H = J = \sigma E = \sigma \nabla u.$$

As a result, recalling the vector calculus identity in (6.6),

$$0 = \operatorname{div}(\operatorname{curl} H) = \operatorname{div}(\sigma \nabla u). \tag{38.1}$$

---

[2]To clarify the jargon, we recall that both the terms "resistance" and "resistivity" describe how difficult it is for electrical current to flow through a material. The only difference is that resistivity is an intrinsic property of the material, while resistance depends on the macroscopic object the material is shaped into. For example, the electrical resistance R of a conductor of uniform cross-section with sectional area A and length ℓ is related to the electrical resistivity ρ through the formula

$$R = \frac{\rho \ell}{A}.$$

That is, the resistivity is a specific property of the material, whereas the resistance depends on the area and length of the conducting wire.

In strict terms, for alternating currents (say, in sinusoidal form) it is also appropriate to replace the notion of resistance with that of "electrical impedance", which is the ratio of the voltage between the terminals of a circuit and the current flowing through it (both in complex notation, so that the impedance is a complex number, with its real part corresponding to the resistance).

To be totally accurate, here we should take into account this notion of impedance (or its reciprocal, which is called "admittance") since, in practice, the electrical impedance tomography that we describe uses either alternating currents at a single frequency or multiple frequencies. Since the basic ideas remain unchanged, we surf over this detail.

The geological problem considered by Calderón was then to apply a boundary voltage potential ϕ to some given region Ω. In this framework, equation (38.1) exists in the domain Ω and is complemented by the boundary condition

$$u = \phi \text{ on } \partial\Omega. \tag{38.2}$$

In this setting, one can measure the energy needed to maintain the potential ϕ along $\partial\Omega$ as

$$Q_\sigma(\phi) := \int_\Omega \sigma(x)|\nabla u(x)|^2 \, dx.$$

Note that, by the divergence theorem, (38.1) and (38.2),

$$Q_\sigma(\phi) = \int_\Omega \Big(\operatorname{div}\left(\sigma(x)u(x)\nabla u(x)\right) - u(x)\operatorname{div}\left(\sigma(x)\nabla u(x)\right) \Big) \, dx$$

$$= \int_\Omega \operatorname{div}\left(\sigma(x)u(x)\nabla u(x)\right) \, dx$$

$$= \int_{\partial\Omega} u(x)\sigma(x)\partial_\nu u(x) \, d\mathcal{H}_x^{n-1}$$

$$= \int_{\partial\Omega} \phi(x)\sigma(x)\partial_\nu u(x) \, d\mathcal{H}_x^{n-1}.$$

We stress that, for all $x \in \partial\Omega$, the quantity $\sigma(x)\partial_\nu u(x)$ accounts for the normal component of the electric current at the boundary.

Calderón's problem thus consisted of determining whether or not one can recover, at least approximately, the electrical conductivity σ in Ω using the knowledge of $Q_\sigma(\phi)$, which can be measured along $\partial\Omega$. Note the "noninvasivity" of these measurements in order to determine the inner structure of the material since the necessary detections can be carried out along the boundary of Ω, i.e. without penetrating into the domain.

Some questions posed explicitly by Calderón in his seminal paper [Cal80] were as follows: is σ uniquely determined by $Q_\sigma(\phi)$? Can one calculate σ in terms of $Q_\sigma(\phi)$?

Though the problem remains largely open in its generality, Calderón already provided several interesting results, for instance about the possibility of determining uniquely the electrical conductivity σ from the knowledge of $Q_\sigma(\phi)$, up to a small error, under the assumption that σ is sufficiently close to a constant.

Quite remarkably, for two-dimensional domains, Ω, the electrical conductivity σ is completely determined by $Q_\sigma(\phi)$, see [AP06]. See also [Uhl08, AIM09, Uhl09] for more information[3] about Calderón's problem.

The idea of determining the internal structure of a body through electrical measurements on its surface is called in jargon "electrical impedance tomography", and it is useful not only to provide an educated guess on whether or not it is worth drilling the soil with the expectation of finding oil or gas. Indeed, given that it is a noninvasive, nondestructive, portable and inexpensive methodology, one of the promising applications of this idea lies in the field of medical diagnostics, as pioneered in [HW78].

From a clinical standpoint, the key fact is that electrical conductivity highly varies among different tissues, and it is strongly influenced by the movement of fluids and gases within a given tissue (for instance, muscle and blood have higher electric conductivities than fat and bones). The use of electrical impedance tomography in this situation is typically realized by attaching electrodes to the skin around the body part to be examined, see Figure 38.3.

The streamlines and equipotential surfaces are detected, and after several repetitions in different configurations, the results are elaborated and reconstructed into a two-dimensional image which shows the bending of streamlines and equipotentials caused by the change in conductivity between different tissues, see Figure 38.4.

---

[3]We stress that Calderón's problem is perhaps the prototypical "inverse problem", in which the objective is not to derive information on the solution of a certain problem given a set of data and physical laws, but rather to obtain information about a physical object from observable measurements.

In other words, typical mathematical problems aim to predicting physical observations given the values of the parameters defining the model and the underlying physical laws. Instead, an inverse problem utilizes the results of the observations to infer the values of the parameters characterizing the system under investigation.

Curiously, the groundbreaking paper [Cal80] has remained the only article by Calderón on the topic of inverse problems.

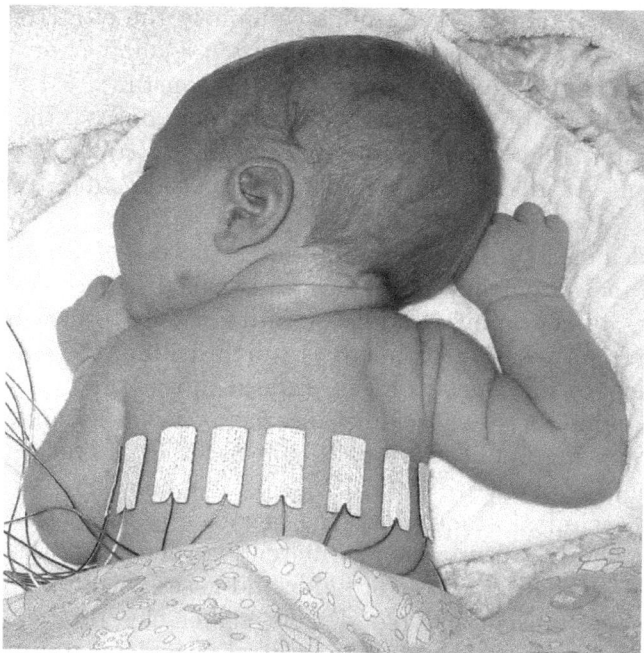

Figure 38.3.   Electrodes attached for electrical impedance tomography lung monitoring on a 10-day old spontaneously breathing neonate (photo by S. Heinrich, H. Schiffmann, A. Frerichs, A. Klockgether-Radke and I. Frerichs; image from Wikipedia, licensed under the Creative Commons Attribution-Share Alike 3.0 Unported license).

A topical clinical application involves, for instance, the monitoring of lung functions. The electrical impedance tomography possesses in this case a very promising potential since the conductivity of lung tissues is about five times lower than the soft tissues in the thorax, thus allowing for a significant contrast. Furthermore, the conductivity highly oscillates depending on the presence of the air in the lungs since air plays the role of an insulating factor, thus detecting vital activities such as inspiration and expiration.

The use of the electrical impedance tomography for lungs monitoring is particularly important for detecting possible injuries caused by mechanical ventilation, thus helping physicians in a bespoke adjustment of ventilators' settings to provide lung ventilation to patients

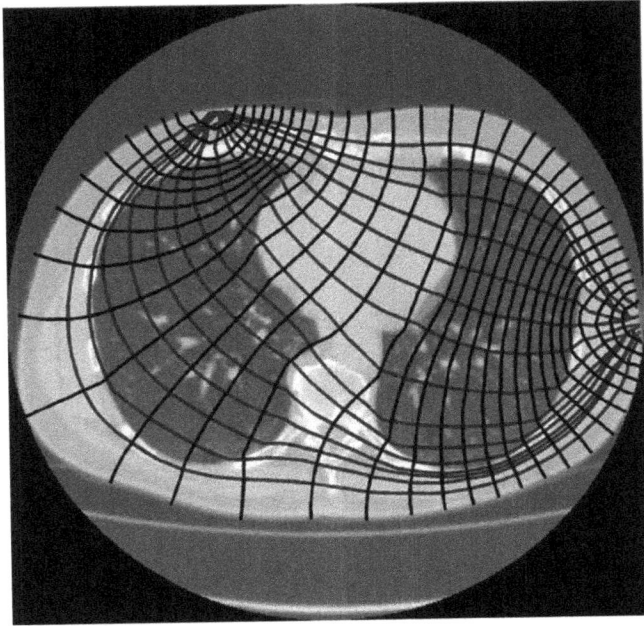

Figure 38.4. A cross-section of a human thorax with electrical current stream-lines and equipotentials (image by Andy Adler from Wikipedia, licensed under the Creative Commons Attribution-Share Alike 3.0 Unported license).

needing it without increasing too much the risk of collateral damages caused by this artificial process.

Another important medical application of the electrical impedance tomography relates to breast imaging for cancer detection, as a possible alternative to mammography, which uses more invasive ionizing radiations, and magnetic resonance imaging, which employs as contrast media chemical substances which can cause problems to patients with kidney failure (also, due to their low specificity, these classical techniques are subject to a relatively high rate of very distressive false positives). In this setting, the electrical impedance tomography shows promising potential due to the different electrical conductivities of normal and malignant breast tissues.

A further application is related to the monitoring of brain function and neuronal activity. In this context, the effectiveness of the method relies on the fact that cerebral hypoglycemia and hypoxemia, ischemia and hemorrhage are typically associated with significant changes in the electrical properties of the brain.

In some cases, the clinical applications require a refinement of the mathematical techniques. For instance, muscle tissues typically present great differences in the electrical conductivities in different directions (e.g. the conductivity of the cardiac muscle changes by a factor of approximately 2.7 between the transverse and the longitudinal directions). This requires mathematicians to take into account also the case of "anisotropic conductivities"; in this situation, rather than considering a scalar function σ in (38.1), it is customary to account for conductivity variations in different directions by taking into consideration a symmetric matrix-valued function $\sigma(x) = \{\sigma_{ij}(x)\}_{i,j\in\{1,\dots,n\}}$. This casts (38.1) into the more general form[4]

$$\sum_{i,j=1}^{n} \partial_i(\sigma_{ij}\partial_j u) = 0. \tag{38.3}$$

The mathematical difficulties of dealing with a nonlinear and ill-posed inverse problem translate into severe practical problems that still need to be overcome for the adoption of the electrical impedance tomography for medical purposes. In particular, the image reconstruction methods involved are difficult to implement since there are typically more than one possible solution corresponding to a three-dimensional region projected onto a two-dimensional surface (see e.g. [GLU03] for a mathematical description of the lack of uniqueness problem). Often, to obtain a reliable reconstruction, detailed knowledge of the exact dimensions and shape of the organ under consideration is required, as well as a very precise electrode location. In addition, sometimes, fluctuations in electrical quantities due to

---

[4]For the reader interested in geometric analysis and differential geometry, let us mention that in dimension $n \geqslant 3$, equation (38.3) corresponds to a geometric equation on a Riemannian manifold with boundary, see [LU89]. The development of this geometric intuition was very fruitful in constructing counterexamples to the uniqueness of the electric conductivity σ in (38.3) (this, in the geometric setting, corresponds to the uniqueness of a Riemann metric related to harmonic functions; the differential operator used for this is the Laplace–Beltrami operator, which will be presented in some detail in Section 2.3 of the companion book [DV23]).

anomalies are still not sufficiently broad to compensate for reconstruction errors.

For all these reasons, the clinical use of the electrical impedance tomography remains mostly experimental. However, some commercial devices relying on this method have been already introduced to monitor lung function in intensive care patients, and hopefully future coordinated progress in the mathematical understanding of the problem, in the precision of the technology used for the measurements and in the algorithms used for image reconstruction will lead to a secure use of the electrical impedance tomography in several areas of critical social and medical importance.

Chapter 39

If You Are Shy, Do as the Romulans Do

The power of invisibility is a very cool superpower to possess, especially if one is shy. This power is actually an evergreen classic. It appeared as early as the year 1100 in Welsh mythology (with a "mantle of invisibility" being one of King Arthur's most valuable possessions), and it also surfaced in Norse mythologies, Celtic stories, Grimm's fairy tales and Japanese traditions (in which the "kakuremino" is a useful raincoat that also provides invisibility).

Invisibility also plays a pivotal role in J. R. R. Tolkien's novels *The Hobbit* and *The Lord of the Rings*, in which the One Ring is presented as a magical device forged by the Dark Lord Sauron, granting the wearer invisibility (see Figure 39.1; as additional benefits, any inherent power of its owner would be automatically amplified, the life of a mortal possessor of this device would extend indefinitely, aging would be stopped and domination over the wills of others would be guaranteed, all with the sole side effect of a malevolent and manipulative agency on the possessor's ego).

Also in cinema, invisibility dates back at least to 1924 (see Figure 39.2) since the main character Ahmed of the silent film *The thief of Bagdad* gains at some point a cloak of invisibility (but no big deal, the movie also presents a crystal ball from the eye of a giant idol that shows whatever one wants to see, a magic powder turning

Figure 39.1.   The One Ring (image from Wikipedia, own work by Peter J. Yost, licensed under the Creative Commons Attribution-Share Alike 4.0 International license).

people into whatever they wish, a flying carpet, a magic rope, and a magic apple which cures any disease, including[1] death).

Invisibility is also one of the cornerstones of science fiction. One of the first and most influential science fiction novels, *The Invisible Man* by H. G. Wells (see Figure 39.2) is about a mysterious man, Griffin, referred to as "the stranger", who happens to be a scientist who manages to render himself invisible (but sadly, failing in his attempt to reverse this condition).

As an example in recent novels and movies, let us mention that cloaks of invisibility are abundantly employed in the Harry Potter series.

With comics, the power of invisibility has maybe reached an extreme. For instance, Susan (Sue) Storm, the Invisible Woman of the Fantastic Four (and the first female hero in the Marvel Universe)

---

[1]Not sure how to eat an apple *after* death, though!

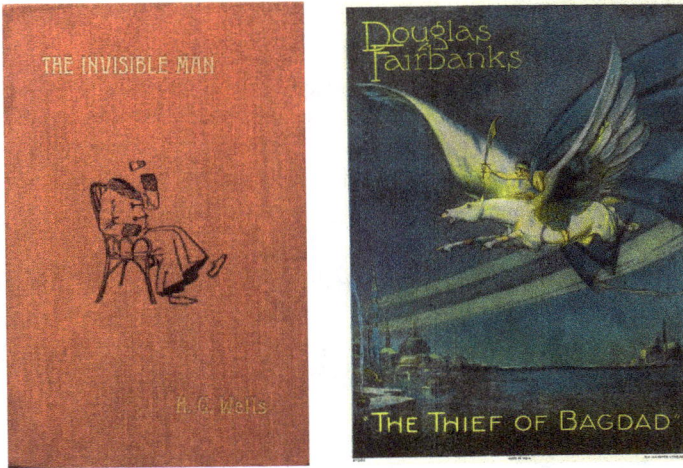

Figure 39.2. First edition cover of The Invisible Man (left, 1897) and a poster for The Thief of Bagdad (right, 1924; Public Domain images from Wikipedia).

possesses extended invisibility powers allowing her to do more than just turn invisible (she can also make the objects and the people around her invisible, as well as creating invisible[2] force fields).

All good, but the main question is: how would you manage to turn yourself invisible?

The idea of Griffin in the novel by H. G. Wells is very ingenuous: using his knowledge of optics, he invents a way to modify the refractive index of his own body, making it precisely equal to that of air (in this way, essentially, he neither absorbs or reflects light, appearing transparent to everybody). Very ingenious indeed, but maybe presenting some side effects since, as pointed out by Yakov Perelman in his book *Physics can be Fun*, this method should have man made Griffin not only invisible but blind as well (because the eye needs to absorb incoming light to transmit visual signals to the brain). On top of that, the fact that Griffin could not restore its refractive index to its normal value suggests that this technique may be too intrusive and not easily reversible, so perhaps not recommendable.

As for Sue Storm, her superhuman powers are the outcome of the exposure to massive cosmic radiation (which happens to be a cliché

---

[2]Not sure what a *visible* force field is, though.

in superheroic plots); again, this doesn't sound like a viable method due to possible side effects.

As for Harry Potter, his cloak of invisibility was passed down to him directly by his father (but the cloak was previously owned by Death), so, once again, unless you come from a wealthy family of sorcerers, this strategy may not be the most practicable one.

Fortunately, however, there is a rather promising strategy that was put forth in the 1966 episode "Balance of Terror" of the science fiction saga of *Star Trek*, see Figure 39.3. Specifically, in that episode, our heroes of the *USS Enterprise* starship had to battle their arch-enemies Romulans, who employed a new type of space vessel which was capable of remaining invisible. The Romulan technology was

Figure 39.3. Mr. Spock and Captain Kirk from the television series Star Trek (with the USS Enterprise starship passing by on the bottom; Public Domain image from Wikipedia).

based[3] on a *selective bending of light* (which was dubbed "cloaking device" in the 1968 episode "The Enterprise Incident").

Since then, the dream of many mathematicians consisted of making this idea work in practice, simply by using a bit of theory of partial differential equations. The gist is that light rays, and in general electromagnetic fields, are described by Maxwell's equations. These equations depend on the specific transmission medium via the permittivity and permeability of the substance under consideration. Therefore, to build an invisibility device according to Romulans, one should find, for instance, a medium with suitable permittivity and permeability, forcing the light rays to bend around a "hole": in this way, a space vessel located inside such a hole would be invisible from the outside, see Figure 39.4.

Some practical questions still remain: how can one determine, at least at a mathematical level, the pointwise values of the permittivity and permeability of the medium in order to guide the lines around the desired hole? And can one do this for real?

---

[3]But after all, the idea of the Romulans guiding light rays to reflect past their spaceship (so that the viewer would have the impression that the spaceship is not there) is a natural one since it is at the basis of various optical illusions, such as mirages in a desert.

Anyway, while the Romulans were created specifically for the episode "Balance of Terror", they have remained in the *Star Trek* saga to embody an aggressive, war-mongering, alien species, having founded an interstellar empire. Not to be confused with the Vulcans, or Vulcanians, who are a peaceful, methodical, logic-oriented species, and perhaps quite unemotional, kind of alien people (Mr. Spock being a prominent representative). Both Romulans and Vulcans are however characterized by pointed ears and slanted eyebrows, so the two populations have very similar physiological traits.

Yet, perhaps the primary antagonists in the classical *Star Trek* saga are the Klingons, an alien species of brutal mood, practicing feudalism and authoritarianism. Likely, the use of the bloodthirsty Klingons over the militaristic but technologically sophisticated Romulans in the original series was motivated by politics and budget reasons. Indeed, the makeup necessary to make the Romulans was at that time rather costly, while the one for the Klingons was cheap and quick (the application of shoe polish being enough to provide Klingons with the bronze skin, which was basically their sole physical difference with respect to humans). Also, in the middle of the Cold War, the physical appearance of Klingons was probably meant to suggest orientalism, at a time when memories of Japanese actions during World War II were still fresh, and their feudalistic society was perhaps alluding to Soviet Union, with the "United Federation of Planets" to which the Enterprise crew belongs, playing the role of the United States.

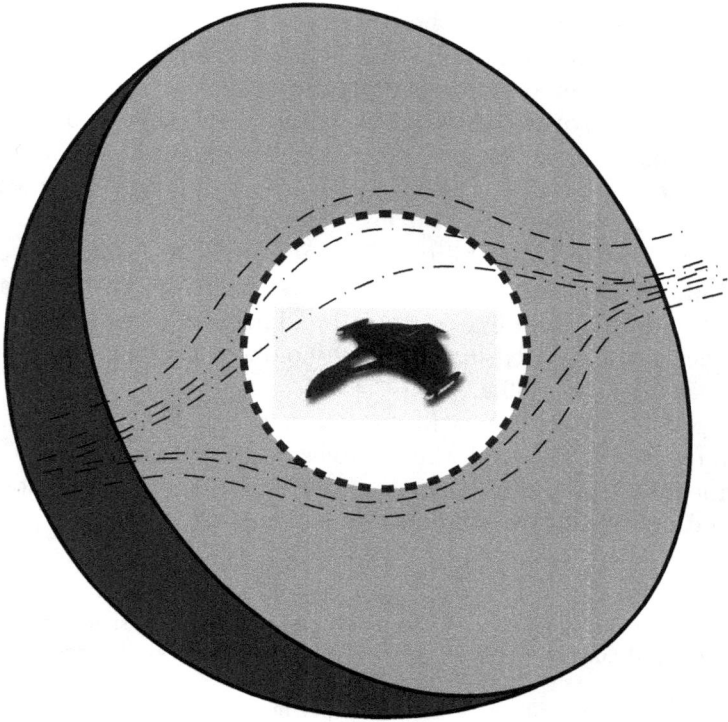

Figure 39.4. Making a Romulan spacecraft invisible by guiding the light rays away via Maxwell's equations. Deflecting light rays around the vessel and restoring them on the other side make the observer believe that light had passed through empty space.

To address these questions, let us start with the mathematical aspect of the problem (see the two relevant articles [Leo06] and [PSS06] for the pioneering aspects of the theory, as well as for links with complex analysis and Riemannian geometry). The calculations to implement these strategies rely on the mathematical property of Maxwell's equations to maintain their structure[4] under

---

[4] The "form invariance" of Maxwell's equations is one of the main building blocks of the so-called "transformation optics", which aims at using the change of coordinate system as a tool to reduce the wave propagation in inhomogeneous materials to that in standard space.

all spatial transformations. In a nutshell, this invariance property allows one to pick a spatial transformation, for example, which maps a ball into an annulus[5] and which coincides with the identity map outside the ball. Therefore, a solution of Maxwell's equations in the usual space would be mapped into a solution of Maxwell's equations in a space with inhomogeneous permittivity and permeability in which the light rays cannot enter the hole that has been produced inside the annulus. At that point, the mathematical task is completed, and it "suffices" to build, at least approximately, a material with those specific values of permittivity and permeability to obtain nice cloaking of any object placed inside the hole.

Let us see in some further detail how to perform these mathematical transformations and how to construct the desired material for the cloaking.

To start with, let us check the invariance of Maxwell's equations with respect to spatial transformations. To this end, consider Maxwell's equations in $\mathbb{R}^3$ for the electric field E and the magnetic field B, without sources or currents, with permittivity ε and permeability μ, namely, up to normalizations:

$$\begin{cases} \operatorname{div}(\varepsilon E) = 0, \\ \operatorname{curl} E = -\mu \partial_t B, \\ \operatorname{div}(\mu B) = 0, \\ \operatorname{curl} B = \varepsilon \partial_t E. \end{cases} \quad (39.1)$$

In this setting, the space variable will be denoted by $x \in \mathbb{R}^3$, and $\mu = \mu(x)$ and $\varepsilon = \varepsilon(x)$ are matrix-valued functions to account for possible inhomogeneity and anisotropy of the medium.

A spatial change of coordinates $\bar{x} = \bar{x}(x)$ maintains the structure in (39.1). Indeed, considering the Jacobian matrix

$$A_{ij} := \frac{\partial \bar{x}_i}{\partial x_j}, \quad (39.2)$$

---

[5]From a technical point of view, such a transformation is singular since the topologies of the initial and final domains are different, and this causes some mathematical difficulty on the notion of (weak) solution to consider, but here we surf over this detail (the interested reader may look at Section 3.5 of [GKLU09]).

with $i, j \in \{1, 2, 3\}$, one can define the transformed quantities as

$$\bar{E}(\bar{x}, t) := A^{-T}(x)\, E(x, t), \quad \bar{B}(\bar{x}, t) := A^{-T}(x)\, B(x, t),$$

$$\bar{\mu}(\bar{x}) := \frac{1}{\mathcal{J}(x)}\, A(x)\, \mu(x)\, A^T(x)$$

$$\text{and} \quad \bar{\varepsilon}(\bar{x}) := \frac{1}{\mathcal{J}(x)}\, A(x)\, \varepsilon(x)\, A^T(x),$$

where $x = x(\bar{x})$, A^{-T} is the inverse of the transpose (which coincides with the transpose of the inverse) of A, and $\mathcal{J} := \det A$.

In this way, for each $i, j \in \{1, 2, 3\}$,

$$\delta_{ij} = \partial_{\bar{x}_i}\bar{x}_j = \sum_{k=1}^{3} \partial_{x_k}\bar{x}_j \partial_{\bar{x}_i} x_k = \sum_{k=1}^{3} A_{jk}\partial_{\bar{x}_i} x_k,$$

showing that the inverse of A is the matrix with entries

$$A_{ij}^{-1} = \partial_{\bar{x}_j} x_i. \tag{39.3}$$

We also calculate[6] that

$$\partial_{x_k} A_{ij} = \partial_{x_k}\partial_{x_j}\bar{x}_i = \partial_{x_j} A_{ik}, \tag{39.4}$$

that

$$\varepsilon E = (\mathcal{J}A^{-1}\bar{\varepsilon}A^{-T})(A^T\bar{E}) = \mathcal{J}A^{-1}\bar{\varepsilon}\bar{E} \tag{39.5}$$

and that

$$\partial_{x_j} = \sum_{i=1}^{3} \partial_{x_j}\bar{x}_i\, \partial_{\bar{x}_i} = \sum_{i=1}^{3} A_{ij}\partial_{\bar{x}_i}. \tag{39.6}$$

---

[6]Some of the tedious computations in these pages could be elegantly streamlined by the use of the Levi–Civita symbols for curls and determinants (see e.g. Chapter 9 in [Gri13] for an introduction to this notation and the appendix in [KFJ08] for a compact proof of the form invariance of Maxwell's equations). We avoid here this quicker, but more sophisticated, approach in order to maintain the calculations in an elementary form.

The last identity also entails that

$$\sum_{j=1}^{3} A_{jk}^{-1} \partial_{x_j} = \sum_{i,j=1}^{3} A_{jk}^{-1} A_{ij} \partial_{\bar{x}_i} = \sum_{i=1}^{3} \delta_{ik} \partial_{\bar{x}_i} = \partial_{\bar{x}_k}. \qquad (39.7)$$

We also recall that

$$A_{ij}^{-1} = \frac{C_{ji}}{\mathcal{J}},$$

where the cofactor matrix has entries

$$C_{\ell m} := (-1)^{\ell+m} M_{\ell m},$$

with $M_{\ell m}$ being the minor of the entry in the ℓth row and mth column (that is, the determinant of the submatrix of A formed by deleting the ℓth row and mth column).

In this way,

$$\mathcal{J} A_{jk}^{-1} = C_{kj} = (-1)^{k+j} M_{kj}. \qquad (39.8)$$

We claim that, for all $k \in \{1, 2, 3\}$,

$$\sum_{j=1}^{3} \partial_{x_j} (\mathcal{J} A_{jk}^{-1}) = 0. \qquad (39.9)$$

We choose $k = 1$ (the cases $k = 2$ and $k = 3$ being similar), and we employ (39.4) and (39.8) to see that

$$\sum_{j=1}^{3} \partial_{x_j} (\mathcal{J} A_{j1}^{-1})$$

$$= -\sum_{j=1}^{3} (-1)^j \partial_{x_j} M_{1j}$$

$$= \partial_{x_1} M_{11} - \partial_{x_2} M_{12} + \partial_{x_3} M_{13}$$

$$= \partial_{x_1} (A_{22} A_{33} - A_{23} A_{32}) - \partial_{x_2} (A_{21} A_{33} - A_{23} A_{31})$$

$$+ \partial_{x_3} (A_{21} A_{32} - A_{22} A_{31})$$

$$= \partial_{x_1} A_{22} A_{33} + A_{22} \partial_{x_1} A_{33} - \partial_{x_1} A_{23} A_{32} - A_{23} \partial_{x_1} A_{32}$$
$$- \partial_{x_2} A_{21} A_{33} - A_{21} \partial_{x_2} A_{33} + \partial_{x_2} A_{23} A_{31} + A_{23} \partial_{x_2} A_{31}$$
$$+ \partial_{x_3} A_{21} A_{32} + A_{21} \partial_{x_3} A_{32} - \partial_{x_3} A_{22} A_{31} - A_{22} \partial_{x_3} A_{31}$$
$$= \partial_{x_1} A_{22} A_{33} + A_{22} \partial_{x_1} A_{33} - \partial_{x_1} A_{23} A_{32} - A_{23} \partial_{x_1} A_{32}$$
$$- \partial_{x_1} A_{22} A_{33} - A_{21} \partial_{x_2} A_{33} + \partial_{x_2} A_{23} A_{31} + A_{23} \partial_{x_1} A_{32}$$
$$+ \partial_{x_1} A_{23} A_{32} + A_{21} \partial_{x_2} A_{33} - \partial_{x_2} A_{23} A_{31} - A_{22} \partial_{x_1} A_{33}$$
$$= 0.$$

The claim in (39.9) is thereby established.

As a result, using (39.5), (39.6), (39.7) and (39.9),

$$0 = \mathrm{div}_x(\varepsilon E)$$
$$= \mathrm{div}_x \left(\mathcal{J} A^{-1} \bar{\varepsilon} \bar{E} \right)$$
$$= \sum_{j=1}^{3} \partial_{x_j} \left(\mathcal{J} A^{-1} \bar{\varepsilon} \bar{E} \right)_j$$
$$= \sum_{j,k=1}^{3} \partial_{x_j} \left(\mathcal{J} A_{jk}^{-1} (\bar{\varepsilon} \bar{E})_k \right)$$
$$= \sum_{j,k=1}^{3} \partial_{x_j} (\mathcal{J} A_{jk}^{-1}) (\bar{\varepsilon} \bar{E})_k + \sum_{j,k=1}^{3} \mathcal{J} A_{jk}^{-1} \partial_{x_j} (\bar{\varepsilon} \bar{E})_k$$
$$= 0 + \sum_{k=1}^{3} \mathcal{J} \partial_{\bar{x}_k} (\bar{\varepsilon} \bar{E})_k$$
$$= \mathcal{J} \, \mathrm{div}_{\bar{x}} (\bar{\varepsilon} \bar{E}).$$

Accordingly,

$$\mathrm{div}_{\bar{x}}(\bar{\varepsilon} \bar{E}) = 0 \qquad (39.10)$$

and similarly

$$\mathrm{div}_{\bar{x}}(\bar{\mu} \bar{B}) = 0. \qquad (39.11)$$

Furthermore, for all i, $j \in \{1, 2, 3\}$, it follows from (39.7) that

$$\partial_{\bar{x}_i} \bar{E}_j = \partial_{\bar{x}_i} (A^{-T} E)_j$$

$$= \sum_{m=1}^{3} A_{mi}^{-1} \partial_{x_m} (A^{-T} E)_j$$

$$= \sum_{m,\ell=1}^{3} A_{mi}^{-1} \partial_{x_m} (A_{\ell j}^{-1} E_\ell)$$

$$= \sum_{m,\ell=1}^{3} A_{mi}^{-1} \partial_{x_m} A_{\ell j}^{-1} E_\ell + \sum_{m,\ell=1}^{3} A_{mi}^{-1} A_{\ell j}^{-1} \partial_{x_m} E_\ell$$

$$= \sum_{\ell=1}^{3} \partial_{\bar{x}_i} A_{\ell j}^{-1} E_\ell + \sum_{m,\ell=1}^{3} A_{mi}^{-1} A_{\ell j}^{-1} \partial_{x_m} E_\ell$$

and, therefore,

$$\partial_{\bar{x}_i} \bar{E}_j - \partial_{\bar{x}_j} \bar{E}_i = \sum_{\ell=1}^{3} (\partial_{\bar{x}_i} A_{\ell j}^{-1} - \partial_{\bar{x}_j} A_{\ell i}^{-1}) E_\ell$$

$$+ \sum_{m,\ell=1}^{3} (A_{mi}^{-1} A_{\ell j}^{-1} - A_{mj}^{-1} A_{\ell i}^{-1}) \partial_{x_m} E_\ell. \qquad (39.12)$$

In addition, by (39.3),

$$\partial_{\bar{x}_i} A_{\ell j}^{-1} = \partial_{\bar{x}_i} \partial_{\bar{x}_j} x_\ell = \partial_{\bar{x}_j} A_{\ell i}^{-1},$$

which reduces (39.12) to

$$\partial_{\bar{x}_i} \bar{E}_j - \partial_{\bar{x}_j} \bar{E}_i$$

$$= 0 + \sum_{m,\ell=1}^{3} (A_{mi}^{-1} A_{\ell j}^{-1} - A_{mj}^{-1} A_{\ell i}^{-1}) \partial_{x_m} E_\ell$$

$$= \sum_{\substack{1 \leqslant m,\ell \leqslant 3 \\ m \neq \ell}} (A_{mi}^{-1} A_{\ell j}^{-1} - A_{mj}^{-1} A_{\ell i}^{-1}) \partial_{x_m} E_\ell$$

$$= \sum_{\substack{1\leqslant m,\ell\leqslant 3\\ m\neq\ell}} A_{mi}^{-1}\, A_{\ell j}^{-1}\, \partial_{x_m} E_\ell - \sum_{\substack{1\leqslant m,\ell\leqslant 3\\ m\neq\ell}} A_{\ell j}^{-1}\, A_{mi}^{-1}\, \partial_{x_\ell} E_m$$

$$= \sum_{\substack{1\leqslant m,\ell\leqslant 3\\ m\neq\ell}} A_{mi}^{-1}\, A_{\ell j}^{-1} (\partial_{x_m} E_\ell - \partial_{x_\ell} E_m)$$

$$= \sum_{\substack{1\leqslant m,\ell\leqslant 3\\ m<\ell}} A_{mi}^{-1}\, A_{\ell j}^{-1} (\partial_{x_m} E_\ell - \partial_{x_\ell} E_m)$$

$$\quad + \sum_{\substack{1\leqslant m,\ell\leqslant 3\\ m>\ell}} A_{mi}^{-1}\, A_{\ell j}^{-1} (\partial_{x_m} E_\ell - \partial_{x_\ell} E_m)$$

$$= \sum_{\substack{1\leqslant m,\ell\leqslant 3\\ m<\ell}} A_{mi}^{-1}\, A_{\ell j}^{-1} (\partial_{x_m} E_\ell - \partial_{x_\ell} E_m)$$

$$\quad - \sum_{\substack{1\leqslant m,\ell\leqslant 3\\ m<\ell}} A_{\ell i}^{-1}\, A_{mj}^{-1} (\partial_{x_m} E_\ell - \partial_{x_\ell} E_m)$$

$$= \sum_{\substack{1\leqslant m,\ell\leqslant 3\\ m<\ell}} (A_{mi}^{-1}\, A_{\ell j}^{-1} - A_{\ell i}^{-1}\, A_{mj}^{-1})(\partial_{x_m} E_\ell - \partial_{x_\ell} E_m). \qquad (39.13)$$

Now, given $q \in \{1,2,3\}$ we let $\alpha_q, \beta_q \in \{1,2,3\}$ be such[7] that $\alpha_q < \beta_q$ and $\{q, \alpha_q, \beta_q\} = \{1,2,3\}$.

Also, let R and $\bar{R}$ be short notation for $\mathrm{curl}_x E$ and $\mathrm{curl}_{\bar{x}}\, \bar{E}$, respectively.

In this way,

$$R_q = (-1)^{q+1}(\partial_{x_{\alpha_q}} E_{\beta_q} - \partial_{x_{\beta_q}} E_{\alpha_q}),$$

and for all $p \in \{1,2,3\}$, we can rewrite (39.13) in the form

$$\bar{R}_p = (-1)^{p+1}(\partial_{\bar{x}_{\alpha_p}} \bar{E}_{\beta_p} - \partial_{\bar{x}_{\beta_p}} \bar{E}_{\alpha_p})$$

$$= (-1)^{p+1} \sum_{\substack{1\leqslant m,\ell\leqslant 3\\ m<\ell}} (A_{m\alpha_p}^{-1}\, A_{\ell\beta_p}^{-1} - A_{\ell\alpha_p}^{-1}\, A_{m\beta_p}^{-1})(\partial_{x_m} E_\ell - \partial_{x_\ell} E_m)$$

---

[7]More explicitly,

$$(\alpha_1, \beta_1) := (2,3), \quad (\alpha_2, \beta_2) := (1,3) \quad \text{and} \quad (\alpha_3, \beta_3) := (1,2).$$

$$= (-1)^{p+1} \sum_{q=1}^{3} (A_{\alpha_q \alpha_p}^{-1} A_{\beta_q \beta_p}^{-1} - A_{\beta_q \alpha_p}^{-1} A_{\alpha_q \beta_p}^{-1})(\partial_{x_{\alpha_q}} E_{\beta_q} - \partial_{x_{\beta_q}} E_{\alpha_q})$$

$$= \sum_{q=1}^{3} (-1)^{p+q} (A_{\alpha_q \alpha_p}^{-1} A_{\beta_q \beta_p}^{-1} - A_{\beta_q \alpha_p}^{-1} A_{\alpha_q \beta_p}^{-1}) R_q.$$

On this account, for all $d \in \{1, 2, 3\}$,

$$(A^{-1} \bar{R})_d = \sum_{p=1}^{3} A_{dp}^{-1} \bar{R}_p$$

$$= \sum_{p,q=1}^{3} (-1)^{p+q} (A_{dp}^{-1} A_{\alpha_q \alpha_p}^{-1} A_{\beta_q \beta_p}^{-1} - A_{dp}^{-1} A_{\beta_q \alpha_p}^{-1} A_{\alpha_q \beta_p}^{-1}) R_q.$$

$$(39.14)$$

We now claim that if $q \neq d$, then

$$\sum_{p=1}^{3} (-1)^p (A_{dp}^{-1} A_{\alpha_q \alpha_p}^{-1} A_{\beta_q \beta_p}^{-1} - A_{dp}^{-1} A_{\beta_q \alpha_p}^{-1} A_{\alpha_q \beta_p}^{-1}) = 0. \qquad (39.15)$$

Indeed, if $q \neq d$, then $d \in \{\alpha_q, \beta_q\}$. Let us assume that $d = \alpha_q$ (the case $d = \beta_q$ being similar). In this situation,

$$\sum_{p=1}^{3} (-1)^p (A_{dp}^{-1} A_{\alpha_q \alpha_p}^{-1} A_{\beta_q \beta_p}^{-1} - A_{dp}^{-1} A_{\beta_q \alpha_p}^{-1} A_{\alpha_q \beta_p}^{-1})$$

$$= \sum_{p=1}^{3} (-1)^p (A_{\alpha_q p}^{-1} A_{\alpha_q \alpha_p}^{-1} A_{\beta_q \beta_p}^{-1} - A_{\alpha_q p}^{-1} A_{\beta_q \alpha_p}^{-1} A_{\alpha_q \beta_p}^{-1})$$

$$= -(A_{\alpha_q 1}^{-1} A_{\alpha_q 2}^{-1} A_{\beta_q 3}^{-1} - A_{\alpha_q 1}^{-1} A_{\beta_q 2}^{-1} A_{\alpha_q 3}^{-1})$$

$$+ (A_{\alpha_q 2}^{-1} A_{\alpha_q 1}^{-1} A_{\beta_q 3}^{-1} - A_{\alpha_q 2}^{-1} A_{\beta_q 1}^{-1} A_{\alpha_q 3}^{-1})$$

$$- (A_{\alpha_q 3}^{-1} A_{\alpha_q 1}^{-1} A_{\beta_q 2}^{-1} - A_{\alpha_q 3}^{-1} A_{\beta_q 1}^{-1} A_{\alpha_q 2}^{-1}),$$

which vanishes (canceling the first term with the third, the second with the fifth, and the fourth with the sixth). This calculation proves (39.15).

Hence, plugging (39.15) into (39.14), we arrive at

$$(A^{-1}\bar{R})_d = \sum_{p=1}^{3}(-1)^{p+d}(A_{dp}^{-1}A_{\alpha_d\alpha_p}^{-1}A_{\beta_d\beta_p}^{-1} - A_{dp}^{-1}A_{\beta_d\alpha_p}^{-1}A_{\alpha_d\beta_p}^{-1})R_d.$$

(39.16)

Now, we claim that for every matrix $M \in \mathrm{Mat}(3\times3)$ and every $d \in \{1,2,3\}$,

$$\sum_{p=1}^{3}(-1)^{p+d}(M_{dp}M_{\alpha_d\alpha_p}M_{\beta_d\beta_p} - M_{dp}M_{\beta_d\alpha_p}M_{\alpha_d\beta_p}) = \det M.$$

(39.17)

To check this identity one can rely on the "axiomatic approach" to determinant functions (see e.g. [LL09, Theorem 8.14]), stating that the determinant is the unique multilinear and alternating function on matrices, which returns 1 when applied to the identity. More explicitly, to prove (39.17), it suffices to check three properties: namely, if we call M_i the ith row of a matrix M and $\mathcal{D}(M_1, M_2, M_3)$ the function on the left-hand side of (39.17) (thinking about it as a function acting on rows), our aim is to check that

if $i \in \{1,2,3\}$ and $M_i = av + bw$ for some $a, b \in \mathbb{R}$ and $v, w \in \mathbb{R}^3$,

$$\text{then } \mathcal{D}(\dots, M_i, \dots) = a\mathcal{D}(\dots, v, \dots) + b\mathcal{D}(\dots, w \dots), \quad (39.18)$$

if $M_i = M_j$ for some $i \neq j \in \{1,2,3\}$,

$$\text{then } \mathcal{D}(M_1, M_2, M_3) = 0 \qquad (39.19)$$

$$\text{and } \mathcal{D}(e_1, e_2, e_3) = 1. \qquad (39.20)$$

To check (39.18), let us suppose that $i = d$ (the cases in which $i = \alpha_d$ and $i = \beta_d$ being completely analogous). In this situation, we have that

$$M_{dp}M_{\alpha_d\alpha_p}M_{\beta_d\beta_p} - M_{dp}M_{\beta_d\alpha_p}M_{\alpha_d\beta_p}$$
$$= (av+bw)_p M_{\alpha_d\alpha_p}M_{\beta_d\beta_p} - (av+bw)_p M_{\beta_d\alpha_p}M_{\alpha_d\beta_p}$$
$$= av_p M_{\alpha_d\alpha_p}M_{\beta_d\beta_p} + bw_p M_{\alpha_d\alpha_p}M_{\beta_d\beta_p}$$
$$\quad - av_p M_{\beta_d\alpha_p}M_{\alpha_d\beta_p} - bw_p M_{\beta_d\alpha_p}M_{\alpha_d\beta_p},$$

from which (39.18) plainly follows.

To establish (39.19), one should consider three cases: $M_d = M_{\alpha_d}$, $M_d = M_{\beta_d}$ and $M_{\alpha_d} = M_{\beta_d}$. The first and second case however are similar, so we focus on the proof of the second and third cases.

Thus, let us suppose that $M_d = M_{\beta_d}$. Then,

$$(-1)^d \mathcal{D}(M_1, M_2, M_3)$$

$$= \sum_{p=1}^{3} (-1)^p (M_{dp}\, M_{\alpha_d \alpha_p}\, M_{d\beta_p} - M_{dp}\, M_{d\alpha_p}\, M_{\alpha_d \beta_p})$$

$$= -(M_{d1}\, M_{\alpha_d 2}\, M_{d3} - M_{d1}\, M_{d2}\, M_{\alpha_d 3})$$

$$+ (M_{d2}\, M_{\alpha_d 1}\, M_{d3} - M_{d2}\, M_{d1}\, M_{\alpha_d 3})$$

$$- (M_{d3}\, M_{\alpha_d 1}\, M_{d2} - M_{d3}\, M_{d1}\, M_{\alpha_d 2}),$$

which vanishes after simplifying the first term with the sixth, the second with the fourth, and the third with the fifth.

This establishes (39.19) in this case, and thus we take now into consideration the case $M_{\alpha_d} = M_{\beta_d}$. In this situation,

$$\mathcal{D}(M_1, M_2, M_3)$$

$$= \sum_{p=1}^{3} (-1)^{p+d} (M_{dp}\, M_{\alpha_d \alpha_p}\, M_{\alpha_d \beta_p} - M_{dp}\, M_{\alpha_d \alpha_p}\, M_{\alpha_d \beta_p}),$$

which clearly vanishes and the proof of (39.19) is thereby complete.

Moreover,

$$\mathcal{D}(e_1, e_2, e_3) = \sum_{p=1}^{3} (-1)^{p+d} (\delta_{dp}\, \delta_{\alpha_d \alpha_p}\, \delta_{\beta_d \beta_p} - \delta_{dp}\, \delta_{\beta_d \alpha_p}\, \delta_{\alpha_d \beta_p})$$

$$= (-1)^{d+d} (\delta_{\alpha_d \alpha_d}\, \delta_{\beta_d \beta_d} - \delta_{\beta_d \alpha_d}\, \delta_{\alpha_d \beta_d})$$

$$= 1 - 0,$$

giving (39.20).

This completes the proof of (39.17).

Thus, we plug (39.17) into (39.16), and we obtain that

$$(A^{-1}\bar{R})_d = \det(A^{-1}) R_d = \frac{R_d}{\mathcal{J}},$$

that is, $A^{-1}\bar{R} = \frac{R}{\mathcal{J}}$.

As a result,

$$\operatorname{curl}_{\bar{x}} \bar{E} = \bar{R} = \frac{AR}{\mathcal{J}} = \frac{A \operatorname{curl}_x E}{\mathcal{J}} = -\frac{A\mu \partial_t B}{\mathcal{J}}$$

$$= -\frac{A(\mathcal{J}A^{-1}\bar{\mu}A^{-T})\partial_t(A^T \bar{B})}{\mathcal{J}} = -\bar{\mu}\partial_t \bar{B}$$

and similarly

$$\operatorname{curl}_{\bar{x}} \bar{B} - \bar{\varepsilon}\partial_t \bar{E} = 0.$$

These observations, together with (39.10) and (39.11), establish the form invariance of Maxwell's equations, i.e. the property that Maxwell's equations (39.1) maintain their structure under every spatial transformation.

As discussed on p. 425, this feature of Maxwell's equations of "looking the same" in every coordinate system can be conveniently interpreted, rather than simply as the representation of the same electromagnetic properties of a material written in different coordinates, as the representation of different electromagnetic properties in the usual space.

Namely, we consider a spatial transformation to map a ball into an annulus, i.e. a ring-shaped object with a hole in the middle. The form invariance of Maxwell's equations entails that light cannot pass through the hole (hence objects can be made invisible by placing them inside the hole). Since the transformation is designed to coincide with the identity map outside the ball, the light rays will behave "as usual" far from the invisible object: the transformed space corresponds to a medium with inhomogeneous permittivity and permeability, and we will discuss in the following how to construct in practice a material with the appropriate structure to respond to electric and magnetic fields in such a prescribed way.

To map a ball, say B_2 (or better to say, a ball minus a point, say $B_2 \setminus \{0\}$), into an annulus, say $B_2 \setminus B_1$, one can consider the transformation

$$\bar{x} := \begin{cases} \frac{|x|+2}{2|x|} x & \text{if } x \in B_2 \setminus \{0\}, \\ x & \text{if } x \in \mathbb{R}^3 \setminus B_2. \end{cases}$$

Since $|\bar{x}| = \frac{|x|+2}{2} \in [1,2]$ for all $x \in B_2 \setminus \{0\}$, we have that $\bar{x}(B_2 \setminus \{0\}) = B_2 \setminus B_1$. Also, $\bar{x} = x$ outside B_2, and hence this change of coordinates possesses the features described above.

In this case, the Jacobian matrix in (39.2) in the relevant region $B_2 \setminus \{0\}$ becomes

$$A_{ij} = \partial_{x_j}\left(\frac{|x|+2}{2|x|}x_i\right) = \frac{|x|+2}{2|x|}\delta_{ij} - \frac{x_i\,x_j}{|x|^3}.$$

Therefore, in this situation, the matrix A is symmetric,

$$J = \det A = \frac{(|x|+2)^2|x|^3}{8|x|^5},$$

and accordingly the new permittivity $\bar{\varepsilon}$ and permeability $\bar{\mu}$ obtained from the homogeneous case $\varepsilon = \mu = 1$ are equal to

$$\frac{1}{J}\,A\,A^T = \frac{1}{J}\,A^2,$$

with A as above.

To appreciate the bending of a light ray induced by this transformation, for a given parameter $h \in (-1, 1)$, we can consider the parallel segments

$$\ell_h := \left\{(t, h, 0),\ \text{with}\ |t| < \sqrt{3}\right\}.$$

Then,

$$\bar{x}(\ell_h) = \frac{\sqrt{t^2+h^2}+2}{2\sqrt{t^2+h^2}}(t, h, 0), \tag{39.21}$$

see Figure 39.5 for a visual representation of such a hole which cannot be entered by light rays.

Now, let us briefly discuss how, in practice, there is a potential to build efficient electromagnetic cloaking devices. The idea is that one can use special media, called "metamaterials", whose physical parameters can be accurately designed to steer light around a hidden region of space. These artificial media are composites, usually assembled from multiple elements, such as metals and special plastics, often arranged in repeating patterns and presenting small structures (such as tiny needles). The metamaterial design can get quite sophisticated, involving thousands of elements with complex geometry, see e.g. [LJM+09, Woo09].

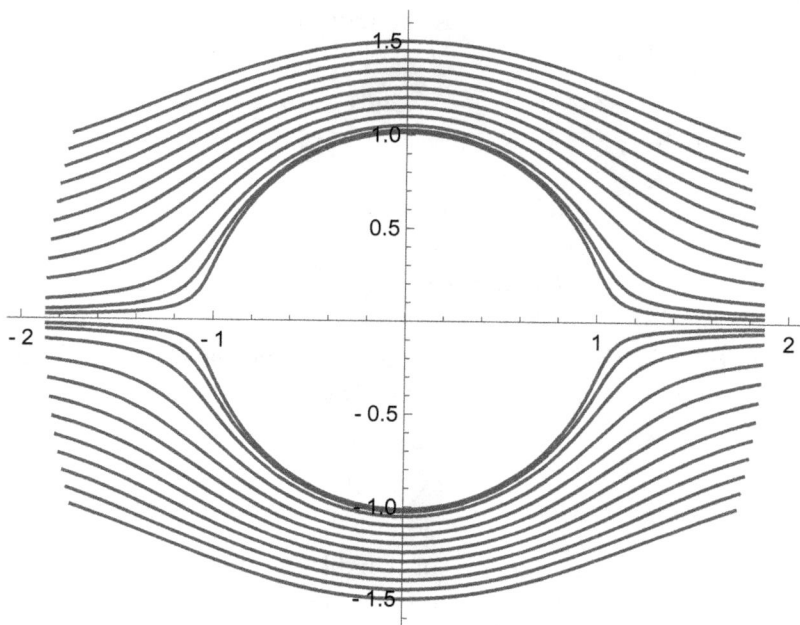

Figure 39.5. Steering light around a hole, as analytically described in (39.21) (with parameters $h \in \{\pm 0.025, \pm 0.05, \pm 0.1, \pm 0.2, \pm 0.3, \pm 0.4, \pm 0.5, \pm 0.6, \pm 0.7, \pm 0.8, \pm 0.9, \pm 1\}$). To be compared with Figure 39.4.

Interestingly, not only is mathematics pivotal in the development of metamaterials, but also metamaterials have been recently used to perform mathematical operations, such as differentiation, integration or convolution, so we can expect a nice interplay between mathematics and technology in this field, see [SMC$^+$14].

As a matter of curiosity, let us mention that the first patent for artificial materials of this type probably dates back to 1880, when Prussian-American inventor Hermann Hammesfahr registered a new material known as fiberglass. The utmost application of this material at that time was the design of a new dress that was very successfully presented at the 1893 World's Exposition in Chicago (the electric light bulb and the Ferris wheel also debuted at the same fair, but allegedly the "glass dress" was far more sensational, see Figure 39.6).

However, returning to the topic of invisibility and cloaking devices, at this point, a natural remark arises. If the light bends around a shell, any object enclosed within this region must be

Figure 39.6. The famous actress Georgia Cayvan posing in her glass dress (Public Domain image from Wikipedia).

necessarily blind since it cannot communicate electromagnetically[8] with the outside world. That is, the observation made by Yakov Perelman about Wells' Invisible Man (recall the discussion on p. 421) also applies to electromagnetic cloaking (though under different physical

---

[8]Interestingly, for the same reason, cloaked objects have no shadow. All in all, not only would observers be incapable of seeing a cloaked object, they would also be unaware that something has been hidden!

circumstances). Several scientists have tried to get around this some-what unavoidable issue, allowing some kind of information exchange between the inside and outside of the cloak, possibly at the cost of losing perfect invisibility, or by placing the invisible outside the cloak, utilizing so-called "complementary media", which can optically "cancel" a certain region of space at a certain frequency to create an "antiobject" see e.g. [HMSQZXJW09, LCZC09, XX19] (one could also try to rely on the use of different frequencies, e.g. cloaking for visible lights and using infrared vision).

It is also worth retaking inspiration from the *Star Trek* series mentioned on p. 422. Probably to make the plot more intriguing, in the *Star Trek* saga, while a space vessel is using its cloaking device, it is not allowed to fire weapons, use defensive shields, or operate transporters. To do any of these things, the vessel must "decloak". Quite interestingly, this has a counterpart in the mathematical theory of cloaking. Namely, what we described here is the cloaking of "passive" objects, i.e. objects without internal currents; in this case, cloaking for Maxwell's equations is indeed mathematically possible. Instead, if an object is "active" (e.g. it presents internal currents, as supposedly should occur for a space vessel firing weapons, employing defensive shields or operating transporters), even the mathematical theory of cloaking poses additional restrictions, and complete invisibility seems more problematic[9] and a "double coating" may be necessary, see [GKLU07]. One may need to line the inner surface of the cloak with a perfectly reflecting material or induce extraordinary electric and magnetic surface voltages at the inner surface of the cloak, see [Zha08].

In rough terms, active electromagnetic objects are more difficult to cloak because they are constantly emitting or interacting with electromagnetic waves (similar difficulties arise when cloaking moving objects); to circumvent issues of this type. one may also consider approximate cloaks, see e.g. [BLZ14].

---

[9]From a mathematical point of view, this has to deal with the fact that "finite energy" solutions to Maxwell's equations do not exist in the presence of active sources inside the cloaked region.

Moreover, we have discussed here some basic features of "passive" ways of cloaking an object, but there are also "active" cloaking methods, corresponding to the recent technology of noise-canceling headphones: broadly, one could try to "cancel" signals by actively producing "opposite" signals (see [NE92, PT07] for the case of acoustic waves and [SE13] for electromagnetic cloaking). Drawbacks of these active cloaking devices are that they may be subject to field instabilities and the design of the cloak also becomes even more difficult and complicated, see e.g. [AT09], thus posing severe challenges in practical applications.

We stress that the use of the technical word "active" in the context of cloaking could be sometimes confusing since this may refer to either the object to be cloaked, in which some electromagnetic activity may take place, or to the cloaking device, which may exploit active sources specifically designed to cancel suitable signals (or both, see [GVMO11, Remark 2]).

Let us also mention that here we considered the cloaking problem in a rather broad generality. Additional assumptions on the object to be cloaked or on the angle of incidence of the light rays may allow more effective cloaking strategies that are specifically tailored to a specific situation, see e.g. [AE08, LJM+09, LS13, ULD+13].

Note that the ideas related to invisibility and cloaking, though at the moment still in their early stages of development, can have fruitful applications beyond their original goals (and hopefully beyond the obvious military scope). For instance, if one improved the practical possibility of manipulating light rays, this would allow the creation of powerful microscopes and telescopes and maybe even faster computers.

Certainly, the mathematical difficulties of it are enormous, and so are the technological hurdles to be overcome. For instance, to effectively obtain invisibility, a cloak should efficiently interact at once with all frequencies of light rays, or at least the ones in the visible spectrum, while the parameters in the equation seem to be sensible to the change in frequency due to dispersive phenomena. The steps toward a possible improvement in this context may include the combination of metamaterials and dielectric materials, see e.g. [VLZ+09].

On the one hand, at the moment, metamaterials seem to be employed fruitfully mostly at microwave[10] frequencies, see [SMJ+06], with additional difficulties in assembling and designing them above this threshold; on the other hand, metamaterials engineering at optical frequencies has also been tested, see [GCPL09].

Another possible hindrance to the development of cloaking devices may be the cost of the materials employed. A promising line of research tries to build cloaks out of ordinary materials, such as calcite, see e.g. [CLZ+11].

The cloaking problem arising from Maxwell's equations is also related to the electrostatic problem presented in Chapter 38: indeed, a natural analogue in that case is to construct examples of objects that cannot be detected by electrostatic measurements performed along the boundary (for examples, electrostatically cloaked complex objects that look the same under electrical impedance tomography as an isotropic material). See [GKLU09] for additional information on this type of problems, as well as for related questions about electromagnetic wormholes.

---

[10]What happens is that, typically, for cloaking devices, one has to engineer the metamaterial at a scale smaller than the wavelength of radiation. For the optical radiation of visible light, the wavelength is less than a micron; therefore, the structure of a metamaterial effective in this situation has to rely on nanotechnology.

For this reason, cloaking devices often work well for a single wavelength or for a small wavelength range.

On a positive, or maybe negative, note, let us mention that for radars, the wavelength is about 3 cm, so one can relatively easily design metamaterials at the scale of millimeters to hide objects from radars, with military-specific applications.

By the way, possible military uses of cloaking devices are not limited to making vehicles and weapons invisible. For example, cloaking may be employed to deactivate structures that interrupt signals since one could cloak these structures to allow signals to pass by freely (in a sense, this leverages the blindness of the cloaked object turning it into an advantage).

Also, steering light rays may make objects larger or smaller (or different).

Chapter 40

Fighting a Pandemic Using Differential Equations

A topical subject nowadays is the spread of a virus among a given population, see Figure 40.1. Mathematics has established a solid reputation in trying to understand the fundamental properties of epidemic disease diffusion and has often helped taking farsighted decisions in the critical moments of global pandemics.

In this set of notes, we certainly do not aim at presenting all possible mathematical models that can be effectively used to describe infectious diseases, nor at giving a fully "realistic" description of an epidemic in the real world. Rather, we briefly recall a classical approach which, despite its simplifications and limitations, can already highlight the great potential of mathematics in dealing with epidemiology and can obviously open the possibility of presenting more sophisticated and bespoke models to address more realistically concrete cases of epidemics.

The setting that we recall here builds on the research (among others) of Sir Ronald Ross, Hilda Phoebe Hudson, William Ogilvy Kermack and Anderson Gray McKendrick and is called the "SIR model", not because of Ross' formal honorific address but because it divides the whole population into three compartments:

- *Susceptible*: individuals who might become infected if exposed to the infectious agent (e.g. a virus).
- *Infected (and Infectious)*: individuals who are currently infected and can transmit the infection to susceptible individuals (e.g. by contact).

441

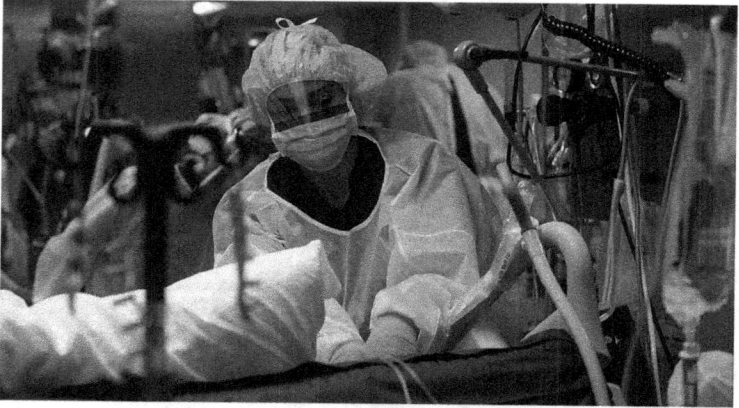

Figure 40.1.   Natasha McClinton, a surgical nurse, prepares a patient for a procedure in a COVID-19 intensive care unit (Public Domain image from Wikipedia).

- *Recovered (or, better to say, Removed)*: individuals who, after being infected, become immune to the infection (in this category, it is also common to place the people who died following infection since, like those recovered, they cannot contribute to the spread of the disease).

This is an example of a compartmental model since it splits a given population into "compartments" (three compartments, in this case, which are labeled with the letters S for susceptible, I for infected and R for recovered), and several types of related models are indeed broadly utilized in epidemiology.

For simplicity, we can consider a large population and denote by S, I and R the proportion of susceptible, infected and recovered individuals, respectively; in this way, $S, I, R \in [0, 1]$ (the value 1 corresponding to 100% of the population, the value $\frac{1}{2}$ to 50%, and so on). Also, by possibly adopting the above-mentioned convention of counting casualties in R, we can suppose that the total population remains constant and hence

$$S + I + R = 1. \tag{40.1}$$

The SIR model then specifies the evolution, $S(t)$, $I(t)$ and $R(t)$, of the different compartments over time according to the transition rates between them. First of all, in units of time, one assumes that some susceptible individuals may become infected. The *ansatz* in this situation is that the number of newly infected people in a unit of time is proportional to the number of susceptible individuals S (the higher this number, the easier it becomes for someone to catch the disease) and to the number of infected I (the more the infected people, the easier for the epidemic to spread). Accordingly, if we denote by $\alpha \geqslant 0$ this proportionality coefficient, in a unit of time, we have that αSI individuals transit from the compartment of susceptible individuals to that of infected (the parameter α can be seen as a "transmission rate").

Concurrently, some infected people can recover. This number is taken to be proportional to the number of infected, e.g. if the medicine is effective in a given proportion of treatments, the higher the number of people receiving the medical treatment, the higher the number of recovered patients (also, in the pessimistic scenario of counting casualties in this compartment, the higher the number of infected individuals, the higher the number of possible deaths). Accordingly, if we denote by $\beta \in [0, 1]$ the proportionality coefficient involved in this transition, in a unit of time, we have that βI individuals move from the compartment of infected to that of recovered (the parameter β thus plays the role of a "recovery rate").

The transition between compartments described in this way is summarized in Figure 40.2.

It is certainly convenient to translate Figure 40.2 into a mathematical formulation; this is done by writing the system of ordinary differential equations corresponding to this compartmental transit:

$$\begin{cases} \dot{S} = -\alpha SI, \\ \dot{I} = \alpha SI - \beta I, \\ \dot{R} = \beta I, \end{cases} \tag{40.2}$$

where the dot represents the derivative with respect to time, $S = S(t)$, $I = I(t)$ and $R = R(t)$. The system in (40.2) is usually taken as the mathematical description of the SIR model. It is interesting to observe that, by (40.2),

$$\frac{d}{dt}(S + I + R) = \dot{S} + \dot{I} + \dot{R} = -\alpha SI + (\alpha SI - \beta I) + \beta I = 0,$$

consistently with (40.1).

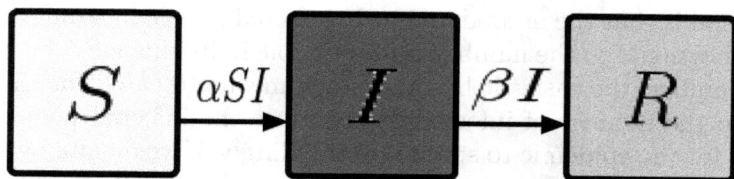

Figure 40.2. States in a SIR epidemic model and transition between compartments.

Similarly, one can also focus only on the first two equations in (40.2) to determine the time evolution of the number of susceptible and infected individuals and then obtain, as a byproduct, the number of recovered individuals by using (40.1). In this way, one can reduce (40.2) to a system of two ordinary differential equations:

$$\begin{cases} \dot{S} = -\alpha SI, \\ \dot{I} = \alpha SI - \beta I. \end{cases} \tag{40.3}$$

Despite its exceptional conceptual simplicity (after all, we are simply stating that susceptible individuals may become infected and infected individuals may recover or possibly die), the SIR model already provides some very interesting information about the spread of a disease.

First of all, the first equation in (40.3) already suggests that the infection occurs through contact between infectious and susceptible people (since the quantity SI is a good model for "random encounters" between S and I). As a consequence, the parameter α in (40.3) can be considered as a "contact rate". For this reason, measures of social distancing and lockdowns aim at reducing contacts between people and hence at reducing the value of α.

The parameter β in the second equation in (40.3) instead reduces the number of infected people, correspondingly [1] raising the number of recovered. To increase the parameter β, one effective way is clearly to have better medicines since more effective treatments facilitate healing and accelerate recovery rates. Another useful measure

---

[1]Beware that since casualties are also counted among recoveries, a tragic way to increase the number of recoveries consists of killing infectious people! This is a classical topic of science fiction horror films, such as "The Crazies" by George A. Romero.

relies on the early detection of infected people and their isolation in hospitals or quarantine areas; in this way, infected people are *de facto* removed from the dynamics of (40.3) and, from a mathematical point of view, transit to the R compartment (even if they are not yet recovered from a medical perspective).

See Figure 40.3 to appreciate how decreasing the transmission rate α and increasing the recovery rate β can help in slowing an epidemic's spread.

In the description of epidemic diseases, it is also interesting to consider the basic reproduction number (often called "R naught" in jargon), e.g. defined by

$$\mathcal{R}_o := \frac{\alpha}{\beta}. \tag{40.4}$$

Its importance lies in the fact that, by (40.3),

$$\dot{I} = \beta I \left(\frac{\alpha}{\beta} S - 1 \right) = \beta I \left(\mathcal{R}_o S - 1 \right) \tag{40.5}$$

and thus, since $S \in [0, 1]$,

$$\dot{I} \leqslant \beta I \left(\mathcal{R}_o - 1 \right).$$

Hence, when

$$\mathcal{R}_o < 1, \tag{40.6}$$

we find that

$$\frac{d}{dt} \ln I(t) = \frac{\dot{I}(t)}{I(t)} \leqslant \beta \left(\mathcal{R}_o - 1 \right) = -\beta \left(1 - \mathcal{R}_o \right),$$

leading to

$$I(t) \leqslant I(0) \, e^{-\beta(1-\mathcal{R}_o)t},$$

which means that the number of infected people decreases exponentially fast, and the epidemic is likely to be overcome sufficiently quickly (conversely, when $\mathcal{R}_o > 1$, one can expect that the number of infected people will grow exponentially, with obvious tragic consequences).

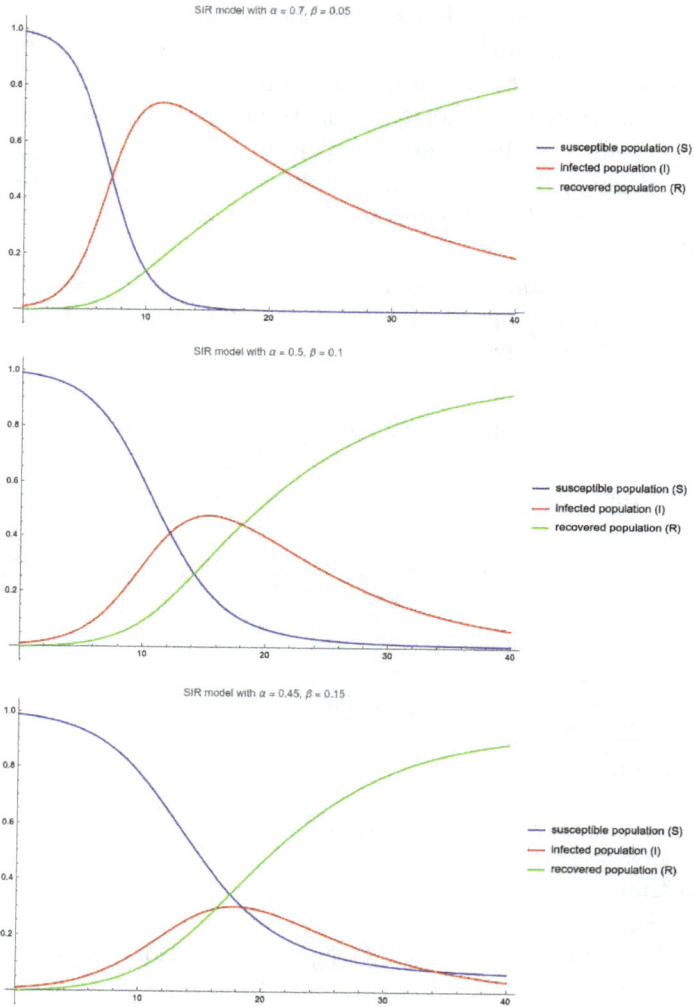

Figure 40.3.   Flattening the curve of infection: the parameters (α, β) are chosen here in $\{(0.7, 0.05),\ (0.5, 0.1),\ (0.45, 0.15)\}$.

The computation in (40.5) also reveals an interesting piece of information about the possibility of embanking the spread of the epidemic by reducing the number of susceptible individuals; indeed, when

$$\mathcal{R}_o\, S < 1, \qquad\qquad (40.7)$$

a similar computation as above would lead to an exponential decrease in the infected individuals. This is related to the notion of "herd immunity", namely by achieving community protection due to a critical proportion of the population having become immune to the disease. Note indeed that condition (40.7) prescribes that, to reach this situation, at least a proportion $\mathcal{H}_o := 1 - \frac{1}{\mathcal{R}_o}$ must have become immune.

If we wish to play with some numbers, an initial estimate of $\mathcal{R}_o$ for COVID-19 by the World Health Organization took into account a possibility of $\mathcal{R}_o \simeq 2.4$, thus largely violating the optimistic threshold in (40.6). Also, with this value of $\mathcal{R}_o$, one would obtain a herd immunity threshold of $\mathcal{H}_o \simeq 1 - \frac{1}{2.4} = 0.58\overline{3}$, meaning that about 60% of the population must be immune to protect the community from an exponential spread of epidemic (but not from the disease[2] in itself).

Reaching such a threshold of immunization without a vaccination campaign is likely extremely dangerous (since it entails that a vast majority of the population must catch the disease). Since, realistically, vaccines are also not 100% effective in preventing the dissemination of the virus, if one aims at reaching a herd immunity via a vaccination campaign, the efficacy of the vaccine is an important parameter to take into account. For instance, if 80% of the population is vaccinated with a vaccine which is effective in 70% of the cases in avoiding new infections, we have that 56% of the population is immune to the disease (and this number is still below the 60% threshold discussed above for herd immunity).

Of course, we are not proposing here concrete measures to deal with COVID-19, and the numbers plugged in for the naive computations above must not be taken seriously. Our aim is to simply stress

---

[2]We stress that reaching herd immunity does not stop the epidemic overnight. It only makes the curve of newly infected people bend down and approach zero exponentially fast. But, according to the SIR model, other individuals will get infected even after having reached the status of herd immunity, and the curve representing the total number of infected people will keep increasing toward a horizontal asymptote.

We also remark that we are not accounting here for important realistic features, such as the possible diminished effectiveness of a vaccine on a rapidly mutating virus, the possibility of repeated infections within the population, and the different personal responses of individuals to virus exposition.

the importance of dealing with epidemics using a scientific approach and how useful mathematics can be for such a life-or-death situation.

Another striking application of the SIR model consists of the prediction of the epidemic peak, namely the maximal number I_{max} of people who are simultaneously infected during the epidemic. Indeed, one can deduce from (40.3) that

$$I_{max} \simeq \frac{\mathcal{R}_o - 1 - \ln \mathcal{R}_o}{\mathcal{R}_o}. \tag{40.8}$$

Note that as $\mathcal{R}_o \to +\infty$, we have $I_{max} \to 1$, which corresponds to the total population ending up being infected at the same time. We also remark that estimating I_{max} is of great strategic importance because, to accommodate patients, one always wants to have a number of available beds in the hospitals commensurate to the number of people in need of hospitalization.

To prove (40.8), one can pick a time $t_\star$ at which we expect the number of infected individuals to be maximal, i.e. such that $I_{max} = I(t_\star)$. Then, by (40.3),

$$0 = \frac{dI}{dt}(t_\star) = \alpha\, S(t_\star)\, I(t_\star) - \beta\, I(t_\star),$$

from which we infer that

$$S(t_\star) = \frac{\beta}{\alpha} = \frac{1}{\mathcal{R}_o}.$$

Note that this is not quite what (40.8) is looking for since $S(t_\star)$ represents only the number of susceptible individuals at the time when the number of infected people is maximum. Nevertheless, the structure of (40.3) comes in handy now and provides that

$$\begin{aligned}
\frac{d(I+S)}{dt} &= (\alpha S I - \beta I) - \alpha S I \\
&= -\beta I \\
&= -\frac{\alpha \beta S I}{\alpha S} \\
&= \frac{\beta}{\alpha S}\frac{dS}{dt} \\
&= \frac{1}{\mathcal{R}_o}\frac{d}{dt}(\ln S).
\end{aligned}$$

Therefore,

$$I(t) + S(t) = I(0) + S(0) + \frac{\ln(S(t)) - \ln(S(0))}{\mathcal{R}_o}$$

and, as a result,

$$I_{\max} = I(t_\star)$$

$$= I(0) + S(0) + \frac{\ln(S(t_\star)) - \ln(S(0))}{\mathcal{R}_o} - S(t_\star)$$

$$= I(0) + S(0) + \frac{-\ln \mathcal{R}_o - \ln(S(0))}{\mathcal{R}_o} - \frac{1}{\mathcal{R}_o}$$

$$= I(0) + S(0) - \frac{1 + \ln \mathcal{R}_o + \ln(S(0))}{\mathcal{R}_o}.$$

If we assume that at the beginning of the epidemic there is nobody recovering, namely[3] that $R(0) = 0$, we deduce from (40.1) that $I(0) + S(0) = 1$ and thus

$$I_{\max} = 1 - \frac{1 + \ln \mathcal{R}_o + \ln(S(0))}{\mathcal{R}_o},$$

which completes the proof of (40.8).

We also mention that explicit solutions of the SIR model can be obtained through an implicit "time reparameterization" procedure. Specifically, one considers the integral of I as a "new time" τ by setting

$$\tau(t) := \int_0^t I(s)\, ds. \tag{40.9}$$

Note that $\frac{\tau}{t}$ represents the average number of infected people at time t. In this new variable, the solution of (40.2) becomes explicit and takes the form

$$\begin{cases} \tilde{S}(\tau) = S(0)\, e^{-\alpha\tau}, \\ \tilde{I}(\tau) = I(0) + S(0)(1 - e^{-\alpha\tau}) - \beta\tau, \\ \tilde{R}(\tau) = R(0) + \beta\tau. \end{cases} \tag{40.10}$$

---

[3]No confusion should arise between the initial number of recovered $R(0)$ and the structural parameter $\mathcal{R}_o$ in (40.4).

Of course, one of the limitations of this explicit formulation is that the new solutions $\tilde{S}$, $\tilde{I}$ and $\tilde{R}$ are functions of an implicitly defined time τ through the relations $\tilde{S}(\tau(t)) = S(t)$, $\tilde{I}(\tau(t)) = I(t)$ and $\tilde{R}(\tau(t)) = R(t)$, whence the explicit representation in (40.10) is not always particularly pleasant for concrete calculations (see, however, the forthcoming equation (40.11) for a reformulation of the time parameter problem).

To check the validity of (40.10), one can proceed with implicit differentiation, observing that from (40.2), it follows that

$$
\begin{cases}
\dfrac{d\tilde{R}}{d\tau} = \beta, \\[4mm]
\dfrac{d\tilde{S}}{d\tau} = -\alpha\tilde{S},
\end{cases}
$$

which are explicitly solvable, leading to $\tilde{S}(\tau)$ and $\tilde{R}(\tau)$ in (40.10). Then, using

$$
\frac{d\tilde{I}}{d\tau} = \alpha\tilde{S} - \beta
$$

and exploiting the previous explicit solution for $\tilde{S}(\tau)$, one obtains the expression of $\tilde{I}(\tau)$ in (40.10).

Having completed the proof of (40.10), we also observe that, by (40.9),

$$
\frac{d\tau}{dt} = \tilde{I}(\tau),
$$

from which one obtains an explicit integral formula relating the old and new times of the type

$$
t = \int_0^\tau \frac{ds}{I(0) + S(0)(1 - e^{-\alpha s}) - \beta s}. \tag{40.11}
$$

Once again, this time reparameterization is explicit, but the integral cannot be usually written in terms of standard elementary mathematical functions.

We stress that the classical SIR model in (40.3) does not take into account the possible mobility of individuals (or, in rough terms, the whole population in (40.3) is concentrated at the same place).

There are of course a number of models available in the literature which account for possible spatial displacements of the population; for instance, if one models the space configuration by a variable x and assumes that the population moves according to a random walk, as in Chapter 2, one may think about the possibility of replacing the ordinary differential equations in (40.3) with a system of partial differential equations of the type

$$\begin{cases} \partial_t S = \mu \Delta S - \alpha SI, \\ \partial_t I = \nu \Delta I + \alpha SI - \beta I. \end{cases} \tag{40.12}$$

In this case, $S = S(x,t)$, $I = I(x,t)$ and the diffusion coefficients μ, $\nu \in [0, +\infty)$ may also be different to account for the possibility of different speeds between the mobility of susceptible and infected people (e.g. in a situation where the diffusivity of infected individuals is limited by the illness itself, or by the responsibility of self-isolating individuals with symptoms).

Restrictions on travel are of course a consequential measure to avoid or limit the spread of an epidemic due to the diffusive nature of the system in (40.12).

See for instance [BvdDW08] for additional information on the SIR models and for many related models utilized in mathematical epidemiology.

Now, for completeness, as well as in view of the discussion presented in Chapter 24, we recall here a variation of the SIR model based on the evolution of the probability density $p(x,y,t)$ corresponding to having, at time t, a proportion $x \in [0,1]$ of susceptible individuals and $y \in [0,1]$ of infected ones, as proposed in [CS11]. The core of this model is to obtain a partial differential equation of the form

$$\partial_t p + \operatorname{div}(pF) = 0, \tag{40.13}$$

where the above divergence is taken with respect to the variables (x,y), and the vector field F corresponds to the right-hand side of the SIR model in (40.3), that is,

$$F(x,y) := (-\alpha xy, \ \alpha xy - \beta y). \tag{40.14}$$

We remark that (40.13) has the form of a transport equation, as in (5.4).

To deduce (40.13) from the principles of the SIR model, one can argue as follows. Given n, $m \in \mathbb{N}$, with $n + m \leqslant N$, we consider the probability $P(n, m, t)$ of having n susceptible and m infected at time t, with N being the total number of individuals (note that, in view of the constancy of the population in (40.1), the number of recovered is thus necessarily $N - n - m$). We argue that, given an "infinitesimal" time increment τ, the probability $P(n, m, t + \tau)$ of having n susceptible and m infected at time $t + \tau$ is built by the superposition of three occurrences:

- either, during the elapsed time τ, an infected individual may have recovered: this corresponds to the probability $P(n, m+1, t)$ of having one more infected at time t, times the probability of recovering, which, inspired by Figure 40.2, we take to be equal to $\frac{\beta(m+1)}{N}$;
- or, during the elapsed time τ, a susceptible individual may have become infected: this corresponds to the probability $P(n + 1, m - 1, t)$ of having one more susceptible and one less infected at time t, times the probability[4] of getting infected, which, inspired by Figure 40.2, we take to be equal to $\frac{\alpha(n+1)(m-1)}{N^2}$;
- or, during the elapsed time τ, no susceptible individual becomes infected and no infected recovers: this corresponds to the probability $P(n, m, t)$ of having the same number of susceptible and infected individuals at time t, times the probability of having no infections or recovery, which is equal to the remaining probability $1 - \frac{\beta m}{N} - \frac{\alpha nm}{N^2}$.

Due to these considerations, we write that

$$
\begin{aligned}
P(n, m, t + \tau) &= \frac{\beta(m + 1)}{N} P(n, m + 1, t) \\
&+ \frac{\alpha(n + 1)(m - 1)}{N^2} P(n + 1, m - 1, t) \\
&+ \left(1 - \frac{\beta m}{N} - \frac{\alpha nm}{N^2}\right) P(n, m, t).
\end{aligned}
\tag{40.15}
$$

---

[4]In all this discussion, we are implicitly assuming that $\frac{\beta(m+1)}{N}$, $\frac{\alpha(n+1)(m-1)}{N^2} \in [0, 1]$, consistently with the standard notion of probability. This hypothesis is justified, for instance, if the number of infected m is sufficiently small compared to the total population N.

Now, we try to achieve a continuum model by taking the limit as $N \to +\infty$, corresponding to a large population. For this, for all $x, y \in [0, 1]$, we define

$$p(x, y, t) := NP(Nx, Ny, t).$$

In this way, we deduce from (40.15) that

$$
\begin{aligned}
\partial_t p(x, y, t) \\
\simeq \frac{p(x, y, t + \tau) - p(x, y, t)}{\tau} \\
= \frac{N}{\tau} \Big[P(Nx, Ny, t + \tau) - P(Nx, Ny, t) \Big] \\
= \frac{N}{\tau} \bigg[\frac{\beta(Ny + 1)}{N} P(Nx, Ny + 1, t) \\
+ \frac{\alpha(Nx + 1)(Ny - 1)}{N^2} P(Nx + 1, Ny - 1, t) \\
+ (1 - \beta y - \alpha xy) P(Nx, Ny, t) - P(Nx, Ny, t) \bigg] \\
= \frac{1}{\tau} \bigg[\beta \left(y + \frac{1}{N} \right) p \left(x, y + \frac{1}{N}, t \right) \\
+ \alpha \left(x + \frac{1}{N} \right) \left(y - \frac{1}{N} \right) p \left(x + \frac{1}{N}, y - \frac{1}{N}, t \right) \\
- (\beta y + \alpha xy) p(x, y, t) \bigg] \\
= \frac{1}{\tau} \bigg[B \left(x, y + \frac{1}{N}, t \right) + A \left(x + \frac{1}{N}, y - \frac{1}{N}, t \right) \\
- B(x, y, t) - A(x, y, t) \bigg] \\
\simeq \frac{1}{\tau} \bigg[\frac{1}{N} \partial_y B(x, y, t) + \frac{1}{N} \left(\partial_x A(x, y, t) - \partial_y A(x, y, t) \right) \bigg],
\end{aligned}
$$

where

$$A(x,y,t) := \alpha xy \, p(x,y,t) \quad \text{and} \quad B(x,y,t) := \beta y \, p(x,y,t).$$

Hence, by choosing a scaling between the time step and the total population such that $\tau N \simeq 1$, we reduce the infectious disease spreading problem to

$$\partial_t p = \partial_x A + \partial_y (B - A) = \partial_x(\alpha xyp) + \partial_y\big((\beta y - \alpha xy)p\big),$$

which, due to the choice of the vector field in (40.14), corresponds to the desired transport equation in (40.13).

For further motivations about elliptic partial differential equations, see e.g. Chapter 2 in [Som49], Chapter 4 in [TS90], Chapter 1 in [Pet91], pp. 160–168 in [HT96], the introduction of [Eva98], Chapter 1 in [FB02], Section 1.4 in [PR05], Chapter 1 in [Str08], Chapters 4 and 5 in [BT09], Section 1.2 in [Sal15], the appendix of [FRRO21] and the references therein.

Having established solid heuristic motivations for the study of partial differential equations and especially the Laplace operator, we now enter the mathematical theory of these objects. From this point forward, we are not allowed to "waive hands", provide formal expansions or make additional simplifying assumptions; instead, we must fully adhere to rigorous mathematical precision.

References

[AV19] N. Abatangelo and E. Valdinoci, *Getting Acquainted with the Fractional Laplacian*, Contemporary research in elliptic PDEs and related topics, Springer INdAM Ser., vol. 33, Springer, Cham, 2019, pp. 1–105. MR3967804.

[AZ79] M. J. Ablowitz and A. Zeppetella, Explicit solutions of Fisher's equation for a special wave speed, *Bull. Math. Biol.* **41**, no. 6, 835–840 (1979), DOI: 10.1016/S0092-8240(79)80020-8. MR639998.

[Ada80] D. Adams, *The Hitchhiker's Guide to the Galaxy*, Hitchhiker series, Harmony Books, 1980.

[AT09] P. Alitalo and S. Tretyakov, Electromagnetic cloaking with metamaterials, *Mater. Today* **12**, no. 3, 22–29 (2009), DOI: 10.1016/S1369-7021(09)70072-0.

[AC75] S. M. Allen and J. W. Cahn, Coherent and incoherent equilibria in iron-rich iron-aluminum alloys, *Acta Metall.* **23**, no. 9, 1017–1026 (1975), DOI: 10.1016/0001-6160(75)90106-6.

[AE08] A. Alù and N. Engheta, Plasmonic and metamaterial cloaking: Physical mechanisms and potentials, *J. Opt. A: Pure Appl. Opt.* **10**, no. 9, 1–55 (2008), DOI: 10.1088/1464-4258/10/9/093002.

[Ami00] Yu. Aminov, Differential geometry and the topology of curves, Gordon and Breach Science Publishers, Amsterdam, 2000. Translated from the Russian by V. Gorkavy. MR1807240.

[And99] B. Andrews, Gauss curvature flow: The fate of the rolling stones, *Invent. Math.* **138**, no. 1, 151–161 (1999), DOI: 10.1007/s002220050344. MR1714339.

[ACGL20] B. Andrews, B. Chow, C. Guenther, and M. Langford, Extrinsic geometric flows, Graduate Studies in Mathematics, vol. 206, American Mathematical Society, Providence, RI, 2020. MR4249616.

[APS06] N. J. Armstrong, K. J. Painter, and J. A. Sherratt, A continuum approach to modelling cell-cell adhesion, *J. Theoret. Biol.* **243**, no. 1, 98–113 (2006), DOI: 10.1016/j.jtbi.2006.05.030. MR2279324.

[AW75] D. G. Aronson and H. F. Weinberger, *Nonlinear Diffusion in Population Genetics, Combustion, and Nerve Pulse Propagation*, Partial differential equations and related topics (Program, Tulane University, New Orleans, La., 1974), Springer, Berlin, 1975, pp. 5–49. Lecture Notes in Math., Vol. 446. MR0427837.

[AW78] D. G. Aronson and H. F. Weinberger, Multidimensional nonlinear diffusion arising in population genetics, *Adv. Math.* **30**, no. 1, 33–76 (1978), DOI: 10.1016/0001-8708(78)90130-5. MR511740.

[AA13] V. Asensi and J. M. Asensi, Euler's Right Eye: The Dark Side of a Bright Scientist, *Clin. Infect. Dis.* **57**, no. 1, 158–159 (2013), DOI: 10.1093/cid/cit170.

[AIM09] K. Astala, T. Iwaniec, and G. Martin, *Elliptic Partial Differential Equations and Quasiconformal Mappings in the Plane*, Princeton Mathematical Series, vol. 48, Princeton University Press, Princeton, NJ, 2009. MR2472875.

[AP06] K. Astala and L. Päivärinta, Calderón's inverse conductivity problem in the plane, *Ann. Math. (2)* **163**, no. 1, 265–299 (2006), DOI: 10.4007/annals.2006.163.265. MR2195135.

[Bal99] P. Ball, *The Self-Made Tapestry. Pattern Formation in Nature*, Oxford University Press, Oxford, 1999.

[BLZ14] G. Bao, H. Liu, and J. Zou, Nearly cloaking the full Maxwell equations: Cloaking active contents with general conducting layers, *J. Math. Pures Appl. (9)* **101**, no. 5, 716–733 (2014), DOI: 10.1016/j.matpur.2013.10.010 (English, with English and French summaries). MR3192429.

[BCMM15] D. Bambusi, A. Carati, A. Maiocchi, and A. Maspero, Some analytic results on the FPU paradox, Hamiltonian partial differential equations and applications, *Fields Inst. Commun.*, vol. 75, *Fields Inst. Res. Math. Sci.*, Toronto, ON, 2015, pp. 235–254, DOI: 10.1007/978-1-4939-2950-4_8. MR3445504.

[BMP08] D. Bambusi, D. Muraro, and T. Penati, Numerical studies on boundary effects on the FPU paradox, *Phys. Lett. A* **372**, no. 12, 2039–2042 (2008), DOI: 10.1016/j.physleta.2007.11.011. MR2398159.

[BJ08] A. Banerjee and J. Jost, On the spectrum of the normalized graph Laplacian, *Linear Algebra Appl.* **428**, no. 11–12, 3015–3022 (2008), DOI: 10.1016/j.laa.2008.01.029. MR2416605.

[BT09] H. T. Banks and H. T. Tran, *Mathematical and Experimental Modeling of Physical and Biological Processes*, Textbooks in Mathematics, CRC Press, Boca Raton, FL, 2009. With 1 CD-ROM (Windows, Macintosh and UNIX). MR2488750.

[BDVRV21] F. Bartolucci, E. De Vito, L. Rosasco, and S. Vigogna, Understanding Neural Networks with Reproducing Kernel Banach Spaces, arXiv e-prints (2021).

[Bas69] F. M. Bass, A new product growth for model consumer durables, *Manag. Sci.* **15**, no. 5, 215–227 (1969).

[Bat99] G. K. Batchelor, *An Introduction to Fluid Dynamics*, Second paperback edition, Cambridge Mathematical Library, Cambridge University Press, Cambridge, 1999. MR1744638.

[BNS06] M. Belkin, P. Niyogi, and V. Sindhwani, Manifold regularization: A geometric framework for learning from labeled and unlabeled examples, *J. Mach. Learn. Res.* **7**, 2399–2434 (2006). MR2274444.

[BD11] N. Bellomo and C. Dogbe, On the modeling of traffic and crowds: A survey of models, speculations, and perspectives, *SIAM Rev.* **53**, no. 3, 409–463 (2011), DOI: 10.1137/090746677. MR2834083.

[Ben05] G. Benettin, Time scale for energy equipartition in a two-dimensional FPU model, *Chaos* **15**, no. 1, 015108, 10 (2005), DOI: 10.1063/1.1854278. MR2133459.

[BP11] G. Benettin and A. Ponno, Time-scales to equipartition in the Fermi-Pasta-Ulam problem: Finite-size effects and thermodynamic limit, *J. Stat. Phys.* **144**, no. 4, 793–812 (2011), DOI: 10.1007/s10955-011-0277-9. MR2826618.

[BP21] G. Benettin and A. Ponno, Understanding the FPU state in FPU-like models, *Math. Eng.* **3**, no. 3, Paper No. 025, 22 (2021), DOI: 10.3934/mine.2021025. MR4146721.

[BG04] L. Berchialla and A. Giorgilli, Exponentially long times to equipartition in the thermodynamic limit, *Phys. Lett. A* **321**, no. 3, 167–172 (2004), DOI: 10.1016/j.physleta.2003.11.052.

[BN10] H. Berestycki and J.-P. Nadal, Self-organised critical hot spots of criminal activity, *Euro. J. Appl. Math.* **21**, no. 4–5, 371–399 (2010), DOI: 10.1017/S0956792510000185. MR2671615.

[BNR22] H. Berestycki, S. Nordmann, and L. Rossi, Modeling the propagation of riots, collective behaviors and epidemics, *Math. Eng.* **4**, no. 1, Paper No. 003, 53 (2022), DOI: 10.3934/mine.2022003. MR4231145.

[BPV86] P. Bergé, Y. Pomeau, and C. Vidal, *Order Within Chaos*, A Wiley-Interscience Publication, John Wiley & Sons, Inc., New York; Hermann, Paris, 1986. Towards a deterministic approach to turbulence; With a preface by David Ruelle; Translated from the French by Laurette Tuckerman. MR882723.

[BCAKCC06] L. M. A. Bettencourt, A. Cintrón-Arias, D. I. Kaiser, and C. Castillo-Chàvez, The power of a good idea: Quantitative modeling of the spread of ideas from epidemiological models, *Phys. A* **364**, no. C, 513–536 (2006).

[BT08]   J. Binney and S. Tremaine, Galactic dynamics, xvi + 885 (2008).

[Bir50]   G. Birkhoff, *Hydrodynamics. A Study in Logic, Fact, and Similitude*, Princeton University Press, Princeton, NJ, 1950. MR0038180.

[Bir60]   G. Birkhoff, *Hydrodynamics: A Study in Logic, Fact and Similitude*, Princeton University Press, Princeton, NJ, 1960. Revised ed. MR0122193.

[BR19]   P. Blanc and J. D. Rossi, *Game Theory and Partial Differential Equations*, De Gruyter Series in Nonlinear Analysis and Applications, vol. 31, De Gruyter, Berlin, 2019. MR4316246.

[Bly14]   S. Blyth, *An Introduction to Quantitative Finance*, Oxford University Press, Oxford, 2014. MR3115237.

[Bou77]   J. Boussinesq, Essai sur la théorie des eaux courantes, *Mém. prés. p. div. sav. de Paris* **XXIII**, 666–680 (1877) (French).

[BS03]   T. J. M. Boyd and J. J. Sanderson, *The Physics of Plasmas*, Cambridge University Press, Cambridge, 2003. MR1960956.

[BvdDW08] F. Brauer, P. van den Driessche, and J. Wu (eds.), *Mathematical Epidemiology*, Lect. Notes Math., vol. 1945, Springer, Berlin, 2008.

[BSF13]   N. J. Brown, A. D. Sokal, and H. L. Friedman, The complex dynamics of wishful thinking: The critical positivity ratio, *Am. Psychol.* **68**, no. 9, 801–813 (2013), DOI: 10.1037/a0032850.

[Bru16]   G. Bruzzaniti, *Enrico Fermi. The Obedient Genius. Translated from the Italian by Ugo Bruzzo*, Springer Biogr., Springer, New York, 2016.

[BV16]   C. Bucur and E. Valdinoci, *Nonlocal Diffusion and Applications*, Lecture Notes of the Unione Matematica Italiana, vol. 20, Springer, [Cham]; Unione Matematica Italiana, Bologna, 2016. MR3469920.

[Bun79]   L. A. Bunimovich, On the ergodic properties of nowhere dispersing billiards, *Comm. Math. Phys.* **65**, no. 3, 295–312 (1979). MR530154.

[CG97]   L. A. Caffarelli and C. E. Gutiérrez, Properties of the solutions of the linearized Monge-Ampère equation, *Am. J. Math.* **119**, no. 2, 423–465 (1997). MR1439555.

[Cal80]   A.-P. Calderón, On an inverse boundary value problem, Seminar on Numerical Analysis and its Applications to Continuum Physics (Rio de Janeiro, 1980), *Soc. Brasil. Mat., Rio de Janeiro*, 1980, pp. 65–73. MR590275.

[CH93]   R. Camassa and D. D. Holm, An integrable shallow water equation with peaked solitons, *Phys. Rev. Lett.* **71**, no. 11, 1661–1664 (1993), DOI: 10.1103/PhysRevLett.71.1661. MR1234453.

[CML07] B. Camenen and Magnus Larson, Predictive formulas for breaker depth index and breaker type, *J. Coast. Res.* **234**, 1028–1041 (2007), DOI: 10.2112/05-0566.1.

[CHJ16] B. Cao, S.-h. Han, and Z. Jin, Modeling of knowledge transmission by considering the level of forgetfulness in complex networks, *Phys. A* **451**, 277–287 (2016), DOI: 10.1016/j.physa.2015.12.137. MR3471879.

[CJW06] J. Carlson, A. Jaffe, and A. Wiles (eds.), *The Millennium Prize Problems*, Clay Mathematics Institute, Cambridge, MA; American Mathematical Society, Providence, RI, 2006. MR2246251.

[CMS90] R. Carmona, W. C. Masters, and B. Simon, Relativistic Schrödinger operators: Asymptotic behavior of the eigenfunctions, *J. Funct. Anal.* **91**, no. 1, 117–142 (1990), DOI: 10.1016/0022-1236(90)90049-Q. MR1054115.

[CS11] F. A. C. C. Chalub and M. O. Souza, The SIR epidemic model from a PDE point of view, *Math. Comput. Modelling* **53**, no. 7–8, 1568–1574 (2011), DOI: 10.1016/j.mcm.2010.05.036. MR2782833.

[CL23] H. A. Chang-Lara, El problema del Barrendero, *Mixba'al Rev. Metropol. Matem.* **14**, no. 1 (2023), DOI: 10.24275/uami/dcbi/mix/v14n1/hchang.

[Che70] J. Cheeger, *A Lower Bound for the Smallest Eigenvalue of the Laplacian*, Problems in analysis (Papers dedicated to Salomon Bochner, 1969), Princeton University Press, Princeton, NJ, 1970, pp. 195–199. MR0402831.

[CLZ$^+$11] X. Chen, Y. Luo, J. Zhang, K. Jiang, J. B. Pendry, and S. Zhang, Macroscopic invisibility cloaking of visible light, *Nat. Comm.* **2**, 1–6 (2011), DOI: 10.1038/ncomms1176.

[CP06] Z.-M. Chen and W. G. Price, On the relation between Rayleigh-Bénard convection and Lorenz system, *Chaos Solitons Fractals* **28**, no. 2, 571–578 (2006), DOI: 10.1016/j.chaos.2005.08.010. MR2178136.

[CM93] A. J. Chorin and J. E. Marsden, *A Mathematical Introduction to Fluid Mechanics*, 3rd ed., Texts in Applied Mathematics, vol. 4, Springer-Verlag, New York, 1993. MR1218879.

[Chu97] F. R. K. Chung, *Spectral Graph Theory*, CBMS Regional Conference Series in Mathematics, vol. 92, Published for the Conference Board of the Mathematical Sciences, Washington, DC; by the American Mathematical Society, Providence, RI, 1997. MR1421568.

[Chu10] F. Chung, *Four Proofs for the Cheeger Inequality and Graph Partition Algorithms*, Fourth International Congress of Chinese

Mathematicians, AMS/IP Stud. Adv. Math., vol. 48, American Mathematical Society, Providence, RI, 2010, pp. 331–349. MR2744229.

[CBV74] R. V. Churchill, J. W. Brown, and R. F. Verhey, *Complex Variables and Applications*, 3rd ed., McGraw-Hill Book Co., New York-Düsseldorf-Johannesburg, 1974. MR0348082.

[CMS12] D. Ciregan, U. Meier, and J. Schmidhuber, *Multi-column Deep Neural Networks for Image Classification*, 2012 IEEE Conference on Computer Vision and Pattern Recognition, 2012, pp. 3642–3649, DOI: 10.1109/CVPR.2012.6248110.

[CKS+01] J. Colliander, M. Keel, G. Staffilani, H. Takaoka, and T. Tao, Global well-posedness for KdV in Sobolev spaces of negative index, *Electron. J. Differ. Eq.*, No. 26, 7 (2001). MR1824796.

[CKS+03] J. Colliander, M. Keel, G. Staffilani, H. Takaoka, and T. Tao, Sharp global well-posedness for KdV and modified KdV on $\mathbb{R}$ and $\mathbb{T}$, *J. Am. Math. Soc.* **16**, no. 3, 705–749 (2003), DOI: 10.1090/S0894-0347-03-00421-1. MR1969209.

[Con93] L. Conlon, *Differentiable Manifolds: A First Course*, Birkhäuser Advanced Texts: Basler Lehrbücher. [Birkhäuser Advanced Texts: Basel Textbooks], Birkhäuser Boston, Inc., Boston, MA, 1993. MR1209437.

[Cor12] I. Corwin, The Kardar-Parisi-Zhang equation and universality class, *Random Matrices Theory Appl.* **1**, no. 1, 1130001, 76 (2012), DOI: 10.1142/S2010326311300014. MR2930377.

[Cse00] J. Cserti, Application of the lattice Green's function for calculating the resistance of an infinite network of resistors, *Am. J. Phys.* **68**, 896–906 (2000), DOI: 10.1119/1.1285881.

[CAC+22] M. Cucchi, S. Abreu, G. Ciccone, D. Brunner, and H. Kleemann, Hands-on reservoir computing: A tutorial for practical implementation, *Neuromorph. Comput. Eng.* **2**, 1–32 (2022), DOI: 10.1088/2634-4386/ac7db7.

[Cyb89] G. Cybenko, Approximation by superpositions of a sigmoidal function, *Math. Control Signals Sys.* **2**, no. 4, 303–314 (1989), DOI: 10.1007/BF02551274. MR1015670.

[Cyb92] G. Cybenko, Correction: "Approximation by superpositions of a sigmoidal function" [*Math. Control Signals Syst.* **2** (1989), no. 4, 303–314; MR1015670 (90m:41033)], *Math. Control Signals Syst.* **5**, no. 4, 455 (1992), DOI: 10.1007/BF02134016. MR1178852.

[DdPDV15] J. Dávila, M. del Pino, S. Dipierro, and E. Valdinoci, Concentration phenomena for the nonlocal Schrödinger equation with Dirichlet datum, *Anal. PDE* **8**, no. 5, 1165–1235 (2015), DOI: 10.2140/apde.2015.8.1165. MR3393677.

[DGF75] E. De Giorgi and T. Franzoni, Su un tipo di convergenza variazionale, *Atti Accad. Naz. Lincei Rend. Cl. Sci. Fis. Mat. Nat. (8)* **58**, no. 6, 842–850 (1975) (Italian). MR448194.

[DS06] F. Delbaen and W. Schachermayer, *The Mathematics of Arbitrage*, Springer Finance, Springer-Verlag, Berlin, 2006. MR2200584.

[Dem97] J. W. Demmel, *Applied Numerical Linear Algebra*, SIAM, Philadelphia, PA, 1997.

[DPF14] G. De Philippis and A. Figalli, The Monge-Ampère equation and its link to optimal transportation, *Bull. Am. Math. Soc. (N.S.)* **51**, no. 4, 527–580 (2014), DOI: 10.1090/S0273-0979-2014-01459-4. MR3237759.

[DiB93] E. DiBenedetto, *Degenerate Parabolic Equations*, Universitext, Springer-Verlag, New York, 1993. MR1230384.

[DV24] S. Dipierro and E. Valdinoci, *Elliptic Partial Differential Equations From an Elementary Viewpoint—A Fresh Glance at the Classical Theory*, World Scientific Publishing Co. Pte. Ltd., Hackensack, NJ, 2024. MR4784613.

[DS84] P. G. Doyle and J. L. Snell, *Random Walks and Electric Networks*, Carus Mathematical Monographs, vol. 22, Mathematical Association of America, Washington, DC, 1984. MR920811.

[DJ89] P. G. Drazin and R. S. Johnson, *Solitons: An Introduction*, Cambridge Texts in Applied Mathematics, Cambridge University Press, Cambridge, 1989. MR985322.

[Edw01] H. M. Edwards, *Riemann's Zeta Function*, Dover Publications, Inc., Mineola, NY, 2001. Reprint of the 1974 original [Academic Press, New York; MR0466039 (57 #5922)]. MR1854455.

[Eft06] S. Efthimiades, Physical meaning and derivation of Schrödinger and Dirac equations, *arXiv e-prints* (2006).

[Ein05] A. Einstein, Über die von der molekularkinetischen Theorie der Wärme geforderte Bewegung von in ruhenden Flüssigkeiten suspendierten Teilchen, *Ann. der Phys.* (4) **17**, 549–560 (1905), DOI: 10.1002/andp.19053220806 (German).

[vEH20] J. E. van Engelen and H. H. Hoos, A survey on semi-supervised learning, *Mach. Learn.* **109**, no. 2, 373–440 (2020), DOI: 10.1007/s10994-019-05855-6. MR4070203.

[Eva98] L. C. Evans, *Partial Differential Equations*, Graduate Studies in Mathematics, vol. 19, American Mathematical Society, Providence, RI, 1998. MR1625845.

[Eva13] L. C. Evans, *An Introduction to Stochastic Differential Equations*, American Mathematical Society, Providence, RI, 2013. MR3154922.

[EG15] L. C. Evans and R. F. Gariepy, *Measure Theory and Fine Properties of Functions*, Revised edition, Textbooks in Mathematics, CRC Press, Boca Raton, FL, 2015. MR3409135.

[ES91] L. C. Evans and J. Spruck, Motion of level sets by mean curvature. I, *J. Differential Geom.* **33**, no. 3, 635–681 (1991). MR1100206.

[Fad16] L. Faddeev, *Fifty Years of Mathematical Physics*, World Scientific Series in 21st Century Mathematics, vol. 2, World Scientific Publishing Co. Pte. Ltd., Hackensack, NJ, 2016. Selected works of Ludwig Faddeev; Edited by Molin Ge and Antti J. Niemi. MR3470076.

[Fef06] C. L. Fefferman, *Existence and Smoothness of the Navier-Stokes Equation*, The millennium prize problems, Clay Math. Inst., Cambridge, MA, 2006, pp. 57–67. MR2238274.

[Fer22] E. Fermi, Sopra i fenomeni che avvengono in vicinanza di una linea oraria, *Atti Accad. Naz. Lincei Rend. Cl. Sci. Fis. Mat. Nat.* **31**, 21–23 (1922).

[Fer23] E. Fermi, Dimostrazione che in generale un sistema meccanico normale è quasi-ergodico, *Nuovo Cimento* **25**, 267–269 (1923).

[FPU55] E. Fermi, J. Pasta, and S. Ulam, *Studies of Nonlinear Problems*, 978–988 (1955).

[FRRO21] X. Fernández-Real and X. Ros-Oton, Regularity Theory for Elliptic PDE, Preprint, 2021.

[LU89] J. M. Lee and G. Uhlmann, Determining anisotropic real-analytic conductivities by boundary measurements, *Comm. Pure Appl. Math.* **42**, no. 8, 1097–1112 (1989), DOI: 10.1002/cpa.3160420804. MR1029119.

[Leo06] U. Leonhardt, Optical conformal mapping, *Science* **312**, no. 5781, 1777–1780 (2006), DOI: 10.1126/science.1126493.

[FLS65] R. P. Feynman, R. B. Leighton, and M. Sands, *The Feynman Lectures on Physics. Vol. 3: Quantum Mechanics*, Addison-Wesley Publishing Co., Inc., Reading, MA-London, 1965. MR0213079.

[Fig17] A. Figalli, *The Monge-Ampère Equation and its Applications*, Zurich Lectures in Advanced Mathematics, European Mathematical Society (EMS), Zürich, 2017. MR3617963.

[Fir74] W. J. Firey, Shapes of worn stones, *Mathematika* **21**, 1–11 (1974), DOI: 10.1112/S0025579300005714. MR362045.

[Fis37a] R. A. Fisher, The wave of advance of advantageous genes, *Ann. Eugenics* **7** (1937).

[Fis37b] R. A. Fisher, *The Design of Experiments*, Oliver and Boyd, ix+252 (1937).

[FNT06] J.-P. Françoise, G. L. Naber, and T. S. Tsun (eds.), *Encyclopedia of Mathematical Physics. Vol. 5*, Academic Press/Elsevier Science, Oxford, 2006. MR2238867.

[Fri71] A. Friedman, *Differential Games*, Pure and Applied Mathematics, Vol. XXV, Wiley-Interscience [A division of John Wiley & Sons, Inc.], New York-London, 1971. MR0421700.

[FP99] G. Friesecke and R. L. Pego, Solitary waves on FPU lattices. I. Qualitative properties, renormalization and continuum limit, *Nonlinearity* **12**, no. 6, 1601–1627 (1999), DOI: 10.1088/0951-7715/12/6/311. MR1726667.

[FP02] G. Friesecke and R. L. Pego, Solitary waves on FPU lattices. II. Linear implies nonlinear stability, *Nonlinearity* **15**, no. 4, 1343–1359 (2002), DOI: 10.1088/0951-7715/15/4/317. MR1912298.

[FP04a] G. Friesecke and R. L. Pego, Solitary waves on Fermi-Pasta-Ulam lattices. III. Howland-type Floquet theory, *Nonlinearity* **17**, no. 1, 207–227 (2004), DOI: 10.1088/0951-7715/17/1/013. MR2023440.

[FP04b] G. Friesecke and R. L. Pego, Solitary waves on Fermi-Pasta-Ulam lattices. IV. Proof of stability at low energy, *Nonlinearity* **17**, no. 1, 229–251 (2004), DOI: 10.1088/0951-7715/17/1/014. MR2023441.

[FB02] G. R. Fulford and P. Broadbridge, *Industrial Mathematics*, Australian Mathematical Society Lecture Series, vol. 16, Cambridge University Press, Cambridge, 2002. Case studies in the diffusion of heat and matter. MR2007714.

[Fun89] K.-I. Funahashi, On the approximate realization of continuous mappings by neural networks, *Neural Net.* **2**, no. 3, 183–192 (1989), DOI: 10.1016/0893-6080(89)90003-8.

[GCPL09] L. H. Gabrielli, J. Cardenas, C. B. Poitras, and M. Lipson, Silicon nanostructure cloak operating at optical frequencies, *Nat. Photon* **3**, no. 8, 461–463 (2009), DOI: 10.1038/nphoton.2009.117.

[Gal08] G. Gallavotti (ed.), *The Fermi-Pasta-Ulam Problem*, Lecture Notes in Physics, vol. 728, Springer, Berlin, 2008. A status report. MR2402016.

[GPR21] M. Gallone, A. Ponno, and B. Rink, Korteweg–de Vries and Fermi-Pasta-Ulam-Tsingou: Asymptotic integrability of quasi unidirectional waves, *J. Phys. A* **54**, no. 30, Paper No. 305701, 29 (2021), DOI: 10.1088/1751-8121/ac0a2e. MR4294860.

[GGMV92] L. Galgani, A. Giorgilli, A. Martinoli, and S. Vanzini, On the problem of energy equipartition for large systems of the Fermi-Pasta-Ulam type: Analytical and numerical estimates, *Phys. D* **59**, no. 4, 334–348 (1992), DOI: 10.1016/0167-2789(92)90074-W. MR1192748.

[GP06]   M. Garavello and B. Piccoli, *Traffic Flow on Networks*, AIMS Series on Applied Mathematics, vol. 1, American Institute of Mathematical Sciences (AIMS), Springfield, MO, 2006. Conservation laws models. MR2328174.

[Gar09]   C. Gardiner, *Stochastic Methods*, 4th ed., Springer Series in Synergetics, Springer-Verlag, Berlin, 2009. A handbook for the natural and social sciences. MR2676235.

[GBGB21]   D. J. Gauthier, E. Bollt, A. Griffith, and W. A. S. Barbosa, Next generation reservoir computing, *Nature Comm.* **12**, 1–8 (2021), DOI: 10.1038/s41467-021-25801-2.

[GJS23]   F. Gazzola, M. Jleli, and B. Samet, A new detailed explanation of the Tacoma collapse and some optimization problems to improve the stability of suspension bridges, *Math. Eng.* **5**, no. 2, Paper No. 045, 35 (2023), DOI: 10.3934/mine.2023045. MR4468373.

[Get98]   A. V. Getling, *Rayleigh-Bénard Convection*, Advanced Series in Nonlinear Dynamics, vol. 11, World Scientific Publishing Co., Inc., River Edge, NJ, 1998. Structures and dynamics. MR1629996.

[GQ18]   A. Geyer and R. Quirchmayr, Shallow water equations for equatorial tsunami waves, *Philos. Trans. Roy. Soc. A* **376**, no. 2111, 20170100, 12 (2018), DOI: 10.1098/rsta.2017.0100. MR3744219.

[GM73]   A. Gierer and H. Meinhardt, A theory of biological pattern formation, *Kybernetik* **12**, 30–39 (1973).

[GK04]   B. H. Gilding and R. Kersner, *Travelling Waves in Nonlinear Diffusion-convection Reaction*, Progress in Nonlinear Differential Equations and their Applications, vol. 60, Birkhäuser Verlag, Basel, 2004. MR2081104.

[God10]   Y. Goda, Reanalysis of regular and random breaking wave statistics, *Coastal Eng. J.* **1**, no. 52, 71–106 (2010).

[GBC16]   I. Goodfellow, Y. Bengio, and A. Courville, *Deep Learning*, Adaptive Computation and Machine Learning, MIT Press, Cambridge, MA, 2016. MR3617773.

[GK09]   L. Graham and J.-M. Kantor, *Naming Infinity*, The Belknap Press of Harvard University Press, Cambridge, MA, 2009. A true story of religious mysticism and mathematical creativity. MR2526973.

[GKLU07]   A. Greenleaf, Y. Kurylev, M. Lassas, and G. Uhlmann, Full-wave invisibility of active devices at all frequencies, *Comm. Math. Phys.* **275**, no. 3, 749–789 (2007), DOI: 10.1007/s00220-007-0311-6. MR2336363.

[GKLU09] A. Greenleaf, Y. Kurylev, M. Lassas, and G. Uhlmann, Cloaking devices, electromagnetic wormholes, and transformation optics, *SIAM Rev.* **51**, no. 1, 3–33 (2009), DOI: 10.1137/080716827. MR2481110.

[GLU03] A. Greenleaf, M. Lassas, and G. Uhlmann, On nonuniqueness for Calderón's inverse problem, *Math. Res. Lett.* **10**, no. 5–6, 685–693 (2003), DOI: 10.4310/MRL.2003.v10.n5.a11. MR2024725.

[Gri59] N. T. Gridgeman, The lady tasting tea and allied topics, *J. Am. Stat. Assoc.* **54**, 776–783 (1959), DOI: 10.2307/2282501.

[GY16] R. Grimshaw and C. Yuan, Depression and elevation tsunami waves in the framework of the Korteweg-de Vries equation, *Nat. Hazards* **84**, no. 2, 493–511 (2016), DOI: 10.1007/s11069-016-2479-6.

[Gri13] P. Grinfeld, *Introduction to Tensor Analysis and the Calculus of Moving Surfaces*, Springer, New York, 2013. MR3136419.

[GVMO11] F. Guevara Vasquez, G. W. Milton, and D. Onofrei, Exterior cloaking with active sources in two dimensional acoustics, *Wave Motion* **48**, no. 6, 515–524 (2011), DOI: 10.1016/j.wavemoti.2011.03.005. MR2811880.

[Gut01] C. E. Gutiérrez, *The Monge-Ampère Equation*, Progress in Nonlinear Differential Equations and their Applications, vol. 44, Birkhäuser Boston, Inc., Boston, MA, 2001. MR1829162.

[Han19] B. Hanin, Universal function approximation by deep neural nets with bounded width and ReLU activations, *Mathematics* **7**, no. 10 (2019), DOI: 10.3390/math7100992.

[HKM06] J. Heinonen, T. Kilpeläinen, and O. Martio, *Nonlinear Potential Theory of Degenerate Elliptic Equations*, Dover Publications, Inc., Mineola, NY, 2006. Unabridged republication of the 1993 original. MR2305115.

[HW78] R. P. Henderson and J. G. Webster, An impedance camera for spatially specific measurements of the thorax, *IEEE Trans. Biomed. Eng.* **BME-25**, no. 3, 250–254 (1978), DOI: 10.1109/TBME.1978.326329.

[HK08] A. Henrici and T. Kappeler, Results on normal forms for FPU chains, *Comm. Math. Phys.* **278**, no. 1, 145–177 (2008), DOI: 10.1007/s00220-007-0387-z. MR2367202.

[HI11] R. B. Hetnarski and J. Ignaczak, *The Mathematical Theory of Elasticity*, 2nd ed., CRC Press, Boca Raton, FL, 2011. MR2779440.

[Hil94] R. C. Hilborn, *Chaos and Nonlinear Dynamics*, The Clarendon Press, Oxford University Press, New York, 1994. An introduction for scientists and engineers. MR1263025.

[HT96]   S. Hildebrandt and A. Tromba, *The Parsimonious Universe*, Copernicus, New York, 1996. Shape and form in the natural world. MR1398883.

[HP09]   T. Hillen and K. J. Painter, A user's guide to PDE models for chemotaxis, *J. Math. Biol.* **58**, no. 1–2, 183–217 (2009), DOI: 10.1007/s00285-008-0201-3. MR2448428.

[HSD13] M. W. Hirsch, S. Smale, and R. L. Devaney, *Differential Equations, Dynamical Systems, and an Introduction to Chaos*, 3rd ed., Elsevier/Academic Press, Amsterdam, 2013. MR3293130.

[HJJ16] J. Hoffman, J. Jansson, and C. Johnson, New theory of flight, *J. Math. Fluid Mech.* **18**, no. 2, 219–241 (2016), DOI: 10.1007/s00021-015-0220-y. MR3503185.

[HJ09]   J. Hoffman and C. Johnson, The mathematical secret of flight, *Normat* **57**, no. 4, 148–169, 192 (2009). MR2650049.

[HW04]   B. Hopkins and R. J. Wilson, The truth about Königsberg, *College Math. J.* **35**, no. 3, 198–207 (2004), DOI: 10.2307/4146895. MR2053120.

[HSW89] K. Hornik, M. Stinchcombe, and H. White, Multilayer feedforward networks are universal approximators, *Neural Netw.* **2**, no. 5, 359–366 (1989), DOI: 10.1016/0893-6080(89)90020-8.

[Hor91] K. Hornik, Approximation capabilities of multilayer feedforward networks, *Neural Netw.* **4**, no. 2, 251–257 (1991), DOI: 10.1016/0893-6080(91)90009-T.

[Hor93] K. Hornik, Some new results on neural network approximation, *Neural Netw.* **6**, no. 8, 1069–1072 (1993), DOI: 10.1016/S0893-6080(09)80018-X.

[HS16]   L. Huo and N. Song, Dynamical interplay between the dissemination of scientific knowledge and rumor spreading in emergency, *Phys. A* **461**, 73–84 (2016), DOI: 10.1016/j.physa.2016.05.028. MR3519855.

[Hur17] V. M. Hur, Wave breaking in the Whitham equation, *Adv. Math.* **317**, 410–437 (2017), DOI: 10.1016/j.aim.2017.07.006. MR3682673.

[HT14]   V. M. Hur and L. Tao, Wave breaking for the Whitham equation with fractional dispersion, *Nonlinearity* **27**, no. 12, 2937–2949 (2014), DOI: 10.1088/0951-7715/27/12/2937. MR3291137.

[HT18]   V. M. Hur and L. Tao, Wave breaking in a shallow water model, *SIAM J. Math. Anal.* **50**, no. 1, 354–380 (2018), DOI: 10.1137/15M1053281. MR3749383.

[Ioa84] H. Ioannidou, Explicit derivation of the relativistic Schrödinger equation, *Nuovo Cimento B* (11) **79**, no. 1, 67–75 (1984), DOI: 10.1007/BF02723838 (English, with Italian and Russian summaries). MR734851.

[IK54] A. T. Ippen and G. Kulin, The shoaling and breaking of the solitary wave, *Coast. Eng. Proc.* **1**, no. 5, 27–49 (1954), DOI: 10.9753/icce.v5.4.

[Isa65] R. Isaacs, *Differential Games. A Mathematical Theory with Applications to Warfare and Pursuit, Control and Optimization*, John Wiley & Sons, Inc., New York-London-Sydney, 1965. MR0210469.

[Isr10] S. Israwi, Variable depth KdV equations and generalizations to more nonlinear regimes, *M2AN Math. Model. Numer. Anal.* **44**, no. 2, 347–370 (2010), DOI: 10.1051/m2an/2010005. MR2655953.

[IC66] F. M. Izraĭlev and B. V. Chirikov, Statistical properties of a nonlinear string, *Sov. Phys., Dokl.* **11**, 30–32 (1966).

[Jar10] S. Jardin, *Computational Methods in Plasma Physics*, Chapman & Hall/CRC Computational Science Series, CRC Press, Boca Raton, FL, 2010. MR2779346.

[JHRW16] R. A. Jeffs, J. Hayward, P. A. Roach, and J. Wyburn, Activist model of political party growth, *Phys. A* **442**, 359–372 (2016), DOI: 10.1016/j.physa.2015.09.002. MR3412972.

[Joh97] R. S. Johnson, *A Modern Introduction to the Mathematical Theory of Water Waves*, Cambridge Texts in Applied Mathematics, Cambridge University Press, Cambridge, 1997. MR1629555.

[JPS10] D. S. Jones, M. J. Plank, and B. D. Sleeman, *Differential Equations and Mathematical Biology*, Chapman & Hall/CRC Mathematical and Computational Biology Series, CRC Press, Boca Raton, FL, 2010. Second edition [of MR1967145]. MR2573923.

[KN86] T. Kano and T. Nishida, A mathematical justification for Korteweg-de Vries equation and Boussinesq equation of water surface waves, *Osaka J. Math.* **23**, no. 2, 389–413 (1986). MR856894.

[Kap15] T. Kapitula, *Ordinary Differential Equations and Linear Algebra*, Society for Industrial and Applied Mathematics, Philadelphia, PA, 2015. A systems approach. MR3450069.

[KP03] T. Kappeler and J. Pöschel, *KdV & KAM*, Ergebnisse der Mathematik und ihrer Grenzgebiete. 3. Folge. A Series of Modern Surveys in Mathematics [Results in Mathematics and Related Areas. 3rd Series. A Series of Modern Surveys in Mathematics], vol. 45, Springer-Verlag, Berlin, 2003. MR1997070.

[KPZ86] M. Kardar, G. Parisi, and Y.-C. Zhang, Dynamic scaling of growing interfaces, *Phys. Rev. Lett.* **56**, no. 9, 889–892 (1986), DOI: 10.1103/PhysRevLett.56.889.

[Kat72] T. Kato, Nonstationary flows of viscous and ideal fluids in $\mathbf{R}^3$, *J. Func. Anal.* **9**, 296–305 (1972), DOI: 10.1016/0022-1236(72)90003-1. MR0481652.

[Kaz85] J. L. Kazdan, *Prescribing the Curvature of a Riemannian Manifold*, CBMS Regional Conference Series in Mathematics, vol. 57, Published for the Conference Board of the Mathematical Sciences, Washington, DC; by the American Mathematical Society, Providence, RI, 1985. MR787227.

[KS70] E. F. Keller and L. A. Segel, Initiation of slime mold aggregation viewed as an instability, *J. Theoret. Biol.* **26**, no. 3, 399–415 (1970), DOI: 10.1016/0022-5193(70)90092-5. MR3925816.

[KL20] P. Kidger and T. Lyons, *Universal Approximation with Deep Narrow Networks*, Proceedings of Thirty Third Conference on Learning Theory, 2020, pp. 2306–2327.

[Kin08] W. D. King, The physics of a stove-top espresso machine, *Am. J. Phys.* **76**, no. 6, 558–565 (2008), DOI: 10.1119/1.2870524.

[Kir76] G. Kirchhoff, *Vorlesungen über mathematische Physik, Mechanik. 1te und 2te Lieferung*, Teubner, Leipzig, 1876 (German).

[KT14] D. Kisacik and P. Troch, The influence of an existing vertical structure on the inception of wave breaking point, *Coastal Eng. Proc.* **1**, no. 34, 1–11 (2014), DOI: 10.9753/icce.v34.structures.54.

[KL03] M. Kleman and O. D. Lavrentovich, *Soft Matter Physics. An Introduction*, Partially Ordered Systems, Springer-Verlag, New York, 2003.

[Kol91] A. N. Kolmogorov, *Selected Works of A. N. Kolmogorov. Vol. I*, Mathematics and its Applications (Soviet Series), vol. 25, Kluwer Academic Publishers Group, Dordrecht, 1991. Mathematics and mechanics; With commentaries by V. I. Arnol'd, V. A. Skvortsov, P. L. Ul'yanov *et al.*; Translated from the Russian original by V. M. Volosov; Edited and with a preface, foreword and brief biography by V. M. Tikhomirov. MR1175399.

[Kon17] S. Kondo, An updated kernel-based Turing model for studying the mechanisms of biological pattern formation, *J. Theoret. Biol.* **414**, 120–127 (2017), DOI: 10.1016/j.jtbi.2016.11.003. MR3614990.

[KA95] S. Kondo and R. Asai, A reaction-diffusion wave on the skin of the marine angelfish Pomacanthus, *Nature* **376**, 765–768 (1995), DOI: 10.1038/376765a0.

[KWM21] S. Kondo, M. Watanabe, and S. Miyazawa, Studies of Turing pattern formation in zebrafish skin, *Philos. Trans. Roy. Soc. A* **379**, no. 2213, 20200274 (2021), DOI: 10.1098/rsta.2020.0274.

[KdV95] D. J. Korteweg and G. de Vries, On the change of form of long waves advancing in a rectangular canal, and on a new type of long stationary waves, *Philos. Mag.* (5) **39**, no. 240, 422–443 (1895), DOI: 10.1080/14786449508620739. MR3363408.

[KFJ08] C. Kottke, A. Farjadpour, and S. G. Johnson, Perturbation theory for anisotropic dielectric interfaces, and application to subpixel smoothing of discretized numerical methods, *Phys. Rev. E* **77**, 036611-1–10 (2008), DOI: 10.1103/PhysRevE.77.036611.

[Kra40] H. A. Kramers, Brownian motion in a field of force and the diffusion model of chemical reactions, *Physica* **7**, 284–304 (1940). MR2962.

[KFW⁺14] J. Krauss, H. G. Frohnhöfer, B. Walderich, H.-M. Maischein, C. Weiler, U. Irion, and C. Nüsslein-Volhard, Endothelin signalling in iridophore development and stripe pattern formation of zebrafish, *Biol. Open* **3**, no. 6, 503–509 (2014), DOI: 10.1242/bio.20148441.

[Kru74] M. D. Kruskal, The Korteweg-de Vries equation and related evolution equations, Nonlinear wave motion (Proc. AMS-SIAM Summer Sem., Clarkson Coll. Tech., Potsdam, N.Y., 1972), American Mathematical Society, Providence, RI, 1974, pp. 61–83. Lectures in Appl. Math., Vol. 15. MR0352741.

[LCZC09] Y. Lai, H. Chen, Z.-Q. Zhang, and C. T. Chan, Complementary media invisibility cloak that cloaks objects at a distance outside the cloaking shell, *Phys. Rev. Lett.* **102**, 093901, 4 (2009), DOI: 10.1103/PhysRevLett.102.093901.

[LL60] L. D. Landau and E. M. Lifshitz, *Mechanics*, Course of Theoretical Physics, Vol. 1, Pergamon Press, Oxford-London-New York-Paris; Addison-Wesley Publishing Co., Inc., Reading, Mass., 1960. Translated from the Russian by J. B. Bell. MR0120782.

[LS13] N. Landy and D. R. Smith, A full-parameter unidirectional metamaterial cloak for microwaves, *Nat. Mater.* **12**, 25–28 (2013), DOI: 10.1038/nmat3476.

[Law10] G. F. Lawler, *Random Walk and the Heat Equation*, Student Mathematical Library, vol. 55, American Mathematical Society, Providence, RI, 2010. MR2732325.

[LC21] K.-H. Lee and Y.-H. Cho, Simple breaker index formula using linear model, *J. Mar. Sci. Eng.* **9**, no. 7, 1–17 (2021), DOI: 10.3390/jmse9070731.

[Lem92] C. M. Lemos, *Wave Breaking. A Numerical Study*, Lect. Notes Eng., vol. 71, Springer-Verlag, Berlin, 1992.

[LLPS93] M. Leshno, V. Ya. Lin, A. Pinkus, and S. Schocken, Multilayer feedforward networks with a nonpolynomial activation function can approximate any function, *Neural Netw.* **6**, no. 6, 861–867 (1993), DOI: 10.1016/S0893-6080(05)80131-5.

[Lin08] P. Lin, *Numerical Modeling of Water Waves*, Taylor & Francis, New York, 2008.

[LL09] S. Lipschutz and M. Lipson, *Schaum's Outline of Linear Algebra*, 4th ed., McGraw-Hill Professional, Sydney, 2009.

[LJM+09] R. Liu, C. Ji, J. J. Mock, J. Y. Chin, T. J. Cui, and D. R. Smith, Broadband ground-plane cloak, *Science* **323**, no. 5912, 366–369 (2009), DOI: 10.1126/science.1166949.

[LKN+22] Z. Liu, O. Kitouni, N. Nolte, E. J. Michaud, M. Tegmark, and M. Williams, Towards understanding grokking: An effective theory of representation learning, *arXiv e-prints* (2022).

[Lor63] E. N. Lorenz, Deterministic Nonperiodic Flow, *J. Atmosph. Sci.* **20**, no. 2, 130–141 (1963), DOI: 10.1175/1520-0469(1963)020¡0130:DNF¿2.0.CO;2.

[LPW+17] Z. Lu, H. Pu, F. Wang, Z. Hu, and L. Wang, The expressive power of neural networks: A view from the width, *31st Conference on Neural Information Processing Systems (NIPS)*, Long Beach, CA, 2017, pp. 1–9.

[HMSQZXJW09] Hua Ma, Shaobo Qu, Zhuo Xu, and Jiafu Wang, The open cloak, *Appl. Phys. Lett.* **94**, 103501, 3 (2009), DOI: 10.1063/1.3095436.

[MS06] N. A. Magnitskii and S. V. Sidorov, *New Methods for Chaotic Dynamics*, World Scientific Series on Nonlinear Science. Series A: Monographs and Treatises, vol. 58, World Scientific Publishing Co. Pte. Ltd., Hackensack, NJ, 2006. MR2310642.

[Mar07] P. A. Markowich, *Applied Partial Differential Equations: A Visual Approach*, Springer, Berlin, 2007. With 1 CD-ROM (Windows, Macintosh and UNIX). MR2309862.

[IYK12] M. Inaba, H. Yamanaka, and S. Kondo, Pigment pattern formation by contact-dependent depolarization, *Science* **335**, no. 6069, 677–677 (2012), DOI: 10.1126/science.1212821.

[Mat07] L. E. Matsona, The Malkus-Lorenz water wheel revisited, *Am. J. Phys.* **75**, 1114–1122 (2007), DOI: 10.1119/1.2785209.

[McC91] J. McCowan, On the solitary wave, *London Edinburgh Dublin Phil. Magaz. J. Sci.* **32**, no. 194, 45–58 (1891), DOI: 10.1080/14786449108621390.

[JM94] J. McCowan, On the highest wave of permanent type, *London Edinburgh Dublin Phil. Magaz. J. Sci.* **38**, no. 233, 351–358 (1894), DOI: 10.1080/14786449408620643.

[Mil81] J. W. Miles, The Korteweg-de Vries equation: a historical essay, *J. Fluid Mech.* **106**, 131–147 (1981), DOI: 10.1017/S0022112081001559.

[MT60] L. M. Milne-Thomson, *Theoretical Hydrodynamics*, The Macmillan Co., New York, 1960. 4th ed. MR0112435.

[Min10] H. Q. Minh, Some properties of Gaussian reproducing kernel Hilbert spaces and their implications for function approximation and learning theory, *Constr. Approx.* **32**, no. 2, 307–338 (2010), DOI: 10.1007/s00365-009-9080-0. MR2677883.

[MS09] M. C. Moh and S. Shen, The roles of cell adhesion molecules in tumor suppression and cell migration: A new paradox, *Cell Adhesion Migr.* **3–4**, 334–336 (2009), DOI: 10.4161/cam.3.4.9246.

[MRT18] M. Mohri, A. Rostamizadeh, and A. Talwalkar, *Foundations of Machine Learning*, Adaptive Computation and Machine Learning, MIT Press, Cambridge, MA, 2018. MR3931734.

[MRSY19] A. Montanari, F. Ruan, Y. Sohn, and J. Yan, The generalization error of max-margin linear classifiers: Benign overfitting and high dimensional asymptotics in the overparametrized regime, *arXiv e-prints* (2019).

[Moy49] J. E. Moyal, Stochastic processes and statistical physics, *J. Roy. Statist. Soc. Ser. B* **11**, 150–210 (1949). MR34975.

[Mun49] W. H. Munk, The Solitary wave theory and its application to surf problems, *Ann. New York Acad. Sci.* **51**, no. 3, 376–424 (1949).

[Mur88] J. D. Murray, How the Leopard Gets its Spots, *Sci. Am.* **258**, no. 3, 80–87 (1988), DOI: 10.1038/scientificamerican0388-80.

[Mur02] J. D. Murray, *Mathematical Biology. I*, 3rd ed., Interdisciplinary Applied Mathematics, vol. 17, Springer-Verlag, New York, 2002. An introduction. MR1908418.

[Mur03] J. D. Murray, *Mathematical Biology. II*, 3rd ed., Interdisciplinary Applied Mathematics, vol. 18, Springer-Verlag, New York, 2003. Spatial models and biomedical applications. MR1952568.

[Nah88] P. J. Nahin, *Oliver Heaviside: Sage in Solitude*, IEEE Press, New York, 1988. The life, work, and times of an electrical genius of the Victorian age. MR939170.

[Nah09] P. J. Nahin, *Mrs. Perkins's Electric Quilt and Other Intriguing Stories of Mathematical Physics*, Princeton University Press, Princeton, NJ, 2009. MR2535945.

[NE92] P. A. Nelson and S. J. Elliott, *Active Control of Sound*, Academic Press, Elsevier Science, New York, 1992.

[New06] M. E. J. Newman, Finding community structure in networks using the eigenvectors of matrices, *Phys. Rev. E (3)* **74**, no. 3, 036104, 19 (2006), DOI: 10.1103/PhysRevE.74.036104. MR2282139.

[NG13] B.-T. Nguyen and D. S. Grebenkov, Localization of Laplacian eigenfunctions in circular, spherical, and elliptical domains, *SIAM J. Appl. Math.* **73**, no. 2, 780–803 (2013), DOI: 10.1137/120869857. MR3038118.

[Nic18] B. Nica, *A Brief Introduction to Spectral Graph Theory*, EMS Textbooks in Mathematics, European Mathematical Society (EMS), Zürich, 2018. MR3821579.

[Öch21] A. Öchsner, *Classical Beam Theories of Structural Mechanics*, Springer, Cham, 2021. MR4298495.

[OPLH04] T. Okegawa, R.-C. Pong, Y. Li, and J.-T. Hsieh, The role of cell adhesion molecule in cancer progression and its application in cancer therapy, *Acta Bioch. Pol.* **51**, no. 2, 445–457 (2004), DOI: 10.18388/abp.2004_3583.

[Pal97] R. S. Palais, The symmetries of solitons, *Bull. Am. Math. Soc. (N.S.)* **34**, no. 4, 339–403 (1997), DOI: 10.1090/S0273-0979-97-00732-5. MR1462745.

[PMSB21] J. Park, S. Moon, J. M. Seo, and J.-J. Baik, Systematic comparison between the generalized Lorenz equations and DNS in the two-dimensional Rayleigh-Bénard convection, *Chaos* **31**, no. 7, Paper No. 073119, 16 (2021), DOI: 10.1063/5.0051482. MR4283024.

[PR16] V. I. Paulsen and M. Raghupathi, *An Introduction to the Theory of Reproducing Kernel Hilbert Spaces*, Cambridge Studies in Advanced Mathematics, vol. 152, Cambridge University Press, Cambridge, 2016. MR3526117.

[PSS06] J. B. Pendry, D. Schurig, and D. R. Smith, Controlling electromagnetic fields, *Science* **312**, 1780–1782 (2006), DOI: 10.1126/science.1125907.

[PM90] P. Perona and J. Malik, Scale-space and edge detection using anisotropic diffusion, *IEEE Trans. Pattern Analysis Mach. Intell.* **12**, no. 7, 629–639 (1990), DOI: 10.1109/34.56205.

[Per01] L. Perko, *Differential Equations and Dynamical Systems*, 3rd ed., Texts in Applied Mathematics, vol. 7, Springer-Verlag, New York, 2001. MR1801796.

[Per15] B. Perthame, *Parabolic Equations in Biology*, Lecture Notes on Mathematical Modelling in the Life Sciences, Springer, Cham, 2015. Growth, reaction, movement and diffusion. MR3408563.

[PT07] A. W. Peterson and S. V. Tsynkov, Active control of sound for composite regions, *SIAM J. Appl. Math.* **67**, no. 6, 1582–1609 (2007), DOI: 10.1137/060662368. MR2349998.

[PS17] V. M. Petkov and L. N. Stoyanov, *Geometry of the Generalized Geodesic Flow and Inverse Spectral Problems*, 2nd ed., John Wiley & Sons, Ltd., Chichester, 2017. MR3617212.

[Pet91] I. G. Petrovsky, *Lectures on Partial Differential Equations*, Dover Publications, Inc., New York, 1991. Translated from the Russian by A. Shenitzer; Reprint of the 1964 English translation. MR1160355.

[PR05] Y. Pinchover and J. Rubinstein, *An Introduction to Partial Differential Equations*, Cambridge University Press, Cambridge, 2005. MR2164768.

[Pin99] A. Pinkus, Approximation theory of the MLP model in neural networks, *Acta Numer.* **8**, 143–195 (1999), DOI: 10.1017/S0962492900002919. MR1819645.

[Pho17] P. Photinos, *Musical Sound, Instruments, and Equipment*, Morgan & Claypool Publishers, 2017.

[PB05] A. Ponno and D. Bambusi, Korteweg-de Vries equation and energy sharing in Fermi-Pasta-Ulam, *Chaos* **15**, no. 1, 015107, 5 (2005), DOI: 10.1063/1.1832772. MR2133458.

[PBE+22] A. Power, Y. Burda, H. Edwards, I. Babuschkin, and V. Misra, Grokking: Generalization Beyond Overfitting on Small Algorithmic Datasets, *arXiv e-prints* (2022).

[Pra05] L. Prandtl, Über Flüssigkeitsbewegung bei sehr kleiner Reibung, *Verh. d. 3. intern. Math.-Kongr. Heidelb.*, 484–491 (1905) (German).

[PM13] E. M. Purcell and D. J. Morin, *Electricity and Magnetism*, 3rd ed., Cambridge University Press, Cambridge, 2013.

[LR10] L. Rayleigh, The problem of the whispering gallery, *London Edinburgh Dublin Phil. Mag. J. Sci.* **20**, no. 120, 1001–1004 (1910), DOI: 10.1080/14786441008636993.

[LR16] L. Rayleigh, On convection currents in a horizontal layer of fluid, when the higher temperature is on the under side, *London Edinburgh Dublin Phil. Mag. J. Sci.* **32**, 529–546 (1916), DOI: 10.1080/14786441608635602.

[Rin01] B. Rink, Symmetry and resonance in periodic FPU chains, *Comm. Math. Phys.* **218**, no. 3, 665–685 (2001), DOI: 10.1007/s002200100428. MR1831098.

[Ris89] H. Risken, *The Fokker-Planck Equation*, 2nd ed., Springer Series in Synergetics, vol. 18, Springer-Verlag, Berlin, 1989. Methods of solution and applications. MR987631.

[RM18]   A. Romano and A. Marasco, *Classical Mechanics with Mathematica®*, Modeling and Simulation in Science, Engineering and Technology, Birkhäuser/Springer, Cham, 2018. Second edition [MR2976921]. MR3837529.

[Rus45]   J. S. Russell, Report on Waves, *Report of the Fourteenth Meeting of the British Association for the Advancement of Science*, 311–390 (1845).

[Sal15]   S. Salsa, *Partial Differential Equations in Action*, 2nd ed., Unitext, vol. 86, Springer, Cham, 2015. From modelling to theory; La Matematica per il 3+2. MR3362185.

[Sal01]   D. Salsburg, *The Lady Tasting Tea*, W. H. Freeman and Company, New York, 2001. How statistics revolutionized science in the twentieth century. MR1815390.

[SMJ⁺06] D Schurig, J. J. Mock, B. J. Justice, S Cummer A., J. B. Pendry, A. F. Starr, and D. R. Smith, Metamaterial electromagnetic cloak at microwave frequencies, *Science* **314**, no. 5801, 977–980 (2006), DOI: 10.1126/science.1133628.

[SF82]   L. W. Schwartz and J. D. Fenton, *Strongly Nonlinear Waves*, Annual review of fluid mechanics, Vol. 14, Annual Reviews, Palo Alto, CA, 1982, pp. 39–60. MR642535.

[SE13]   M. Selvanayagam and G. V. Eleftheriades, Experimental demonstration of active electromagnetic cloaking, *Phys. Rev. X* **3**, 041011, 13 (2013), DOI: 10.1103/PhysRevX.3.041011.

[Sen14]   R. Sentis, *Mathematical Models and Methods for Plasma Physics. Vol. 1*, Modeling and Simulation in Science, Engineering and Technology, Birkhäuser/Springer, Cham, 2014. Fluid models. MR3184808.

[Ser71]   J. Serrin, A symmetry problem in potential theory, *Arch. Rational Mech. Anal.* **43**, 304–318 (1971), DOI: 10.1007/BF00250468. MR333220.

[Sho05]   A. D. Short, *Beaches of the Western Australian Coast: Eucla to Roebuck Bay. A Guide to Their Nature, Characteristics, Surf and Safety*, Sydney University Press, Sydney, 2005.

[SDP⁺08] M. B. Short, M. R. D'Orsogna, V. B. Pasour, G. E. Tita, P. J. Brantingham, A. L. Bertozzi, and L. B. Chayes, A statistical model of criminal behavior, *Math. Models Methods Appl. Sci.* **18**, no. suppl., 1249–1267 (2008), DOI: 10.1142/S0218202508003029. MR2438215.

[SKSC21] I. Shugan, S. Kuznetsov, Y. Saprykina, and Y.-Y. Chen, Physics of traveling waves in shallow water environment, *Water* **13**, no. 21 (2021), DOI: 10.3390/w13212990.

[SMC⁺14] A. Silva, F. Monticone, G. Castaldi, V. Galdi, A. Alù, and N. Engheta, Performing mathematical operations with metamaterials, *Science* **343**, no. 6167, 160–163 (2014), DOI: 10.1126/science.1242818. MR3156078.

[SM95] A. Sitenko and V. Malnev, *Plasma Physics Theory*, Applied Mathematics and Mathematical Computation, vol. 10, Chapman & Hall, London, 1995. MR1368631.

[Smo83] J. Smoller, *Shock Waves and Reaction-Diffusion Equations*, Grundlehren der Mathematischen Wissenschaften [Fundamental Principles of Mathematical Sciences], vol. 258, Springer-Verlag, New York-Berlin, 1983. MR688146.

[Som49] A. Sommerfeld, *Partial Differential Equations in Physics*, Academic Press, Inc., New York, NY, 1949. Translated by Ernst G. Straus. MR0029463.

[Spa82] C. Sparrow, *The Lorenz Equations: Bifurcations, Chaos, and Strange Attractors*, Applied Mathematical Sciences, vol. 41, Springer-Verlag, New York-Berlin, 1982. MR681294.

[Ste02] I. Steinwart, On the influence of the kernel on the consistency of support vector machines, *J. Mach. Learn. Res.* **2**, no. 1, 67–93 (2002), DOI: 10.1162/153244302760185252. MR1883281.

[ST18] I. Stewart and D. Tall, *Complex Analysis*, Cambridge University Press, Cambridge, 2018. The hitch hiker's guide to the plane; Second edition of [MR0698076]. MR3839273.

[Ste81] K. Stewartson, d'Alembert's paradox, *SIAM Rev.* **23**, no. 3, 308–343 (1981), DOI: 10.1137/1023063. MR631832.

[Sto51] J. J. Stoker, Book Review: Hydrodynamics, a study in logic, fact, and similitude, *Bull. Am. Math. Soc.* **57**, no. 6, 497–499 (1951), DOI: 10.1090/S0002-9904-1951-09552-X. MR1565341.

[Str08] W. A. Strauss, *Partial Differential Equations*, 2nd ed., John Wiley & Sons, Ltd., Chichester, 2008. An introduction. MR2398759.

[Str15] S. H. Strogatz, *Nonlinear Dynamics and Chaos*, 2nd ed., Westview Press, Boulder, CO, 2015. With applications to physics, biology, chemistry, and engineering. MR3837141.

[SG69] C. H. Su and C. S. Gardner, Korteweg-de Vries equation and generalizations. III. Derivation of the Korteweg-de Vries equation and Burgers equation, *J. Math. Phys.* **10**, 536–539 (1969), DOI: 10.1063/1.1664873. MR271526.

[Tay20] M. E. Taylor, *Linear Algebra*, Pure Appl. Undergrad. Texts, vol. 45, American Mathematical Society, Providence, RI, 2020.

[Tho42] D. W. Thompson, *On Growth and Form*, New edition, Cambridge University Press, Cambridge, England, 1942. MR0006348.

[TS90] A. N. Tikhonov and A. A. Samarskiĭ, *Equations of Mathematical Physics*, Dover Publications, Inc., New York, 1990. Translated from the Russian by A. R. M. Robson and P. Basu; Reprint of the 1963 translation. MR1149383.

[Tol78] J. F. Toland, On the existence of a wave of greatest height and Stokes's conjecture, *Proc. Roy. Soc. London Ser. A* **363**, no. 1715, 469–485 (1978), DOI: 10.1098/rspa.1978.0178. MR513927.

[TW08] N. S. Trudinger and X.-J. Wang, *The Monge-Ampère Equation and its Geometric Applications*, Handbook of geometric analysis. No. 1, Adv. Lect. Math. (ALM), vol. 7, Int. Press, Somerville, MA, 2008, pp. 467–524. MR2483373.

[Tuc02] W. Tucker, A rigorous ODE solver and Smale's 14th problem, *Found. Comput. Math.* **2**, no. 1, 53–117 (2002), DOI: 10.1007/s002080010018. MR1870856.

[Tur52] A. M. Turing, The chemical basis of morphogenesis, *Philos. Trans. Roy. Soc. London Ser. B* **237**, no. 641, 37–72 (1952). MR3363444.

[Uhl08] G. Uhlmann, *Commentary on Calderón's paper (29), on an Inverse Boundary Value Problem*, Selected papers of Alberto P. Calderón, American Mathematical Society, Providence, RI, 2008, pp. 623–636. MR2435340.

[Uhl09] G. Uhlmann, Electrical impedance tomography and Calderón's problem, *Inverse Prob.* **25**, no. 12, 123011, 39 (2009), DOI: 10.1088/0266-5611/25/12/123011. MR3460047.

[Ula76] S. M. Ulam, *Adventures of a Mathematician*, Charles Scribner's Sons, New York, 1976. MR0485098.

[ULD⁺13] Y. Urzhumov, N. Landy, T. Driscoll, D. Basov, and D. R. Smith, Thin low-loss dielectric coatings for free-space cloaking, *Opt. Lett.* **38**, no. 10, 1606–1608 (2013), DOI: 10.1364/OL.38.001606.

[VLZ⁺09] J. Valentine, J. Li, T. Zentgraf, G. Bartal, and X. Zhang, An optical cloak made of dielectrics, *Nat. Mater.* **8**, 568–571 (2009), DOI: 10.1038/nmat2461.

[VL00] N. Vasconcelos and A. Lippman, *A Unifying View of Image Similarity*, Proceedings 15th International Conference on Pattern Recognition (2000), 2000, pp. 38–41, DOI: 10.1109/ICPR.2000.905271.

[Váz07] J. L. Vázquez, *The Porous Medium Equation*, Oxford Mathematical Monographs, The Clarendon Press, Oxford University Press, Oxford, 2007. Mathematical theory. MR2286292.

[Vis04] D. Viswanath, The fractal property of the Lorenz attractor, *Phys. D* **190**, no. 1–2, 115–128 (2004), DOI: 10.1016/j.physd. 2003.10.006. MR2043795.

[XX19] Y. Xuan and X. Xu, Three-dimensional reciprocal invisibility cloak with multilayered structure, *J. Microw. Optoelectron. Electromagn. Appl.* **18**, 184–195 (2019), DOI: 10.1590/2179-10742019v18i21551.

[YKO68] H. Yamada, G. Kimura, and J. Okabe, Precise determination of the solitary waves of extreme height on water of a uniform depth, *Rep. Res. Inst. Applied Mech., Kyushu Univ.* **16**, no. 52, 15–32 (1968).

[YYK07] M. Yamaguchi, E. Yoshimoto, and S. Kondo, Pattern regulation in the stripe of zebrafish suggests an underlying dynamic and autonomous mechanism, *Proc. Natl. Acad. Sci. USA* **104**, 4790–4793 (2007), DOI: 10.1073/pnas.0607790104.

[Yan21] X.-B. Yan, The physical origin of Schrödinger equation, *Europ. J. Phys.* **42**, no. 4, 045402 (2021), DOI: 10.1088/1361-6404/abf6a0.

[YKD22] M. Yun, J. Kim, and K. Do, Estimation of wave-breaking index by learning nonlinear relation using multilayer neural network, *J. Marine Sci. Eng.* **10**, no. 1, 1–16 (2022), DOI: 10.3390/jmse10010050.

[Wan09] X.-J. Wang, *The k-Hessian Equation*, Geometric analysis and PDEs, Lecture Notes in Math., vol. 1977, Springer, Dordrecht, 2009, pp. 177–252, DOI: 10.1007/978-3-642-01674-5_5. MR2500526.

[WV06] D. W. Ward and S. M. Volkmer, How to Derive the Schrödinger Equation, *arXiv e-prints* (2006).

[Wat95] G. N. Watson, *A Treatise on the Theory of Bessel Functions*, Cambridge Mathematical Library, Cambridge University Press, Cambridge, 1995. Reprint of the second (1944) edition. MR1349110.

[Whe10] N. Wheatley, On the dimensionality of the Avogadro constant and the definition of the mole, *Nature Precedings*, 1–20, posted on 2010, DOI: 10.1038/npre.2010.5138.1.

[WZ15] R. L. Wheeden and A. Zygmund, Measure and Integral, 2nd ed., *Pure and Applied Mathematics* (Boca Raton), CRC Press, Boca Raton, FL, 2015. An introduction to real analysis. MR3381284.

[WWMS15] V. Wheeler, G. E. Wheeler, J. A. McCoy, and J. J. Sharples, Modelling dynamic bushfire spread: perspectives from the theory of curvature flow, *International Congress on Modelling and Simulation*, 319–325 (2015).

[Whi67] G. B. Whitham, Non-linear dispersion of water waves, *J. Fluid Mech.* **27**, 399–412 (1967), DOI: 10.1017/S0022112067000424. MR208903.

[GBW99] G. B. Whitham, *Linear and Nonlinear Waves*, Pure and Applied Mathematics (New York), John Wiley & Sons, Inc., New York, 1999. Reprint of the 1974 original; A Wiley-Interscience Publication. MR1699025.

[Woo09] B. Wood, Metamaterials and invisibility, *C. R. Phys.* **10**, no. 5, 379–390 (2009), DOI: 10.1016/j.crhy.2009.01.002.

[ZK65] N. J. Zabusky and M. D. Kruskal, Interaction of 'solitons' in a collisionless plasma and the recurrence of initial states, *Phys. Rev. Lett.* **15**, 240–243 (1965), DOI: 10.1103/PhysRevLett.15.240.

[ZBLM85] Ya. B. Zel'dovich, G. I. Barenblatt, V. B. Librovich, and G. M. Makhviladze, *Mathematical Theory of Combustion and Explosions*, Consultants Bureau [Plenum], New York, 1985. Translated from the Russian by Donald H. McNeill. MR781350.

[Zha08] B. a. C. Zhang Hongsheng and Wu, Extraordinary surface voltage effect in the invisibility cloak with an active device inside, *Phys. Rev. Lett.* **100**, 063904 (2008), DOI: 10.1103/PhysRevLett.100.063904.

[DXZ20] Ding-Xuan Zhou, Universality of deep convolutional neural networks, *Appl. Comput. Harm. Anal.* **48**, no. 2, 787–794 (2020), DOI: 10.1016/j.acha.2019.06.004.

Index